Solutions Manual

SAXON Math™
HOMESCHOOL
7/6

Stephen Hake
John Saxon

SAXON™
PUBLISHERS

Saxon Publishers gratefully acknowledges the contributions of the following individuals in the completion of this project:

Authors: Stephen Hake, John Saxon

Editorial: Chris Braun, Bo Björn Johnson, Dana Nixon, Brian E. Rice

Editorial Support Services: Christopher Davey, Jay Allman, Jenifer Sparks, Shelley Turner, Jean Van Vleck, Darlene Terry

Production: Alicia Britt, Karen Hammond, Donna Jarrel, Brenda Lopez, Adriana Maxwell, Cristi D. Whiddon

Project Management: Angela Johnson, Becky Cavnar

© 2005 Saxon Publishers, Inc., and Stephen Hake

Printed in the United States of America

ISBN: 978-1-59-141327-1

21 22 23 24 25 0982 20 19 18 17

4500649270

Solutions for

Lessons and Investigations

LESSON 1, WARM-UP

a. 60

b. 600

c. 120

d. 1200

e. 90

f. 900

Problem Solving

2's: 10, 12, 14, 16, 18, 20
3's: 12, 15, 18
4's: 12, 16, 20
18

LESSON 1, LESSON PRACTICE

a.
$$\begin{array}{r} \overset{1\,1\,1}{3675} \\ 426 \\ +\ \ 1357 \\ \hline \mathbf{5458} \end{array}$$

b.
$$\begin{array}{r} \$6.25 \\ \$_18.23 \\ +\ \$12.00 \\ \hline \mathbf{\$26.48} \end{array}$$

c.
$$\begin{array}{r} 5\,3\,\overset{6}{\cancel{7}}{}^1 4 \\ -\ \ \ 1\,6\,8 \\ \hline \mathbf{5\,2\,0\,6} \end{array}$$

d.
$$\begin{array}{r} \$\overset{4}{\cancel{5}}.\,\overset{9}{\cancel{0}}{}^1 0 \\ -\ \$1.\,3\,5 \\ \hline \mathbf{\$3.\,6\,5} \end{array}$$

e.
$$\begin{array}{rrrr} 6 & 8 & 14 & 14 \\ +\ 8 & +\ 6 & -\ 6 & -\ 8 \\ \hline 14 & 14 & 8 & 6 \end{array}$$

f.
$25 - 15 = 10$
$10 + 15 = 25$
$15 + 10 = 25$

LESSON 1, MIXED PRACTICE

1.
$$\begin{array}{r} 25 \\ +\ 40 \\ \hline \mathbf{65} \end{array}$$

2.
$$\begin{array}{r} \overset{1\,2}{137} \\ 89 \\ +\ \ \ 9 \\ \hline \mathbf{235}\ \text{seeds} \end{array}$$

3.
$$\begin{array}{r} \overset{2}{\cancel{3}}\overset{1}{8}\,7 \\ -\ \ \ 9\,3 \\ \hline \mathbf{2\ 9\ 4} \end{array}$$

4.
$$\begin{array}{r} \$\overset{4}{\cancel{5}}.\,\overset{9}{\cancel{0}}{}^1 0 \\ -\ \$3.\,7\,5 \\ \hline \mathbf{\$1.\,2\,5} \end{array}$$

5.
$$\begin{array}{r} \$1.15 \\ +\ \$5.22 \\ \hline \mathbf{\$6.37} \end{array}$$

6.
$$\begin{array}{r} \overset{1}{\$1.25} \\ \$0.70 \\ +\ \$0.60 \\ \hline \mathbf{\$2.55} \end{array}$$

7.
$$\begin{array}{r} \overset{1}{63} \\ 47 \\ +\ 50 \\ \hline \mathbf{160} \end{array}$$

8.
$$\begin{array}{r} \overset{1\,1}{632} \\ 57 \\ +\ 198 \\ \hline \mathbf{887} \end{array}$$

9.
$$\begin{array}{r} \overset{2}{78} \\ \overset{1}{9} \\ +\ 967 \\ \hline \mathbf{1054} \end{array}$$

10.
$$\begin{array}{r} \overset{1\,1}{432} \\ {}_1579 \\ +\ 3604 \\ \hline \mathbf{4615} \end{array}$$

11. $\begin{array}{r} \overset{2}{\cancel{3}}\overset{13}{\cancel{4}}5 \\ -67 \\ \hline 278 \end{array}$

12. $\begin{array}{r} 678 \\ -416 \\ \hline 262 \end{array}$

13. $\begin{array}{r} 37\overset{6}{\cancel{7}}\overset{15}{\cancel{6}}4 \\ -96 \\ \hline 3668 \end{array}$

14. $\begin{array}{r} \overset{21}{}875 \\ 1086 \\ +980 \\ \hline 2941 \end{array}$

15. $\begin{array}{r} \overset{2}{}10 \\ {}_1156 \\ 8 \\ +27 \\ \hline 201 \end{array}$

16. $\begin{array}{r} \$\overset{2}{\cancel{3}}.\overset{1}{\cancel{4}}7 \\ -\$0.\,9\,2 \\ \hline \$2.\,5\,5 \end{array}$

17. $\begin{array}{r} \$2\,\overset{3}{\cancel{4}}.\overset{1}{\cancel{1}}5 \\ -\$1.\,4\,5 \\ \hline \$2\,2.\,7\,0 \end{array}$

18. $\begin{array}{r} \overset{11}{}\$0.75 \\ +\$0.75 \\ \hline \$1.50 \end{array}$

19. $\begin{array}{r} \overset{1}{}\$0.12 \\ \$0.46 \\ +\$0.50 \\ \hline \$1.08 \end{array}$

20. Sum

21. Difference

22. $\begin{array}{r} 5 \\ +\,6 \\ \hline 11 \end{array}$ $\quad$ $\begin{array}{r} 6 \\ +\,5 \\ \hline 11 \end{array}$ $\quad$ $\begin{array}{r} 11 \\ -\,6 \\ \hline 5 \end{array}$ $\quad$ $\begin{array}{r} 11 \\ -\,5 \\ \hline 6 \end{array}$

23. $16 + 27 = 43$

$43 - 16 = 27$

$43 - 27 = 16$

24. $50 - 29 = 21$

$29 + 21 = 50$

$21 + 29 = 50$

25. One way to check is to add the answer (difference) to the amount subtracted. The total should equal the starting amount.

LESSON 2, WARM-UP

a. **540**

b. **260**

c. **270**

d. **770**

e. **480**

f. **480**

Problem Solving

$1 + 2 + 3 + 4 + 5 =$ **15 coins**

$1 + 2 + 3 + 4 + 5 + 6 =$ **21 coins**

LESSON 2, LESSON PRACTICE

a. $\begin{array}{r} \overset{1}{}37¢ \\ \times20 \\ \hline 740¢ \text{ or } \$7.40 \end{array}$

b. $\begin{array}{r} 37 \\ \times0 \\ \hline 0 \end{array}$

c. $\begin{array}{r} 407 \\ \times37 \\ \hline 2\,849 \\ 12\,210 \\ \hline 15{,}059 \end{array}$

d.
$$5)\overline{\$8.40} \quad \$1.68$$
$$\underline{5}$$
$$3\,4$$
$$\underline{3\,0}$$
$$40$$
$$\underline{40}$$
$$0$$

e.
16 R 8
$$12)\overline{200}$$
$$\underline{12}$$
$$80$$
$$\underline{72}$$
$$8$$

f.
78
$$3)\overline{234}$$
$$\underline{21}$$
$$24$$
$$\underline{24}$$
$$0$$

g. **5; 12; 3**

h.
$8 \times 9 = 72$
$9 \times 8 = 72$
$72 \div 9 = 8$
$72 \div 8 = 9$

LESSON 2, MIXED PRACTICE

1.
$$\begin{array}{r} 11 \\ \times\ 7 \\ \hline 77 \end{array}$$

2.
$$\begin{array}{r} {}^{8}\!\!\!\not{9}{}^{1}7 \\ -\ 79 \\ \hline 1\,8 \end{array}$$

3.
$$\begin{array}{r} {}^{1}170 \\ +\ 130 \\ \hline 300 \end{array}$$

4.
$$4)\overline{36} \quad 9$$
$$\underline{36}$$
$$0$$

5.
$$\begin{array}{r} {}^{2\,1}386 \\ {}^{1}\ 98 \\ +\ 1734 \\ \hline 2218 \end{array}$$

6.
$$\begin{array}{r} \$\not{3}.{}^{4}\not{0}{}^{9}\!{}^{1}0 \\ -\ \$2.25 \\ \hline \$2.75 \end{array}$$

7.
$$\begin{array}{r} \$\not{7}{}^{6}\!\not{0}.{}^{9}\!{}^{1}00 \\ -\ \$47.50 \\ \hline \$22.50 \end{array}$$

8.
$$\begin{array}{r} 75\cent \\ \times\ 12 \\ \hline 150 \\ 750 \\ \hline 900\cent \end{array} \text{ or } \mathbf{\$9.00}$$

9.
$$\begin{array}{r} \not{3}{}^{2}\!\not{1}{}^{1}2 \\ -\ 86 \\ \hline 2\,2\,6 \end{array}$$

10.
$$\begin{array}{r} {}^{1\,1}4106 \\ +\ 1398 \\ \hline 5504 \end{array}$$

11.
$$\begin{array}{r} \not{4}{}^{3}\!\not{0}{}^{9}\!\not{0}{}^{9}\!{}^{1}0 \\ -\ 1357 \\ \hline 2\,6\,4\,3 \end{array}$$

12.
$$\begin{array}{r} \$\not{1}{}^{0}\!\not{0}.{}^{9}\!\not{0}{}^{9}\!{}^{1}0 \\ -\ \$2.83 \\ \hline \$7.17 \end{array}$$

13.
$$\begin{array}{r} 405 \\ \times\ 8 \\ \hline 3240 \end{array}$$

14.
$$\begin{array}{r} 25 \\ \times\ 25 \\ \hline 125 \\ 500 \\ \hline 625 \end{array}$$

15.
$$\begin{array}{r} 48 \\ 6\overline{)288} \\ \underline{24} \\ 48 \\ \underline{48} \\ 0 \end{array}$$

16.
$$\begin{array}{r} 15 \\ 15\overline{)225} \\ \underline{15} \\ 75 \\ \underline{75} \\ 0 \end{array}$$

17.
$$\begin{array}{r} \$1.25 \\ \times \quad 8 \\ \hline \$10.00 \end{array}$$

18.
$$\begin{array}{r} 400 \\ \times \quad 50 \\ \hline 20,000 \end{array}$$

19.
$$\begin{array}{r} 125 \\ 8\overline{)1000} \\ \underline{8} \\ 20 \\ \underline{16} \\ 40 \\ \underline{40} \\ 0 \end{array}$$

20.
$$\begin{array}{r} \$2.25 \\ 20\overline{)\$45.00} \\ \underline{40} \\ 5\ 0 \\ \underline{4\ 0} \\ 1\ 00 \\ \underline{1\ 00} \\ 0 \end{array}$$

21. $6 \times 8 = 48$

$8 \times 6 = 48$

$48 \div 6 = 8$

$48 \div 8 = 6$

22. $36 \div 9 = 4$

$4 \times 9 = 36$

$9 \times 4 = 36$

23. $24 + 12 = 36$

$36 - 24 = 12$

$36 - 12 = 24$

24. (a)
$$\begin{array}{r} 9 \\ + \ 6 \\ \hline 15 \end{array}$$
(b)
$$\begin{array}{r} 9 \\ - \ 6 \\ \hline 3 \end{array}$$

25. $\text{divisor}\overline{)\overset{\text{quotient}}{\text{dividend}}}$

26.
$$\begin{array}{r} 39¢ \\ \times \quad 6 \\ \hline 23.4¢ \end{array} \text{ or } \$2.34$$

27.
$$\begin{array}{r} 365 \\ \times \quad 0 \\ \hline 0 \end{array}$$

28. $50\overline{)\overset{0}{0}}$

29.
$$\begin{array}{r} 1 \\ 365\overline{)365} \\ \underline{365} \\ 0 \end{array}$$

30. One way to check is to multiply the divisor by the quotient. The answer should equal the dividend.

LESSON 3, WARM-UP

a. **7000**

b. **2600**

c. **3020**

d. **920**

e. **4500**

f. **4370**

Problem Solving

top, bottom, front, back, left, right: **6 surfaces**

LESSON 3, LESSON PRACTICE

a.
$$\begin{array}{r} 45 \\ -\,12 \\ \hline 33 \end{array}$$
check:
$$\begin{array}{r} 33 \\ +\,12 \\ \hline 45 \end{array}$$
$A = 33$

b.
$$\begin{array}{r} \overset{5}{\cancel{6}}{}^{1}0 \\ -\,3\,2 \\ \hline 2\,8 \end{array}$$
check:
$$\begin{array}{r} \overset{1}{3}2 \\ +\,28 \\ \hline 60 \end{array}$$
$B = 28$

c.
$$\begin{array}{r} 15 \\ +\,24 \\ \hline 39 \end{array}$$
check:
$$\begin{array}{r} 39 \\ -\,15 \\ \hline 24 \end{array}$$
$C = 39$

d.
$$\begin{array}{r} \overset{2}{\cancel{3}}{}^{1}8 \\ -\,2\,9 \\ \hline 9 \end{array}$$
check:
$$\begin{array}{r} \overset{2}{\cancel{3}}{}^{1}8 \\ -\,9 \\ \hline 2\,9 \end{array}$$
$D = 9$

e.
$$\begin{array}{r} \overset{4}{\cancel{5}}{}^{1}2 \\ -\,2\,4 \\ \hline 2\,8 \end{array}$$
check:
$$\begin{array}{r} \overset{1}{2}8 \\ +\,24 \\ \hline 52 \end{array}$$
$e = 28$

f.
$$\begin{array}{r} \overset{6}{\cancel{7}}{}^{1}0 \\ -\,2\,9 \\ \hline 4\,1 \end{array}$$
check:
$$\begin{array}{r} \overset{1}{2}9 \\ +\,41 \\ \hline 70 \end{array}$$
$f = 41$

g.
$$\begin{array}{r} 67 \\ +\,43 \\ \hline 110 \end{array}$$
check:
$$\begin{array}{r} \overset{0}{\cancel{1}}\overset{10}{\cancel{1}}0 \\ -\,6\,7 \\ \hline 4\,3 \end{array}$$
$g = 110$

h.
$$\begin{array}{r} \overset{7}{\cancel{8}}{}^{1}0 \\ -\,3\,6 \\ \hline 4\,4 \end{array}$$
check:
$$\begin{array}{r} \overset{7}{\cancel{8}}{}^{1}0 \\ -\,4\,4 \\ \hline 3\,6 \end{array}$$
$h = 44$

i.
$$\underbrace{36 + 14 + 8}\, + n = 75$$
$$58 + n = 75$$
$$75 - 58 = 17$$
$$n = 17$$
check: $\quad 36 + 14 + 17 + 8 = 75$

LESSON 3, MIXED PRACTICE

1.
$$\begin{array}{r} 25 \\ \times\,12 \\ \hline 50 \\ 250 \\ \hline 300 \end{array}$$

2.
$$\begin{array}{r} 25 \\ +\,12 \\ \hline 37 \end{array}$$

3.
$$\begin{array}{r} 25 \\ -\,12 \\ \hline 13 \end{array}$$

4.
$$\begin{array}{r} 75 \\ \times\,31 \\ \hline 75 \\ 2250 \\ \hline 2325 \text{ cans} \end{array}$$

5.
$$\begin{array}{r} \$7.85 \\ \times\,12 \\ \hline 1570 \\ 7850 \\ \hline \$94.20 \end{array}$$

6.
$$\begin{array}{r} \overset{0}{\cancel{1}}\overset{9}{\cancel{0}}{}^{1}2 \\ -\,6\,3 \\ \hline 3\,9 \text{ points} \end{array}$$

7.
$$\begin{array}{r} \$3.68 \\ \times\,9 \\ \hline \$33.12 \end{array}$$

8.
$$\begin{array}{r} 407 \\ \times\,80 \\ \hline 32{,}560 \end{array}$$

9.
$$\begin{array}{r} 28¢ \\ \times\,14 \\ \hline 112 \\ 280 \\ \hline 392¢ \text{ or } \$3.92 \end{array}$$

10.
$$\begin{array}{r} 370 \\ \times\,140 \\ \hline 14800 \\ 3700 \\ \hline 51{,}800 \end{array}$$

11. $\begin{array}{r} 100 \\ \times\ \ 100 \\ \hline 10{,}000 \end{array}$

12. $\begin{array}{r} 12 \\ 12\overline{)144} \\ \underline{12} \\ 24 \\ \underline{24} \\ 0 \end{array}$

13. $\begin{array}{r} 12 \\ \times\ \ 5 \\ \hline 60 \end{array}$

14. $\begin{array}{r} {}^{2\,2\,1} \\ 3627 \\ 598 \\ +\ 4881 \\ \hline 9106 \end{array}$

15. $\begin{array}{r} {}^{4}\,{}^{9}\,{}^{10} \\ \cancel{5}\,\cancel{0}\,\cancel{1}\,{}^{1}0 \\ -\ 1\,3\,7\,6 \\ \hline 3\,6\,3\,4 \end{array}$

16. $\begin{array}{r} {}^{0}\,{}^{9}\,{}^{9} \\ \$\cancel{1}\,\cancel{0}.\,\cancel{0}\,{}^{1}0 \\ -\ \ \$0.\,2\,6 \\ \hline \$9.\,7\,4 \end{array}$

17. $\begin{array}{r} 48 \\ -\ 16 \\ \hline 32 \end{array}$
    $A\ =\ 32$

18. $\begin{array}{r} {}^{4} \\ \cancel{5}\,{}^{1}2 \\ -\ 2\,3 \\ \hline 2\,9 \end{array}$
    $B\ =\ 29$

19. $\begin{array}{r} 31 \\ +\ 17 \\ \hline 48 \end{array}$
    $C\ =\ 48$

20. $\begin{array}{r} {}^{3} \\ \cancel{4}\,{}^{1}2 \\ -\ 2\,5 \\ \hline 1\,7 \end{array}$
    $D\ =\ 17$

21. $75\ -\ 38\ =\ 37$
    $x\ =\ 37$

22. $38\ +\ 75\ =\ 113$
    $x\ =\ 113$

23. $75\ -\ 38\ =\ 37$
    $y\ =\ 37$

24. $\underline{6\ +\ 8\ +\ 5}\ +\ w\ =\ 32$
    $\qquad 19\ +\ w\ =\ 32$
    check: $32\ -\ 19\ =\ 13$
    $\qquad\qquad\qquad\qquad w\ =\ 13$

25. $48\ +\ 24\ =\ 72$
    $72\ -\ 24\ =\ 48$
    $72\ -\ 48\ =\ 24$

26. $15\ \times\ 6\ =\ 90$
    $90\ \div\ 6\ =\ 15$
    $90\ \div\ 15\ =\ 6$

27. $\begin{array}{r} 10 \\ 20\overline{)200} \\ \underline{20} \\ 00 \\ \underline{00} \\ 0 \end{array}$

28. $\begin{array}{r} 15 \\ \times\ \ 8 \\ \hline 120 \end{array}$

29. $\begin{array}{r} 1 \\ 144\overline{)144} \\ \underline{144} \\ 0 \end{array}$

30. **To find a missing addend, subtract the known addend(s) from the sum.**

LESSON 4, WARM-UP

a. **2920**

b. **8420**

c. **7740**

d. **2850**

e. 1490

f. 9050

Problem Solving
315, 351, 513, 531

LESSON 4, LESSON PRACTICE

a. $7\overline{)9^21}$ check: $\begin{array}{r} 13 \\ \times \ 7 \\ \hline 91 \end{array}$
$A = 13$ (quotient 13)

b. $20\overline{)44^40}$ check: $\begin{array}{r} 22 \\ \times \ 20 \\ \hline 00 \\ 440 \\ \hline 440 \end{array}$
$B = 22$ (quotient 22)

c. $\begin{array}{r} 15 \\ \times \ 7 \\ \hline 105 \end{array}$ check: $7\overline{)10^35}$ (quotient 15)
$C = 105$

d. $8\overline{)14^64}$ check: $18\overline{)144}$ (quotient 8)
$D = 18$ (quotient 18)

e. $7\overline{)8^14}$ check: $\begin{array}{r} 12 \\ \times \ 7 \\ \hline 84 \end{array}$
$w = 12$ (quotient 12)

f. $8\overline{)11^32}$ check: $\begin{array}{r} 14 \\ \times \ 8 \\ \hline 112 \end{array}$
$m = 14$ (quotient 14)

g. $30\overline{)36^60}$ check: $12\overline{)36^00}$ (quotient 30)
$x = 12$ (quotient 12)

h. $\begin{array}{r} 60 \\ \times \ 5 \\ \hline 300 \end{array}$ check: $5\overline{)30^00}$ (quotient 60)
$n = 300$

LESSON 4, MIXED PRACTICE

1. 4 carrot sticks
$$15\overline{)60} \quad \begin{array}{r} 60 \\ \hline 0 \end{array}$$

2. 25 pennies
$$4\overline{)100} \quad \begin{array}{r} 8 \\ \hline 20 \\ 20 \\ \hline 0 \end{array}$$

3. 20 stacks
$$5\overline{)100} \quad \begin{array}{r} 10 \\ \hline 00 \\ 00 \\ \hline 0 \end{array}$$

4. 21 teams
$$14\overline{)294} \quad \begin{array}{r} 28 \\ \hline 14 \\ 14 \\ \hline 0 \end{array}$$

5. $\begin{array}{r} 2\ 8^{7}0 \\ -\ 1\ 5\ 6 \\ \hline 1\ 2\ 4 \end{array}$ pages

6. $\begin{array}{r} \$0.75 \\ \times \quad 42 \\ \hline 150 \\ 3000 \\ \hline \$31.50 \end{array}$

7. $5\overline{)60}$ check: $\begin{array}{r} 12 \\ \times \ 5 \\ \hline 60 \end{array}$
quotient 12; $\begin{array}{r} 5 \\ \hline 10 \\ 10 \\ \hline 0 \end{array}$
$J = 12$

8. $\begin{array}{r} 7^{6}{}^{1}2 \\ -\ 2\ 7 \\ \hline 4\ 5 \end{array}$ check: $\begin{array}{r} 27 \\ +\ 45 \\ \hline 72 \end{array}$
$K = 45$

9. $\begin{array}{r} 37 \\ -\ 36 \\ \hline 1 \end{array}$ check: $\begin{array}{r} 1 \\ +\ 36 \\ \hline 37 \end{array}$
$L = 1$

SOLUTIONS

10.
$$\begin{array}{r} \overset{5}{\cancel{6}}{}^{1}4 \\ -\ 4\ 6 \\ \hline 1\ 8 \end{array}$$
check:
$$\begin{array}{r} \overset{5}{\cancel{6}}{}^{1}4 \\ -\ 1\ 8 \\ \hline 4\ 6 \end{array}$$

$M = 18$

11.
$$\begin{array}{r} \overset{1}{4}8 \\ +\ 84 \\ \hline 132 \end{array}$$
check:
$$\begin{array}{r} \overset{0}{\cancel{1}}\overset{12}{\cancel{3}}{}^{1}2 \\ -\ \ 4\ 8 \\ \hline 8\ 4 \end{array}$$

$n = 132$

12.
$$\begin{array}{r} 13 \\ 7\overline{)91} \\ \underline{7} \\ 21 \\ \underline{21} \\ 0 \end{array}$$
check:
$$\begin{array}{r} 13 \\ \times\ 7 \\ \hline 91 \end{array}$$

$p = 13$

13.
$$\begin{array}{r} 7 \\ \times\ 0 \\ \hline 0 \end{array}$$
check:
$$\begin{array}{r} 0 \\ 7\overline{)0} \\ \underline{0} \\ 0 \end{array}$$

$q = 0$

14.
$$\begin{array}{r} 24 \\ 6\overline{)144} \\ \underline{12} \\ 24 \\ \underline{24} \\ 0 \end{array}$$
check:
$$\begin{array}{r} 6 \\ 24\overline{)144} \\ \underline{144} \\ 0 \end{array}$$

$r = 24$

15.
$$\begin{array}{r} \$2.06 \\ 6\overline{)\$12.36} \\ \underline{12} \\ 0\ 3 \\ \underline{0\ 0} \\ 36 \\ \underline{36} \\ 0 \end{array}$$

16.
$$\begin{array}{r} 720 \\ 8\overline{)5760} \\ \underline{56} \\ 16 \\ \underline{16} \\ 00 \\ \underline{00} \\ 0 \end{array}$$

17.
$$\begin{array}{r} 29\ \mathbf{R\ 4} \\ 18\overline{)526} \\ \underline{36} \\ 166 \\ \underline{162} \\ 4 \end{array}$$

18.
$$\begin{array}{r} \overset{2}{5}\overset{1}{6}3 \\ 563 \\ 563 \\ +\ 563 \\ \hline \mathbf{2252} \end{array}$$

19.
$$\begin{array}{r} \$3.75 \\ \times\ \ \ 16 \\ \hline 2250 \\ 3750 \\ \hline \mathbf{\$60.00} \end{array}$$

20.
$$\begin{array}{r} \overset{1}{\ }\overset{1}{\ } \\ \$3.00 \\ \$2.86 \\ +\ \$0.98 \\ \hline \mathbf{\$6.84} \end{array}$$

21.
$$\begin{array}{r} \$\overset{0}{\cancel{1}}\overset{9}{\cancel{0}}.\overset{9}{\cancel{0}}{}^{1}0 \\ -\ \ \$6.\ 4\ 3 \\ \hline \mathbf{\$3.\ 5\ 7} \end{array}$$

22.
$$\begin{array}{r} 12 \\ 3\overline{)n} \end{array} \qquad \begin{array}{r} 12 \\ \times\ 3 \\ \hline 36 \end{array}$$

The dividend is 36.

23.
$$5 \times m = 100 \qquad \begin{array}{r} 20 \\ 5\overline{)100} \\ \underline{10} \\ 00 \\ \underline{00} \\ 0 \end{array}$$

The other factor is 20.

24. $17 - 8 = 9$
    $9 + 8 = 17$
    $8 + 9 = 17$

25. $72 \div 9 = 8$
    $8 \times 9 = 72$
    $9 \times 8 = 72$

26. $w + \underbrace{6 + 8 + 10} = 40$

$$w + 24 = 40$$
$$40 - 24 = \mathbf{16}$$
$$w = \mathbf{16}$$

27.
$$\overset{2}{23}¢$$
$$\underline{\times \ 7}$$
$$\mathbf{161¢ \ or \ \$1.61}$$

28.
$$25\overline{)25}\overset{1}{}$$
$$\underline{25}$$
$$0$$
$$m = \mathbf{1}$$

29.
$$15\overline{)0}\overset{0}{}$$
$$\underline{0}$$
$$0$$
$$n = \mathbf{0}$$

30. To find a missing factor, divide the product by the known factor.

LESSON 5, WARM-UP

a. 760

b. 870

c. 7200

d. 790

e. 5800

f. 640

Problem Solving

First think, "5 plus what number equals 13?" (8). Then think, "What number plus 3 plus 1 (from regrouping) equals 9?" (5). Next think, "1 plus what one-digit number equals a two-digit number?" (9). What is the two-digit number? (10).

$$155$$
$$\underline{+ \ 938}$$
$$1093$$

LESSON 5, LESSON PRACTICE

a. $\underline{16 - 3} + 4$
$$13 + 4$$
$$\mathbf{17}$$

b. $16 - \underline{(3 + 4)}$
$$16 - 7$$
$$\mathbf{9}$$

c. $24 \div \underline{(4 \times 3)}$
$$24 \div 12$$
$$\mathbf{2}$$

d. $\underline{24 \div 4} \times 3$
$$6 \times 3$$
$$\mathbf{18}$$

e. $\underline{24 \div 6} \div 2$
$$4 \div 2$$
$$\mathbf{2}$$

f. $24 \div \underline{(6 \div 2)}$
$$24 \div 3$$
$$\mathbf{8}$$

g. $\dfrac{6 + 9}{3} = \dfrac{15}{3} = \mathbf{5}$

h. $\dfrac{12 + 8}{12 - 8} = \dfrac{20}{4} = \mathbf{5}$

LESSON 5, MIXED PRACTICE

1.
$$\$1.25 \qquad \$\overset{4}{\cancel{3}}.\overset{9}{\cancel{0}}{}^{1}0$$
$$\underline{+ \ \$0.60} \qquad \underline{- \ \$1.\ 8\ 5}$$
$$\$1.85 \qquad \mathbf{\$3.\ 1\ 5}$$

2.
$$\overset{1}{82}$$
$$8$$
$$\underline{+ \ 12}$$
$$\mathbf{102 \ pounds}$$

3.
$$\overset{0}{\cancel{1}} \ \overset{1}{\cancel{1}}0$$
$$\underline{- \quad 2\ 5}$$
$$\mathbf{8\ 5}$$

4.
$$
\begin{array}{r}
25¢ \\
\times\ 12 \\
\hline
50 \\
250 \\
\hline
300¢\ \text{ or }\ \$3.00
\end{array}
$$

5.
$$
\begin{array}{r}
{}^{4}\cancel{5}{}^{1}\cancel{1}6 \\
-\ 1\ 4\ 9 \\
\hline
3\ 6\ 7
\end{array}
$$

6. To find the average number of pages she needs to read each day, divide 235 pages by 5.

7.
$$
5 + \underline{(3 \times 4)} \\
5 + 12 \\
\mathbf{17}
$$

8.
$$
\underline{(5 + 3)} \times 4 \\
8 \times 4 \\
\mathbf{32}
$$

9.
$$
800 - \underline{(450 - 125)} \\
800 - 325 \\
\mathbf{475}
$$

10.
$$
600 \div \underline{(20 \div 5)} \\
600 \div 4 \\
\mathbf{150}
$$

11.
$$
\underline{800 - 450} - 125 \\
350 - 125 \\
\mathbf{225}
$$

12.
$$
\underline{600 \div 20} \div 5 \\
30 \div 5 \\
\mathbf{6}
$$

13.
$$
144 \div \underline{(8 \times 6)} \\
144 \div 48 \\
\mathbf{3}
$$

14.
$$
\underline{144 \div 8} \times 6 \\
18 \times 6 \\
\mathbf{108}
$$

15.
$$
\$5 - \underline{(\$1.25 + \$0.60)} \\
\$5 - \$1.85 \\
\mathbf{\$3.15}
$$

16.
$$
7 \times 9 = 63 \\
9 \times 7 = 63 \\
63 \div 7 = 9 \\
63 \div 9 = 7
$$

17.
$$
\begin{array}{r}
24 \\
12\overline{)288} \\
\underline{24} \\
48 \\
\underline{48} \\
0
\end{array}
$$

18.
$$
\begin{array}{r}
\$0.40 \\
25\overline{)\$10.00} \\
\underline{10\ 0} \\
00 \\
\underline{00} \\
0
\end{array}
$$

19.
$$
\begin{array}{r}
378 \\
\times\ \ 64 \\
\hline
1512 \\
2268 \\
\hline
\mathbf{24{,}192}
\end{array}
$$

20.
$$
\begin{array}{r}
506 \\
\times\ 370 \\
\hline
35420 \\
15180 \\
\hline
\mathbf{187{,}220}
\end{array}
$$

21.
$$
\begin{array}{r}
\$\overset{0}{\cancel{1}}\overset{9}{\cancel{0}}.\overset{10}{\cancel{1}}0 \\
-\ \ \$9.\,8\,9 \\
\hline
\$0.\,2\,1
\end{array}
$$

22.
$$
\begin{array}{r}
63 \\
+\ 36 \\
\hline
99
\end{array}
\qquad
\text{check:}
\qquad
\begin{array}{r}
99 \\
-\ 63 \\
\hline
36
\end{array}
$$

$$n = \mathbf{99}$$

23.
$$
\begin{array}{r}
{}^{5}\cancel{6}{}^{1}3 \\
-\ 3\ 6 \\
\hline
2\ 7
\end{array}
\qquad
\text{check:}
\qquad
\begin{array}{r}
{}^{5}\cancel{6}{}^{1}3 \\
-\ 2\ 7 \\
\hline
3\ 6
\end{array}
$$

$$p = \mathbf{27}$$

24.
$$
\begin{array}{r}
\overset{3}{\cancel{4}}\,\overset{12}{\cancel{3}}{}^{1}2 \\
-\ \ 5\ 6 \\
\hline
3\ 7\ 6
\end{array}
\qquad
\text{check:}
\qquad
\begin{array}{r}
{}^{1}\ {}^{1} \\
376 \\
+\ \ 56 \\
\hline
432
\end{array}
$$

$$m = \mathbf{376}$$

25.
$$8\overline{)480}$$
$$\underline{48}$$
$$00$$
$$\underline{00}$$
$$0$$

check:
$$60$$
$$\underline{\times\ 8}$$
$$480$$

$$w\ =\ 60$$

26. $\underline{5\ +\ 12\ +\ 27}\ +\ y\ =\ 50$
$$44\ +\ y\ =\ 50$$
$$50\ -\ 44\ =\ 6$$
$$y\ =\ 6$$

check:
$$_2 5$$
$$12$$
$$27$$
$$\underline{+\ \ 6}$$
$$50$$

27. $4\overline{)36}$ (= 9)

check: $9\overline{)36}$ (= 4)
$$\underline{36}$$
$$0$$

$$\underline{36}$$
$$0$$

$$a\ =\ 9$$

28.
$$8$$
$$\underline{\times\ 4}$$
$$32$$

check: $4\overline{)32}$ (= 8)
$$\underline{32}$$
$$0$$

$$x\ =\ 32$$

29. $7\ +\ 11\ =\ 18$
$$11\ +\ 7\ =\ 18$$
$$18\ -\ 11\ =\ 7$$
$$18\ -\ 7\ =\ 11$$

30. $\underline{3\ \cdot\ 4}\ \cdot\ 5$
$$12\ \cdot\ 5$$
$$60$$

LESSON 6, WARM-UP

a. 2900

b. 8400

c. 770

d. 9740

e. 1560

f. 2980

Problem Solving

$$3Q\ =\ 75¢$$
$$1D\ =\ 10¢$$
$$\underline{+\ \ 3N\ =\ 15¢}$$
$$100¢\ \text{or}\ \$1.00$$

$$2Q\ =\ 50¢$$
$$\underline{+\ \ 5D\ =\ 50¢}$$
$$100¢\ \text{or}\ \$1.00$$

$$1HD\ =\ 50¢$$
$$1Q\ =\ 25¢$$
$$\underline{+\ \ 5N\ =\ 25¢}$$
$$100¢\ \text{or}\ \$1.00$$

$$1HD\ =\ 50¢$$
$$4D\ =\ 40¢$$
$$\underline{+\ \ 2N\ =\ 10¢}$$
$$100¢\ \text{or}\ \$1.00$$

4 combinations are possible: 3Q, 1D, 3N; 2Q, 5D; 1HD, 1Q, 5N; 1HD, 4D, 2N.

LESSON 6, LESSON PRACTICE

a. Three fourths; $\frac{3}{4}$

b. Two fifths; $\frac{2}{5}$

c. Three eighths; $\frac{3}{8}$

d. $2\overline{)72}$ (= 36) $\frac{1}{2}$ of 72 is **36.**
$$\underline{6}$$
$$12$$
$$\underline{12}$$
$$0$$

e. $2\overline{)1000}$ (= 500) $\frac{1}{2}$ of 1000 is **500.**
$$\underline{10}$$
$$00$$
$$\underline{00}$$
$$00$$
$$\underline{00}$$
$$0$$

f.
$$3\overline{)180}$$ 60
$$\frac{18}{00}$$
$$\frac{00}{0}$$
$\frac{1}{3}$ of 180 is **60.**

g.
$$3\overline{)\$3.60}$$ $1.20
$$\frac{3}{06}$$
$$\frac{06}{00}$$
$$\frac{00}{0}$$
$\frac{1}{3}$ of $3.60 is **$1.20.**

h.

LESSON 6, MIXED PRACTICE

1.
$$2\overline{)540}$$ 270
$$\frac{4}{14}$$
$$\frac{14}{00}$$
$$\frac{00}{0}$$
$\frac{1}{2}$ of 540 is **270.**

2.
$$3\overline{)540}$$ 180
$$\frac{3}{24}$$
$$\frac{24}{00}$$
$$\frac{00}{0}$$
$\frac{1}{3}$ of 540 is **180.**

3.
$$\begin{array}{r} 346 \text{ miles} \\ 417 \text{ miles} \\ 289 \text{ miles} \\ + \ 360 \text{ miles} \\ \hline \mathbf{1412 \text{ miles}} \end{array}$$

4.
$$\begin{array}{r} \$20.00 \\ - \ \$12.08 \\ \hline \mathbf{\$7.92} \end{array}$$

5.
$$\begin{array}{r} 52 \\ \times \ \ 7 \\ \hline \mathbf{364 \text{ days}} \end{array}$$

6.
$$\$20\overline{)\$1000}$$ 50 bills
$$\frac{100}{00}$$
$$\frac{00}{0}$$

7. Five sixths; $\frac{5}{6}$

8.
$$\begin{array}{r} 3604 \\ 5186 \\ + \ 7145 \\ \hline \mathbf{15{,}935} \end{array}$$

9.
$$\begin{array}{r} \$30.01 \\ - \ \$15.76 \\ \hline \mathbf{\$14.25} \end{array}$$

10.
$$\begin{array}{r} 376 \\ \times \ \ 87 \\ \hline 2632 \\ 30080 \\ \hline \mathbf{32{,}712} \end{array}$$

11.
$$\begin{array}{r} 470 \\ \times \ 203 \\ \hline 1410 \\ 0000 \\ 94000 \\ \hline \mathbf{95{,}410} \end{array}$$

12.
$$\begin{array}{r} \$20.00 \\ - \ \$11.98 \\ \hline \mathbf{\$8.02} \end{array}$$

13. $596 - (400 - 129)$
$596 - 271$
325

14. $32 \div (8 \times 4)$
$32 \div 32$
1

15.
$$
\begin{array}{r}
502 \\
8\overline{)4016} \\
\underline{40} \\
01 \\
\underline{00} \\
16 \\
\underline{16} \\
0
\end{array}
$$

16.
$$
\begin{array}{r}
400\ R\ 9 \\
15\overline{)6009} \\
\underline{60} \\
00 \\
\underline{00} \\
09 \\
\underline{00} \\
9
\end{array}
$$

17.
$$
\begin{array}{r}
250 \\
36\overline{)9000} \\
\underline{72} \\
180 \\
\underline{180} \\
00 \\
\underline{00} \\
0
\end{array}
$$

18.
$$
\begin{array}{r}
60 \\
8\overline{)480} \\
\underline{48} \\
00 \\
\underline{00} \\
0
\end{array}
\qquad
\text{check:}
\qquad
\begin{array}{r}
60 \\
\times\ 8 \\
\hline
480
\end{array}
$$

$w = 60$

19.
$$
\begin{array}{r}
64 \\
+\ 46 \\
\hline
110
\end{array}
\qquad
\text{check:}
\qquad
\begin{array}{r}
^0\cancel{1}\ ^{1}0{}^{1}0 \\
-\ 6\ 4 \\
\hline
4\ 6
\end{array}
$$

$x = 110$

20.
$$
\begin{array}{r}
7 \\
7\overline{)49} \\
\underline{49} \\
0
\end{array}
\qquad
\text{check:}
\qquad
\begin{array}{r}
7 \\
7\overline{)49} \\
\underline{49} \\
0
\end{array}
$$

$N = 7$

21.
$$
\begin{array}{r}
^3 \\
15 \\
\times\ 7 \\
\hline
105
\end{array}
\qquad
\text{check:}
\qquad
\begin{array}{r}
15 \\
7\overline{)105} \\
\underline{7} \\
35 \\
\underline{35} \\
0
\end{array}
$$

$M = 105$

22.
$$
\begin{array}{r}
^5\cancel{6}\ ^{1}\cancel{5}\ {}^{1}3 \\
-\ 3\ 6\ 5 \\
\hline
2\ 8\ 8
\end{array}
\qquad
\text{check:}
\qquad
\begin{array}{r}
^{1\ 1} \\
365 \\
+\ 288 \\
\hline
653
\end{array}
$$

$P = 288$

23. $\underbrace{36¢\ +\ 25¢}\ +\ m\ =\ 99¢$

$61¢\ +\ m\ \ =\ 99¢$

$99¢\ -\ 61¢\ =\ 38¢$

$m\ =\ 38¢$

check:
$$
\begin{array}{r}
^1 \\
36¢ \\
25¢ \\
+\ 38¢ \\
\hline
99¢
\end{array}
$$

24. $\dfrac{1}{4}$

25.

26.
$$
\begin{array}{r}
\$6.35 \\
\times\ \ \ 12 \\
\hline
1270 \\
6350 \\
\hline
\$76.20
\end{array}
$$

27. $2 + 4 = 6$

$4 + 2 = 6$

$6 - 4 = 2$

$6 - 2 = 4$

28. $2 \times 4 = 8$

$4 \times 2 = 8$

$8 \div 2 = 4$

$8 \div 4 = 2$

29.
$$
\begin{array}{r}
38 \\
\times\ 10 \\
\hline
380
\end{array}
$$

30. **Sample answer: How much money is $\frac{1}{2}$ of \$3.60?**

$$
\begin{array}{r}
\$1.80 \\
2\overline{)\$3.60} \\
\underline{2} \\
1\ 6 \\
\underline{1\ 6} \\
00 \\
\underline{00} \\
0
\end{array}
$$
$\frac{1}{2}$ of \$3.60 is **\$1.80.**

LESSON 7, WARM-UP

a. 500

b. 1000

c. 350

d. 2200

e. 400

f. 500

Problem Solving

$5 \times 5 = $ **25 coins**

$6 \times 6 = $ **36 coins**

LESSON 7, LESSON PRACTICE

a. $1\frac{3}{4}$ in.

b. 25 mm

c. 2 in.; 5 cm

d. Ray

e. Line

f. Segment

LESSON 7, MIXED PRACTICE

1.
$$
\begin{array}{r}
\$0.25 \\
\times \quad 100 \\
\hline
\$25.00
\end{array}
$$

2.
$$
\begin{array}{r}
{}^{2}\cancel{3}{}^{1}6\,5 \text{ days} \\
- \quad 9\,1 \text{ days} \\
\hline
2\,7\,4 \text{ days}
\end{array}
$$

3.
$$
\begin{array}{r}
{}^{1}596 \\
+ \quad 612 \\
\hline
1208 \text{ miles}
\end{array}
\qquad
\begin{array}{r}
1\,8\,\cancel{9}{}^{8}{}^{1}0 \\
- \quad 1\,2\,0\,8 \\
\hline
6\,8\,2 \text{ miles}
\end{array}
$$

4.
$$
\begin{array}{r}
117 \\
2\overline{)234} \\
\underline{2} \\
03 \\
\underline{2} \\
14 \\
\underline{14} \\
0
\end{array}
$$

5.
$$
\begin{array}{r}
\$0.78 \\
3\overline{)\$2.34} \\
\underline{21} \\
24 \\
\underline{24} \\
0
\end{array}
$$

6. Three eighths; $\frac{3}{8}$

7.
$$
\begin{array}{r}
{}^{2\,1\,1} \\
3654 \\
2893 \\
+ \quad 5614 \\
\hline
12{,}161
\end{array}
$$

8.
$$
\begin{array}{r}
{}^{3}\cancel{4}\,{}^{1}\cancel{1}.\,{}^{0}\cancel{0}{}^{1}1 \\
- \quad \$1\,5.\,7\,6 \\
\hline
\$2\,5.\,2\,5
\end{array}
$$

9.
$$
\begin{array}{r}
28¢ \\
\times \quad 74 \\
\hline
112 \\
1960 \\
\hline
2072¢ \ \text{ or } \ \$20.72
\end{array}
$$

10.
$$
\begin{array}{r}
906 \\
\times \quad 47 \\
\hline
6342 \\
36240 \\
\hline
42{,}582
\end{array}
$$

11.
$$
\begin{array}{r}
833 \text{ R } 2 \\
6\overline{)5000} \\
\underline{48} \\
20 \\
\underline{18} \\
20 \\
\underline{18} \\
2
\end{array}
$$

12.
$$
\begin{array}{r}
50 \\
16\overline{)800} \\
\underline{80} \\
00 \\
\underline{00} \\
0
\end{array}
$$

13.
$$\begin{array}{r} 52 \text{ R } 54 \\ 60\overline{)3174} \\ \underline{300} \\ 174 \\ \underline{120} \\ 54 \end{array}$$

14. $\underline{3 + 6 + 5 + 4} + w = 30$

$$18 + w = 30$$
$$30 - 18 = 12$$
$$w = 12$$

15. $\underline{300 - 30} + 3$

$$270 + 3$$
$$\mathbf{273}$$

16. $300 - \underline{(30 + 3)}$

$$300 - 33$$
$$\mathbf{267}$$

17.
$$\begin{array}{r} \$4.32 \\ \times \quad 20 \\ \hline \mathbf{\$86.40} \end{array}$$

18.
$$\begin{array}{r} 48¢ \\ \times \; 24 \\ \hline 192 \\ 960 \\ \hline \mathbf{1152¢} \text{ or } \mathbf{\$11.52} \end{array}$$

19.
$$\begin{array}{r} \$0.35 \\ 25\overline{)\$8.75} \\ \underline{7\,5} \\ 1\,25 \\ \underline{1\,25} \\ 0 \end{array}$$

20.
$$\begin{array}{r} 7 \\ \times\; 6 \\ \hline 42 \end{array} \qquad \text{check: } \begin{array}{r} 7 \\ 6\overline{)42} \\ \underline{42} \\ 0 \end{array}$$

$$W = 42$$

21. $\begin{array}{r} 16 \\ 6\overline{)96} \\ \underline{6} \\ 36 \\ \underline{36} \\ 0 \end{array}$ check: $\begin{array}{r} \;\;3 \\ 16 \\ \times\; 6 \\ \hline 96 \end{array}$

$$n = 16$$

22.
$$\begin{array}{r} 2\overset{1}{\cancel{2}}\overset{0}{\cancel{1}}{}^{1}3 \\ -\quad 5\,8 \\ \hline 1\,5\,5 \end{array} \qquad \text{check: } \begin{array}{r} \overset{1\,1}{155} \\ +\quad 58 \\ \hline 213 \end{array}$$

$$r = 155$$

23.
$$60 - 36 = 24$$
$$36 + 24 = 60$$
$$24 + 36 = 60$$

24. $1\frac{1}{2}$ in.

25. 3 cm; 30 mm

26.
$$9 \times 10 = 90$$
$$10 \times 9 = 90$$
$$90 \div 9 = 10$$
$$90 \div 10 = 9$$

27. **To find a missing dividend, multiply the quotient by the divisor.**

28.
$$\begin{array}{r} \overset{1}{12} \\ +\;\; 8 \\ \hline 20 \end{array} \qquad \text{check: } \begin{array}{r} \overset{1}{2}0 \\ -\; 1\,2 \\ \hline 8 \end{array}$$

$$w = 20$$

29.
$$\begin{array}{r} \overset{0}{\cancel{1}}{}^{1}2 \\ -\;\; 8 \\ \hline 4 \end{array} \qquad \text{check: } \begin{array}{r} \overset{0}{\cancel{1}}{}^{1}2 \\ -\;\; 4 \\ \hline 8 \end{array}$$

$$x = 4$$

30. **1000 millimeters**

LESSON 8, WARM-UP

a. **2800**

b. **786**

c. **8920**

d. **920**

e. **2400**

f. **360**

Problem Solving
Answers will vary.

LESSON 8, ACTIVITY

a. Probably not. Different-size steps will result in different counts.

b. No; the perimeter does not change. A uniform unit of measure is needed for consistent measurements.

c. 2. The molding along the base of the wall.

LESSON 8, LESSON PRACTICE

a. 12 mm + 12 mm + 12 mm + 12 mm
= **48 mm**

b. 15 mm + 20 mm + 15 mm + 20 mm
= **70 mm**

c. 1 cm + 1 cm + 1 cm + 1 cm + 1 cm
= **5 cm**

d. 2 cm + 2 cm + 2 cm = **6 cm**

e. 10 mm + 15 mm + 10 mm + 20 mm
= **55 mm**

f.
$$\begin{array}{r} \textbf{15 cm} \\ 4\overline{)60\ cm} \\ \underline{4} \\ 20 \\ \underline{20} \\ 0 \end{array}$$

LESSON 8, MIXED PRACTICE

1.
$$\begin{array}{r} 25 \\ \times\ 18 \\ \hline 200 \\ 250 \\ \hline \textbf{450 chairs} \end{array}$$

2.
$$\begin{array}{r} {}^{0}\cancel{1}\ {}^{16}\cancel{7}\ {}^{14}\cancel{5}{}^{1}0 \\ -\quad 7\ 6\ 5 \\ \hline \textbf{9 8 5 fewer horses} \end{array}$$

3.
$$\begin{array}{r} \textbf{28 suffragettes} \\ 5\overline{)140} \\ \underline{10} \\ 40 \\ \underline{40} \\ 0 \end{array}$$

4. 20 mm + 15 mm + 25 mm = **60 mm**

5.
$$\begin{array}{r} \textbf{\$3.27} \\ 2\overline{)\$6.54} \\ \underline{6} \\ 0\ 5 \\ \underline{4} \\ 14 \\ \underline{14} \\ 0 \end{array}$$

6.
$$\begin{array}{r} \textbf{218} \\ 3\overline{)654} \\ \underline{6} \\ 05 \\ \underline{3} \\ 24 \\ \underline{24} \\ 0 \end{array}$$

7. $\dfrac{\textbf{3}}{\textbf{10}}$

8.
$$\begin{array}{r} \textbf{\$2.25} \\ 4\overline{)\$9.00} \\ \underline{8} \\ 1\ 0 \\ \underline{8} \\ 20 \\ \underline{20} \\ 0 \end{array}$$

9.
$$\begin{array}{r} \textbf{37 R 3} \\ 10\overline{)373} \\ \underline{30} \\ 73 \\ \underline{70} \\ 3 \end{array}$$

10.
$$\begin{array}{r} \textbf{125} \\ 12\overline{)1500} \\ \underline{12} \\ 30 \\ \underline{24} \\ 60 \\ \underline{60} \\ 0 \end{array}$$

11.
$$\begin{array}{r} \textbf{20 R 20} \\ 39\overline{)800} \\ 78 \\ \hline 20 \\ 00 \\ \hline 20 \end{array}$$

12. $\underbrace{400 \div 20}_{} \div 4$
$\qquad 20 \div 4$
$\qquad \textbf{5}$

13. $400 \div \underbrace{(20 \div 4)}_{}$
$\qquad 400 \div 5$
$\qquad \textbf{80}$

14. $20 \times 12 = 240$
$12 \times 20 = 240$
$240 \div 20 = 12$
$240 \div 12 = 20$

15. $80 \times 60 = 140$
$140 - 80 = 60$
$140 - 60 = 80$

16. $12 \text{ in.} + 12 \text{ in.} + 12 \text{ in.} + 12 \text{ in.} = \textbf{48 in.}$

17. (a) $\begin{array}{r} 6 \\ + 4 \\ \hline 10 \end{array}$ (b) $\begin{array}{r} 6 \\ \times 4 \\ \hline 24 \end{array}$

18.
$$\begin{array}{r} \overset{4}{\$}\overset{9}{5}.\,\overset{}{\cancel{0}}{}^1 0 \\ - \$1.\,4\,8 \\ \hline \$3.\,5\,2 \end{array}$$
$M = \textbf{\$3.52}$

19. $\underbrace{10 \times 20}_{} \times 30$
$\qquad 200 \times 30$
$\qquad \textbf{6000}$

20.
$$\begin{array}{r} \textbf{103 R 1} \\ 8\overline{)825} \\ 8 \\ \hline 02 \\ 0 \\ \hline 25 \\ 24 \\ \hline 1 \end{array}$$

21. $\begin{array}{r} 63 \\ + 36 \\ \hline 99 \end{array}$ check: $\begin{array}{r} 99 \\ - 63 \\ \hline 36 \end{array}$
$w = \textbf{99}$

22. $\underbrace{150 + 165}_{} + a = 397$
$\qquad 315 + a = 397$
$\quad 397 - 315 = 82$
$\qquad a = \textbf{82}$
check: $\begin{array}{r} 150 \\ 165 \\ + 82 \\ \hline 397 \end{array}$

23.
$$\begin{array}{r} \overset{10}{} \\ 12\overline{)120} \\ 12 \\ \hline 00 \\ 00 \\ \hline 0 \end{array}$$ check: $\begin{array}{r} 12 \\ \times 10 \\ \hline 120 \end{array}$
$w = \textbf{10}$

24. $\begin{array}{r} 24 \\ \times 8 \\ \hline \textbf{192} \end{array}$

25. (a) **About 3 centimeters**
(b) **28 millimeters**

26. _____

27. $\begin{array}{r} \overset{1}{2}7 \\ + 18 \\ \hline 45 \end{array}$
$w = \textbf{45}$

28. $\begin{array}{r} \overset{1}{\cancel{2}}7 \\ - 18 \\ \hline 9 \end{array}$
$x = \textbf{9}$

29. $\begin{array}{r} \overset{2}{3}5 \\ \times 4 \\ \hline 140 \end{array}$

30. **One way to calculate the perimeter of a rectangle is to add the lengths of the four sides.**

LESSON 9, WARM-UP

a. 168

b. 89

c. 7720

d. 810

e. 360

f. 165

Problem Solving
246, 264, 426, 462, 624, 642

LESSON 9, LESSON PRACTICE

a. 12¢, $1.20, $12

b. $\dfrac{16 - 8 - 2}{6}$ $\bigcirc\!\!<$ $\dfrac{16 - (8 - 2)}{10}$

c. $\dfrac{8 \div 4 \times 2}{4}$ $\bigcirc\!\!>$ $\dfrac{8 \div (4 \times 2)}{1}$

d. $\dfrac{2 \times 3}{6}$ $\bigcirc\!\!>$ $\dfrac{2 + 3}{5}$

e. $\dfrac{1 \times 1 \times 1}{1}$ $\bigcirc\!\!<$ $\dfrac{1 + 1 + 1}{3}$

f. $\dfrac{1}{2} > \dfrac{1}{4}$

LESSON 9, MIXED PRACTICE

1.
$$\begin{array}{r} 18 \text{ books} \\ 8\overline{)144} \\ \underline{8} \\ 64 \\ \underline{64} \\ 0 \end{array}$$

2.
$$\begin{array}{r} {}^{5} \\ 1\,\cancel{6}^{1}03 \\ -\ 1492 \\ \hline 1\,1\,1 \text{ years} \end{array}$$

3.
$$\begin{array}{r} 4 \text{ R1} \\ 2\overline{)9} \\ \underline{8} \\ 1 \end{array}$$
5 trips

4. length = 2 cm
width = 1 cm
perimeter = 2 cm + 2 cm + 1 cm + 1 cm
= 6 cm

5.
$$\begin{array}{r} \$2.90 \\ 2\overline{)\$5.80} \\ \underline{4} \\ 1\,8 \\ \underline{1\,8} \\ 00 \\ \underline{00} \\ 0 \end{array}$$

6.
$$\begin{array}{r} \$0.25 \longrightarrow 25¢ \\ 4\overline{)\$1.00} \\ \underline{0} \\ 1\,0 \\ \underline{8} \\ 20 \\ \underline{20} \\ 0 \end{array}$$

7. One fourth; $\dfrac{1}{4}$

8. 5012 $\bigcirc\!\!<$ 5120

9. $0, \dfrac{1}{2}, 1$

10. $\dfrac{100 - 50 - 25}{25}$ $\bigcirc\!\!<$ $\dfrac{100 - (50 - 25)}{75}$

11.
$$\begin{array}{r} {}^{2\,1} \\ {}_{1}478 \\ 3692 \\ +\ \ 45 \\ \hline 4215 \end{array}$$

12.
$$\begin{array}{r} {}^{4\ 9\ 9} \\ \$\cancel{5}\,\cancel{0}.\,\cancel{0}^{1}0 \\ -\ \$3\,1.\,7\,6 \\ \hline \$1\,8.\,2\,4 \end{array}$$

13.
$$\begin{array}{r} \$4.20 \\ \times\ \ \ \ 60 \\ \hline \$252.00 \end{array}$$

14.
$$\begin{array}{r} 78 \\ \times\ 36 \\ \hline 468 \\ 2340 \\ \hline \mathbf{2808} \end{array}$$

15.
$$\begin{array}{r} \mathbf{803} \\ 9\overline{)7227} \\ \underline{72} \\ 02 \\ \underline{00} \\ 27 \\ \underline{27} \\ 0 \end{array}$$

16.
$$\begin{array}{r} \mathbf{304} \\ 25\overline{)7600} \\ \underline{75} \\ 10 \\ \underline{00} \\ 100 \\ \underline{100} \\ 0 \end{array}$$

17.
$$\begin{array}{r} \mathbf{400\ R\ 14} \\ 20\overline{)8014} \\ \underline{80} \\ 01 \\ \underline{00} \\ 14 \\ \underline{00} \\ 14 \end{array}$$

18.
$$\begin{array}{r} \mathbf{71\ R\ 36} \\ 100\overline{)7136} \\ \underline{700} \\ 136 \\ \underline{100} \\ 36 \end{array}$$

19.
$$\begin{array}{r} \mathbf{1} \\ 736\overline{)736} \\ \underline{736} \\ 0 \end{array}$$

20.
$$\begin{array}{r} {}^{2}\ {}^{9} \\ \cancel{3}\ \cancel{\emptyset}{}^{1}0 \\ -\ 1\ 6\ 5 \\ \hline 1\ 3\ 5 \end{array}$$
check:
$$\begin{array}{r} 165 \\ +\ 135 \\ \hline 300 \end{array}$$
$a\ =\ 135$

21.
$$\begin{array}{r} {}^{1} \\ 68 \\ +\ 86 \\ \hline 154 \end{array}$$
check:
$$\begin{array}{r} {}^{0}\ {}^{1}4 \\ \cancel{1}\ \cancel{5}{}^{1}4 \\ -\ \ 6\ 8 \\ \hline 8\ 6 \end{array}$$
$b\ =\ 154$

22.
$$\begin{array}{r} 16 \\ 9\overline{)144} \\ \underline{9} \\ 54 \\ \underline{54} \\ 0 \end{array}$$
check:
$$\begin{array}{r} {}^{5} \\ 16 \\ \times\ 9 \\ \hline 144 \end{array}$$
$c\ =\ 16$

23.
$$\begin{array}{r} {}^{3} \\ 15 \\ \times\ 7 \\ \hline 105 \end{array}$$
check:
$$\begin{array}{r} 7 \\ 15\overline{)105} \\ \underline{105} \\ 0 \end{array}$$
$d\ =\ 105$

24. _____

5 cm

25. C. $\longrightarrow$

26. $\dfrac{1}{2} > \dfrac{1}{3}$

27. $9 \times 11 = 99$
$11 \times 9 = 99$
$99 \div 11 = 9$
$99 \div 9 = 11$

28. $\dfrac{25 + 0}{25}$ $\textcircled{>}$ $\dfrac{25 \times 0}{0}$

29. $100 = \underbrace{20 + 30 + 40} + x$
$100 = 90 + x$
$100 - 90 = x$
$x = 10$

30. **Since 5012 is less than 5120, point the small end of the symbol to the smaller number, 5012.**

_____

LESSON 10, WARM-UP

a. 68

b. 870

c. 279

d. 50

e. 250

f. 3200

Problem Solving

There are two possible solutions. Begin with the ones column. Fill the column with digits from the set 5, 6, 7, and 8 that when added to 9 produce numbers that end with another digit from the set. Use all the digits 5, 6, 7, and 8 once each.

$$\begin{array}{r} 58 \\ +\ 9 \\ \hline 67 \end{array} \qquad \begin{array}{r} 76 \\ +\ 9 \\ \hline 85 \end{array}$$

LESSON 10, LESSON PRACTICE

a. ..., <u>54</u>, <u>63</u>, <u>72</u>, ... Addition sequence. Add 9 to the value of a term to find the next term.

b. ..., <u>16</u>, <u>32</u>, <u>64</u>, ... Multiplication sequence. Multiply the value of a term by 2 to find the next term.

c. Odd

d. 72°F; 22°C

LESSON 10, MIXED PRACTICE

1. Add 8 to the value of a term to find the next term. ..., <u>40</u>, <u>48</u>, <u>56</u>, ...

2. $$\begin{array}{r} 1776 \\ -\ 1620 \\ \hline \textbf{156}\ \text{years} \end{array}$$

3. The number 1492 is even because the last digit, 2, is even.

4. **154 pounds**

5. $$\begin{array}{r} \textbf{10 mm} \\ 4\overline{)40} \\ \underline{4} \\ 00 \\ \underline{00} \\ 0 \end{array}$$

6. $$\begin{array}{r} \$3.25 \\ 2\overline{)\$6.50} \\ \underline{6} \\ 0\ 5 \\ \underline{0\ 4} \\ 10 \\ \underline{10} \\ 0 \end{array}$$

7. $$\underset{14}{\underline{4 \times 3 + 2}} \ \textcircled{<}\ \underset{20}{\underline{4 \times (3 + 2)}}$$

8. Three fourths; $\frac{3}{4}$

9. (a) $$\begin{array}{r} 100 \\ \times\ \ 100 \\ \hline \textbf{10,000} \end{array}$$

    (b) $$\begin{array}{r} 100 \\ +\ 100 \\ \hline \textbf{200} \end{array}$$

10. $$\begin{array}{r} 365 \\ \times\ \ 100 \\ \hline \textbf{36,500} \end{array}$$

11. $$\begin{array}{r} 146 \\ \times\ \ 240 \\ \hline 5840 \\ 29200 \\ \hline \textbf{35,040} \end{array}$$

12. $$\begin{array}{r} 78¢ \\ \times\ \ 48 \\ \hline 624 \\ 3120 \\ \hline \textbf{3744¢}\ \ \text{or}\ \ \textbf{\$37.44} \end{array}$$

13. $$\begin{array}{r} 907 \\ \times\ \ 36 \\ \hline 5442 \\ 27210 \\ \hline \textbf{32,652} \end{array}$$

14. $$\begin{array}{r} \textbf{426} \\ 10\overline{)4260} \\ \underline{40} \\ 26 \\ \underline{20} \\ 60 \\ \underline{60} \\ 0 \end{array}$$

15.
$$\begin{array}{r} 213 \\ 20\overline{)4260} \\ \underline{40} \\ 26 \\ \underline{20} \\ 60 \\ \underline{60} \\ 0 \end{array}$$

16.
$$\begin{array}{r} 284 \\ 15\overline{)4260} \\ \underline{30} \\ 126 \\ \underline{120} \\ 60 \\ \underline{60} \\ 0 \end{array}$$

17.
$$\begin{array}{r} {}^{1}2\,{}^{1}8,{}^{1}3\,4\,7 \\ -\quad 9,6\,3\,7 \\ \hline 1\,8,7\,1\,0 \end{array}$$

18.
$$\begin{array}{r} {}^{0}\$\!{}^{1}1.\,4\,9 \\ -\quad \$8.\,0\,0 \\ \hline \$3.\,4\,9 \end{array}$$
$w = \$3.49$

19.
$$\begin{array}{r} \$\,{}^{0}1\,{}^{9}\!\emptyset.\,{}^{9}\!\emptyset\,{}^{1}0 \\ -\quad \$0.\,7\,5 \\ \hline \$9.\,2\,5 \end{array}$$

20.
$$\begin{array}{r} \$0.56 \\ \times\quad 60 \\ \hline \$33.60 \end{array}$$

21.
$$\begin{array}{r} \$1.55 \\ 4\overline{)\$6.20} \\ \underline{4} \\ 2\,2 \\ \underline{2\,0} \\ 20 \\ \underline{20} \\ 0 \end{array}$$

22.
$$\underbrace{56 + 28 + 37}_{} + n = 200$$
$$121 + n = 200$$
$$200 - 121 = 79$$
$$n = 79$$
check: $\quad 56 + 28 + 37 + 79 = 200$

23.
$$\begin{array}{r} {}^{1}67 \\ +\ 49 \\ \hline 116 \end{array}$$
check:
$$\begin{array}{r} {}^{0}\!1\,{}^{1}0\!1\,6 \\ -\quad 6\,7 \\ \hline 4\,9 \end{array}$$
$a = 116$

24.
$$\begin{array}{r} {}^{5}\!\emptyset\,{}^{1}7 \\ -\ 4\,9 \\ \hline 1\,8 \end{array}$$
check:
$$\begin{array}{r} {}^{5}\!\emptyset\,{}^{1}7 \\ -\ 1\,8 \\ \hline 4\,9 \end{array}$$
$b = 18$

25.
$$\begin{array}{r} 15 \\ 8\overline{)120} \\ \underline{8} \\ 40 \\ \underline{40} \\ 0 \end{array}$$
check:
$$\begin{array}{r} {}^{4}15 \\ \times\ 8 \\ \hline 120 \end{array}$$
$c = 15$

26.
$$\begin{array}{r} {}^{3}24 \\ \times\ 8 \\ \hline 192 \end{array}$$
check:
$$\begin{array}{r} 24 \\ 8\overline{)192} \\ \underline{16} \\ 32 \\ \underline{32} \\ 0 \end{array}$$
$d = 192$

27. $5\overline{)20}$; $20 \div 5$; $\dfrac{20}{5}$

28.
$$\begin{array}{r} 12 \\ 3\overline{)36} \\ \underline{3} \\ 06 \\ \underline{6} \\ 0 \end{array}$$

29. $346 + 463 = 809$
$463 + 346 = 809$
$809 - 463 = 346$
$809 - 346 = 463$

30. $32°F$

INVESTIGATION 1

1. **No. In the frequency table, the number of tests with 20 correct answers is combined with the number of tests with 18 and 19 correct answers.**

2. **Each interval is 2 or 3 scores wide. One reason he might have arranged the scores in these intervals is to group the scores by A's, B's, C's, and D's.**

3. |||| |||| ||

4.   Answers will vary.

5.   12–13

6.   18–20

7.   14–15

8.   **Frequency Table**

Number Correct	Tally	Frequency
90–99	||||	4
80–89	|||| ||	7
70–79	||||	4
60–69	|||	3
50–59	||	2

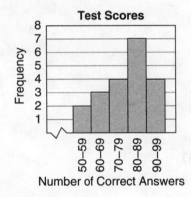

Test Scores

Frequency / Number of Correct Answers

9.   Softball

10. Since girls' preferences are not separated from boys' preferences, the answer **D. cannot be determined from information provided.**

11. **Different sample groups can have different preferences. Therefore, changing the sample group might produce different survey results.**

12. **The percents of the responses might change.**

Extension

   The size of the sample may vary. Students should choose an appropriate method of display (for example, a table or a graph).

24

LESSON 11, WARM-UP

a.   120

b.   1200

c. $5.75

d.   691

e.   4100

f. $3.50

g.   2

Problem Solving

   Begin with the largest stamp denomination and work down.

$$3(37¢) = 111¢$$
$$1(10¢) = 10¢$$
$$2(3¢) = 6¢$$
$$+ \ 1(2¢) = 2¢$$
$$129¢ \text{ or } \$1.29$$

7 stamps total

LESSON 11, LESSON PRACTICE

a.   Step 1:   Subtraction pattern
     Step 2: $B - A = R$
     314 pages − 129 pages = remaining pages
     Step 3:   **314 − 129 = R**

$$\overset{2}{\cancel{3}} \ \overset{1}{\cancel{1}}4 \text{ pages}$$
$$- \ 1 \ 2 \ 9 \text{ pages}$$
$$1 \ 8 \ 5 \text{ pages}$$

     Step 4:   **Tim has 185 pages to read.**

b.   Step 1:   Addition pattern
     Step 2: $S + M = T$
     19 points + M points = 42 points
     Step 3:   **19 + M = 42**

$$\overset{3}{\cancel{4}}{}^{1}2 \text{ points}$$
$$- \ 1 \ 9 \text{ points}$$
$$2 \ 3 \text{ points}$$

     check:
$$\overset{1}{1}9 \text{ points}$$
$$+ \ 23 \text{ points}$$
$$42 \text{ points}$$

     Step 4:   **The team scored 23 points in the second half.**

LESSON 11, MIXED PRACTICE

1. $8 + L = 21$

$$\overset{1}{2}\!{}^{1}1$$
$$-\quad 8$$
$$\overline{1\ 3}\ \textbf{laps}$$

2. (a) $\ \ \ 8$
$$\underline{\times\ 4}$$
$$\textbf{32}$$

(b) $\ \ \ 8$
$$\underline{+\ 4}$$
$$\textbf{12}$$

3. $\underline{(6 \times 4)} \div \underline{(8 - 5)}$
$$\quad 24\ \ \div\ \ 3$$
$$\textbf{8}$$

4. $\$20.00 - M = \$7.75$

$$\$\ \overset{1}{2}\,\overset{9}{\cancel{0}}.\,\overset{9}{\cancel{0}}{}^{1}0$$
$$-\qquad 7.\,7\,5$$
$$\overline{\$\,1\,2.\,2\,5}$$

5. $1 + B = 1000$

$$\overset{0}{\cancel{1}}\,\overset{9}{\cancel{0}}\,\overset{9}{\cancel{0}}{}^{1}0$$
$$-\qquad\quad 1$$
$$\overline{9\ 9\ 9}\ \textbf{meters}$$

6. $\ \ \overset{1}{\ }\$0.65$
$$\underline{+\ \$0.40}$$
$$\textbf{\$1.05}$$

7. $\overset{0}{\cancel{1}}\,\overset{14}{\cancel{5}}{}^{1}5$
$$-\ \ 8\,7$$
$$\overline{\ \ 6\,8}$$
$$w = \textbf{68}$$

check: $\ \ \ \ \overset{1}{\ }87$
$$\underline{+\ 68}$$
$$155$$

8. $\overset{0}{\cancel{1}}\,\overset{9}{\cancel{0}}\,\overset{9}{\cancel{0}}{}^{1}0$
$$-\ \ 3\,8\,6$$
$$\overline{\ \ 6\,1\,4}$$
$$x = \textbf{614}$$

check: $\overset{0}{\cancel{1}}\,\overset{9}{\cancel{0}}\,\overset{9}{\cancel{0}}{}^{1}0$
$$-\ \ 6\,1\,4$$
$$\overline{\ \ 3\,8\,6}$$

9. $\ \ \ 1000$
$$\underline{+\ \ \ 386}$$
$$1386$$
$$y = \textbf{1386}$$

check: $\ \ 1386$
$$\underline{-\ 1000}$$
$$386$$

10. $\underline{42 + 596} + m = 700$
$$638 + m = 700$$
$$700 - 638 = 62$$
$$\boldsymbol{m = 62}$$
check: $\quad 42 + 596 + 62 = 700$

11. $\underline{1000 - (100 - 10)} \ \gtrless\ 1000 - 100 - 10$
$$\qquad 910 \qquad\qquad\qquad\quad 890$$

12. $8\overline{)1000}\ \ \ ^{125}$
$$\ \ \underline{8}$$
$$\ \ 20$$
$$\ \ \underline{16}$$
$$\ \ \ 40$$
$$\ \ \ \underline{40}$$
$$\ \ \ \ 0$$

13. $10\overline{)987}\ \ \ ^{98\ \textbf{R}\ 7}$
$$\ \ \underline{90}$$
$$\ \ 87$$
$$\ \ \underline{80}$$
$$\ \ \ 7$$

14. $\ \ \ \ 35$
$$\underline{\times\ 12}$$
$$\ \ \ 70$$
$$\ 350$$
$$\overline{\textbf{420}}$$

15. $\ \ \ \ 600$
$$\underline{\times\ \ \ 300}$$
$$\textbf{180,000}$$

16. $365\overline{)365}\ \ \ ^{1}$
$$\ \ \underline{365}$$
$$\ \ \ \ 0$$
$$w = \textbf{1}$$

17. Add 4 to the value of a term to find the next term.
2, 6, 10, 14, 18, 22, . . .

18. $\underline{2 \times 3} \times 4 \times 5$
$$\underline{6 \times 4} \times 5$$
$$24 \times 5$$
$$\textbf{120}$$

19. $2\overline{)360}\ \ \ ^{180}$
$$\ \underline{2}$$
$$\ 16$$
$$\ \underline{16}$$
$$\ \ 00$$
$$\ \ \underline{00}$$
$$\ \ \ 0$$

20.
$$\begin{array}{r} 90 \\ 4\overline{)360} \\ \underline{36} \\ 00 \\ \underline{00} \\ 0 \end{array}$$

21.
$$\begin{array}{r} {}^{2\,4}125 \\ \times \quad 8 \\ \hline 1000 \end{array}$$

22. $2\dfrac{1}{4}$ in.

23. $\dfrac{5}{8}$

24. $9\,\text{mm} + 9\,\text{mm} + 9\,\text{mm} + 9\,\text{mm} = \textbf{36 mm}$

25. $1 + 3 + 5 + 7 + 9 = \textbf{25}$

26. $6\overline{)30}$, $30 \div 6$, $\dfrac{30}{6}$

27.
$$\begin{array}{r} {}^{2}\cancel{3}{}^{1}0 \\ -\ 1\,7 \\ \hline 1\,3 \text{ coupes} \end{array} \qquad \dfrac{13}{30}$$

28. 0°C

29.
$6 \times 4 = 24$
$4 \times 6 = 24$
$24 \div 4 = 6$
$24 \div 6 = 4$

30. Answers may vary. Sample answer: Before he went to work, Pham had $24.50. He earned some money putting up a fence at work. Then Pham had $37.00. How much money did Pham earn putting up the fence?

LESSON 12, WARM-UP

a. 240

b. 2400

c. $17.50

d. 475

e. 2500

f. $7.50

g. 0

Problem Solving
$1 + 2 + 3 + 4 + 5 + 6 = \textbf{21 dots}$

LESSON 12, LESSON PRACTICE

a. 3

b. Ten billions

c. Twenty-one million, three hundred fifty thousand, six hundred eight

d. 4,520,000,000

e. $\dfrac{(6 \times 4)}{(6 - 4)} = \dfrac{24}{2} = \textbf{12}$

LESSON 12, MIXED PRACTICE

1. $\underbrace{(1 \times 2 \times 3)}_{6} - \underbrace{(1 + 2 + 3)}_{6}$
$\mathbf{0}$

2. 93,000,000 miles

3. $167 + K = 342$
$$\begin{array}{r} {}^{2}\cancel{3}{}^{13}\cancel{4}{}^{1}2 \\ -\ 1\,6\,7 \\ \hline 1\,7\,5 \text{ pancakes} \end{array}$$

4. $59 + L = 102$
$$\begin{array}{r} {}^{0}\cancel{1}{}^{9}\cancel{0}{}^{1}2 \\ -\ \ 5\,9 \\ \hline 4\,3 \text{ points} \end{array}$$

5. $10\,\text{mm} + 10\,\text{mm} + 18\,\text{mm} + 18\,\text{mm}$
$= \textbf{56 mm}$

6.
$$\begin{array}{r} 10 \\ 6\overline{)60} \\ \underline{6} \\ 00 \\ \underline{00} \\ 0 \end{array}$$
$m = \textbf{10}$

7. (a) $2\overline{)100}$ **50**

       $\underline{10}$

       00

       $\underline{00}$

       0

   (b) $4\overline{)100}$ **25**

       $\underline{8}$

       20

       $\underline{20}$

       0

8. $\dfrac{300 \times 1}{300}$ ⊜ $\dfrac{300 \div 1}{300}$

9. $(3 \times 3) - (3 + 3)$

     $9 \quad - \quad 6$

         3

10. Multiply the value of the previous term by 2 to find the next term.

    2, 4, 8, 16, 32, 64, . . .

11. $\underbrace{1 + 456} + m = 480$

     $457 + m = 480$

     $480 - 457 = 23$

         $m = 23$

12. $\begin{array}{r} \overset{0}{\cancel{1}}{}^{1}0 \ \overset{0}{\cancel{1}}{}^{1}0 \\ - \quad 1\ 0\ 1 \\ \hline 9\ 0\ 9 \end{array}$

    $n = 909$

13. $10\overline{)1234}$ **123 R 4**

     $\underline{10}$

     23

     $\underline{20}$

     34

     $\underline{30}$

     4

14. $12\overline{)1234}$ **102 R 10**

     $\underline{12}$

     03

     $\underline{00}$

     34

     $\underline{24}$

     10

15. $2 + 4 + 6 + 8 + 10 = $ **30**

16. **32 mm**

17. **2**

18. **Millions**

19. **6**

20. $1 \times 10 \times 100 \times 1000 = $ **1,000,000**

21. $\begin{array}{r} {}^{2}\ {}^{1}\ \\ \$3.75 \\ \times \qquad 3 \\ \hline \$11.25 \end{array}$

22. $22\overline{)0}$ 0

     $\underline{0}$

     0

    $y = 0$

23. $\underbrace{100 + 200 + 300 + 400} + w = 2000$

     $1000 + w = 2000$

     $2000 - 1000 = 1000$

         $w = $ **1000**

24. $\begin{array}{r} 24 \\ \times 26 \\ \hline 144 \\ 480 \\ \hline 624 \end{array}$

25. $25\overline{)625}$ 25

     $\underline{50}$

     125

     $\underline{125}$

     0

    $m = 25$

26. $\begin{array}{r} 8 \\ \times 4 \\ \hline 32 \end{array}$

27. $3\overline{)27}, \ 27 \div 3, \ \dfrac{27}{3}$

28. $\begin{array}{r} 10 \\ - 7 \\ \hline 3 \end{array}$ $\dfrac{3}{10}$

29. **4,000,000,000,000**

SOLUTIONS

30. Answers may vary. Sample answer: What is
the difference between the product of 2 and 5
and the sum of 2 and 5?
$(2 \times 5) - (2 + 5)$
  10   −   7
       3

LESSON 13, WARM-UP

a. 1500

b. 15,000

c. $9.25

d. 3830

e. 4000

f. $15.00

Problem Solving

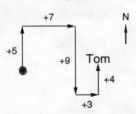

Tom was facing north. He was 10 steps
directly east of the big tree.

LESSON 13, LESSON PRACTICE

a. Step 1: Subtraction pattern
   Step 2: $C - W = D$
   Step 3: $26{,}290 - 18{,}962 = D$

   2 6, 2 9 0
   − 1 8, 9 6 2
   7, 3 2 8 people

   Step 4: **7,328 people**

b. Step 1: Subtraction pattern
   Step 2: $L - E = D$
   Step 3: $1215 - 1066 = D$

   1 2 1 5
   − 1 0 6 6
   1 4 9 years

   Step 4: **149 years**

LESSON 13, MIXED PRACTICE

1. $(8 \times 5) - (8 + 5)$
   40   −   13
        27

2. 250,000 miles

3. Five hundred twenty-one billion

4. 5,200,000

5.   20
   ×   7
   140 merry men

6. $1000 - 487 = R$
   1 0 0 0
   −   4 8 7
   5 1 3 meters

7. $503 - 478 = d$
   5 0 3
   − 4 7 8
   2 5 more girls

8.   99
     100
   + 101
     300

9. $\underline{9 \times 10} \times 11$
     90 × 11
     990

10. 4

11. Billions

12. $18\,\text{mm} + 18\,\text{mm} + 18\,\text{mm} = \mathbf{54\,mm}$

13.   54 R 32
   100)5432
        500
        432
        400
         32

Saxon Math 7/6—Homeschool

14.
$$
\begin{array}{r}
2,000 \\
30\overline{)60,000} \\
\underline{60} \\
0\,0 \\
\underline{0\,0} \\
00 \\
\underline{00} \\
00 \\
\underline{00} \\
0
\end{array}
$$

15.
$$
\begin{array}{r}
142 \ \ R\ 6 \\
7\overline{)1000} \\
\underline{7} \\
30 \\
\underline{28} \\
20 \\
\underline{14} \\
6
\end{array}
$$

16.
$$
\begin{array}{r}
\$1.52 \\
3\overline{)\$4.56} \\
\underline{3} \\
1\,5 \\
\underline{1\,5} \\
06 \\
\underline{6} \\
0
\end{array}
$$

17. $3 + 2 + 1 + 0 \ \bigodot\!> \ 3 \times 2 \times 1 \times 0$

    $\qquad 6 \qquad\qquad\qquad 0$

18. 7

19.
$$
\begin{array}{r}
2640 \\
2\overline{)5280} \\
\underline{4} \\
12 \\
\underline{12} \\
08 \\
\underline{08} \\
00 \\
\underline{00} \\
0
\end{array}
$$

20.
$$
\begin{array}{r}
1 \\
365\overline{)365} \\
\underline{365} \\
0
\end{array}
$$

$w = 1$

21. $\dfrac{(5 + 6 + 7)}{18 \div 3} \div 3$

    $\qquad\qquad 6$

22. $1\dfrac{3}{4}$ in.

23. To find the perimeter of a square, either add the lengths of the four sides or multiply the length of one side by four.

24.
$$
\begin{array}{r}
{}^{1\,3} \\
125 \\
\times \quad 6 \\
\hline
750
\end{array}
$$

25. 212°F

26. $7\overline{)21},\ 21 \div 7,\ \dfrac{21}{7}$

27.
$$
\begin{array}{r}
102 \\
8\overline{)816} \\
\underline{8} \\
01 \\
\underline{0} \\
16 \\
\underline{16} \\
0
\end{array}
$$
check:
$$
\begin{array}{r}
102 \\
\times \quad 8 \\
\hline
816
\end{array}
$$

$a = 102$

28.
$$
\begin{array}{r}
12 \\
\times \quad 4 \\
\hline
48
\end{array}
$$
$b = 48$

check:
$$
\begin{array}{r}
12 \\
4\overline{)48} \\
\underline{4} \\
08 \\
\underline{8} \\
0
\end{array}
$$

29.
$$
\begin{array}{r}
3 \\
4\overline{)12} \\
\underline{12} \\
0
\end{array}
$$
check:
$$
\begin{array}{r}
4 \\
3\overline{)12} \\
\underline{12} \\
0
\end{array}
$$

$c = 3$

30.
$$
\begin{array}{r}
61 \\
+ \ 16 \\
\hline
77
\end{array}
$$
check:
$$
\begin{array}{r}
77 \\
- \ 16 \\
\hline
61
\end{array}
$$

$d = 77$

LESSON 14, WARM-UP

a. 3200

b. 18,000

c. $15.00

d. 590

e. 250

f. $2.50

g. 11

Problem Solving
Andy, Bob, Carol; Andy, Carol, Bob;
Bob, Andy, Carol; Bob, Carol, Andy;
Carol, Andy, Bob; Carol, Bob, Andy

LESSON 14, LESSON PRACTICE

a. -8 $\bigcirc<$ -6

b. Negative eight

c. -3

d. $-3, -1, 0, 2$

e.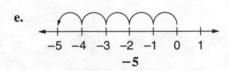
-5

f.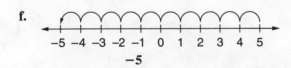
-5

g.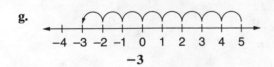
-3

h.
-4

i. True

j. $-12°F$

k. -186 ft

l.
$$\begin{array}{r} \$1\,8.\overset{4}{\cancel{5}}{}^{1}0 \\ -\ \$1\,6.\,2\,5 \\ \hline 2.\,2\,5 \end{array}$$ or $-\$2.25$

LESSON 14, MIXED PRACTICE

1. $\dfrac{(15 + 12)}{(15 - 12)} = \dfrac{27}{3} = \mathbf{9}$

2. Billions

3. 186,000 miles per second

4. -1

5. $-3, -2, 0, 1, 5$

6.
-2

7. $140 - a = 72$
$$\begin{array}{r} \overset{0}{\cancel{1}}\,\overset{13}{\cancel{4}}0 \\ -\ \ 7\,2 \\ \hline 6\,8 \end{array}$$ merry men

8. $1 + 2 + 3 + 4$ $\bigcirc<$ $1 \times 2 \times 3 \times 4$
$\qquad$ **10** $\qquad\qquad\qquad$ **24**

9. $25\,\text{mm} + 15\,\text{mm} + 20\,\text{mm} = \mathbf{60\ mm}$

10. Divide the previous term by 2 to find the next term.
$\mathbf{16, 8, 4, \underline{2}, \underline{1}, \ldots}$

11. $500 - 365 = d$
$$\begin{array}{r} \overset{4}{\cancel{5}}\,\overset{9}{\cancel{0}}{}^{1}0 \\ -\ 3\,6\,5 \\ \hline 1\,3\,5 \end{array}$$
$d = \mathbf{135}$

12.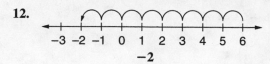

-2

13.

$$\begin{array}{r} 10 \text{ R } 20 \\ 100\overline{)1020} \\ \underline{100} \\ 20 \\ \underline{00} \\ 20 \end{array}$$

14.

$$\begin{array}{r} 3{,}015 \\ 12\overline{)36{,}180} \\ \underline{36} \\ 0\,1 \\ \underline{0\,0} \\ 18 \\ \underline{12} \\ 60 \\ \underline{60} \\ 0 \end{array}$$

15.

$$\begin{array}{r} 31 \text{ R } 6 \\ 18\overline{)564} \\ \underline{54} \\ 24 \\ \underline{18} \\ 6 \end{array}$$

16.

$$\begin{array}{r} {}^{1\,2} \\ 1234 \\ 567 \\ +89 \\ \hline 1890 \end{array}$$

17.

$$\begin{array}{r} 310 \\ +\ 186 \\ \hline 496 \end{array}$$

$n = 496$

18.

$$\begin{array}{r} 11 \\ \times\ 10 \\ \hline 110 \end{array} \qquad \begin{array}{r} 110 \\ \times\ 12 \\ \hline 220 \\ 1100 \\ \hline 1320 \end{array}$$

19.

$$\begin{array}{r} {}^{2}{}^{9} \\ \$\cancel{3}.\,\cancel{0}^{1}5 \\ -\ \$2.\,9\,8 \\ \hline \$0.\,0\,7 \end{array}$$

$m = \$0.07$

20. 4 cm; 40 mm

21.

$$\begin{array}{r} 100 \\ \times\ \ 100 \\ \hline 10{,}000 \end{array} \qquad \begin{array}{r} 10{,}000 \\ \times\ \ \ 100 \\ \hline 1{,}000{,}000 \end{array}$$

22. 5

23. To find the length of the object in millimeters, multiply its length in centimeters by 10.

24.
$19 \times 21 = 399$
$21 \times 19 = 399$
$399 \div 19 = 21$
$399 \div 21 = 19$

25.
$$\underline{12 \div 6} \times 2 \;\bigcirc\!\!>\; 12 \div \underline{(6 \times 2)}$$
$$\underbrace{2 \times 2}_{4} \qquad\qquad \underbrace{12 \div 12}_{1}$$

26. $6\overline{)60}$, $60 \div 6$, $\dfrac{60}{6}$

27. 9,000,000,000,000 nerve cells

28.

$$\begin{array}{r} 4 \text{ eggs} \\ 3\overline{)12} \\ \underline{12} \\ 0 \end{array}$$

29. -10

30. $-1, 0, \dfrac{1}{2}, 1$

LESSON 15, WARM-UP

a. 28,000

b. 2400

c. $25.00

d. 92

e. 6100

f. $17.50

g. 0

Problem Solving

First think, "7 minus what number equals 1?" (6). Then think, "What number minus 9 equals 2?" (11). Next think, "3 (after regrouping the 4) minus what number equals 0?" (3).

$$\begin{array}{r} 417 \\ -\ 396 \\ \hline 21 \end{array}$$

LESSON 15, LESSON PRACTICE

a. Step 1: Equal groups

Step 2: $N \times G = T$

Step 3: $N \times 25¢ = 450¢$

$$\begin{array}{r} 18 \text{ cups} \\ 25¢\overline{)450¢} \\ \underline{25} \\ 200 \\ \underline{200} \\ 0 \end{array}$$

Step 4: **18 cups**

b. Step 1: Equal groups

Step 2: $N \times G = T$

Step 3: $18 \times 12 = T$

$$\begin{array}{r} 18 \\ \times\ 12 \\ \hline 36 \\ 180 \\ \hline 216 \text{ parking spaces} \end{array}$$

Step 4: **216 parking spaces**

LESSON 15, MIXED PRACTICE

1. **Answers may vary. Sample answer: In the auditorium there were 15 rows of chairs with 20 chairs in each row. How many chairs were there in the auditorium?**

2. $212° - 32° = d$

$$\begin{array}{r} \overset{1}{2}^{1}1\ 2 \\ -\ \ \ 3\ 2 \\ \hline 1\ 8\ 0°\text{F} \end{array}$$

3. $16 \cdot 320 = t$

$$\begin{array}{r} 320 \\ \times\ \ 16 \\ \hline 1920 \\ 3200 \\ \hline 5120 \text{ little O's} \end{array}$$

4. $31 - 3 = d$

$$\begin{array}{r} \overset{2}{3}^{1}1 \\ -\ \ \ 3 \\ \hline 2\ 8 \text{ days} \end{array}$$

5. $\begin{array}{c} 3 - 1 \\ 2 \end{array} \gt \begin{array}{c} 1 - 3 \\ -2 \end{array}$

6. $2 - 5 = -3$

Negative three

7.

$$\begin{array}{r} \$ 2\,\overset{7}{8}.^{1}0\ 0 \\ -\ \$ 2\,5.5\ 0 \\ \hline 2.5\ 0 \ \text{ or } \ -\$2.50 \end{array}$$

8. Subtract 2 from a term to get the next term.

6, 4, 2, 0, -2, -4, -6, . . .

9. **$-6°$F; negative six degrees Fahrenheit or six degrees below zero Fahrenheit**

10.

$$\begin{array}{r} \$\ \overset{0}{1}\overset{9}{0}.^{1}0\ 0 \\ -\ \ \$0.1\ 0 \\ \hline \$9.9\ 0 \end{array}$$

11.

$$\begin{array}{r} \$1.75 \\ 2\overline{)\$3.50} \\ \underline{2} \\ 1\ 5 \\ \underline{1\ 4} \\ 10 \\ \underline{10} \\ 0 \end{array}$$

12. **600**

13.

$$\begin{array}{r} \overset{2}{}9 \\ \overset{1}{}87 \\ 654 \\ +\ 3210 \\ \hline 3960 \end{array}$$

14.

$$\begin{array}{r} 574 \\ \times\ \ 76 \\ \hline 3444 \\ 40180 \\ \hline 43,624 \end{array}$$

15.
$$
\begin{array}{r}
480 \\
9\overline{)4320} \\
36 \\
\hline
72 \\
72 \\
\hline
00 \\
00 \\
\hline
0
\end{array}
$$

16.
$$
\begin{array}{r}
13 \text{ R } 25 \\
36\overline{)493} \\
36 \\
\hline
133 \\
108 \\
\hline
25
\end{array}
$$

17.
$$
\begin{array}{r}
4 \\
300\overline{)1200} \\
1200 \\
\hline
0
\end{array}
$$
$w = 4$

check:
$$
\begin{array}{r}
300 \\
4\overline{)1200} \\
12 \\
\hline
00 \\
0 \\
\hline
00 \\
0 \\
\hline
0
\end{array}
$$

18.
$$
\begin{array}{r}
1 \\
63\overline{)63} \\
63 \\
\hline
0
\end{array}
$$
$w = 1$

check:
$$
\begin{array}{r}
63 \\
\times\ 1 \\
\hline
63
\end{array}
$$

19.
$$
\begin{array}{r}
76 \\
\times\ 1 \\
\hline
76
\end{array}
$$
$m = 76$

check:
$$
\begin{array}{r}
1 \\
76\overline{)76} \\
76 \\
\hline
0
\end{array}
$$

20.
$$
\begin{array}{r}
\overset{0\ \ 9\ \ 9}{\$\,\cancel{1}\,\cancel{0}\,\cancel{0}{}^{1}0} \\
-\quad \$\ 6\ 5 \\
\hline
\$\ 9\ 3\ 5
\end{array}
$$
$w = \$935$

check:
$$
\begin{array}{r}
1\ 1 \\
\$935 \\
+\ \$65 \\
\hline
\$1000
\end{array}
$$

21. $\underbrace{3\ +\ 12\ +\ 27}\ +\ n\ =\ 50$
$$42\ +\ n\ =\ 50$$
$$50\ -\ 42\ =\ 8$$
$$n\ =\ 8$$

22. 30 mm

23. $\underbrace{(8\ +\ 9\ +\ 16)}\ \div\ 3$
$$33\ \div\ 3$$
$$11$$

24. Thousands

25. 2

26. $19\ +\ 21\ =\ 40$
$21\ +\ 19\ =\ 40$
$40\ -\ 19\ =\ 21$
$40\ -\ 21\ =\ 19$

27. $-3, -1, 0, 2$

28. $\dfrac{7}{17}$

29.
$$
\begin{array}{r}
\overset{3}{75}¢ \\
\times\ 7 \\
\hline
525¢\ \text{ or }\ \$5.25
\end{array}
$$

30. 0

LESSON 16, WARM-UP

a. 96

b. 92

c. 170

d. 84

e. 750

f. 7500

g. 1

Problem Solving
$$
\begin{array}{r}
2Q\ =\ 50¢ \\
4D\ =\ 40¢ \\
+\ 2N\ =\ 10¢ \\
\hline
100¢
\end{array}
$$
or
$$
\begin{array}{r}
1HD\ =\ 50¢ \\
3D\ =\ 30¢ \\
+\ 4N\ =\ 20¢ \\
\hline
100¢
\end{array}
$$
2Q, 4D, 2N (or 1HD, 3D, 4N)

LESSON 16, LESSON PRACTICE

a. 60

b. 60

c. 50

d. 300

e. 400

f. 400

g. 4000

h. 8000

i. 7000

j.
$$\begin{array}{r} 400 \\ + \ 200 \\ \hline 600 \end{array}$$

k.
$$\begin{array}{r} 700 \\ - \ 600 \\ \hline 100 \end{array}$$

l.
$$\begin{array}{r} 30 \\ \times \ 30 \\ \hline 900 \end{array}$$

m.
$$\begin{array}{r} 20 \\ 30)\overline{600} \\ \underline{60} \\ 00 \\ \underline{00} \\ 0 \end{array}$$

n.
$$\begin{array}{r} 5000 \\ - \ 4000 \\ \hline 1000 \end{array} \text{ fewer people}$$

o.
$$\begin{array}{r} 7000 \\ - \ 5000 \\ \hline 2000 \end{array} \qquad \begin{array}{r} 7000 \\ + \ 2000 \\ \hline 9000 \end{array} \text{ people}$$

LESSON 16, MIXED PRACTICE

1.
$$\underbrace{(20 \times 5)}_{100} - \underbrace{(20 + 5)}_{25}$$
$$75$$

2. $1620 - 1492 = d$

$$\begin{array}{r} 1\ \overset{5}{\cancel{6}}\ \overset{1}{\cancel{2}}{}^{1}0 \\ - \ 1\ 4\ 9\ 2 \\ \hline 1\ 2\ 8 \text{ years} \end{array}$$

3. $5 \cdot g = 140$

$$\begin{array}{r} 28 \text{ merry men} \\ 5)\overline{140} \\ \underline{10} \\ 40 \\ \underline{40} \\ 0 \end{array}$$

4. 3

5. One hundred five million, three hundred ninety-six thousand, six hundred forty-one votes

6.
8

7. 57,000

8. 600

9.
$$\begin{array}{r} 300 \\ \times \ \ \ 400 \\ \hline 120{,}000 \end{array}$$

10.
$$\begin{array}{r} \overset{1}{}\overset{1}{4}5 \\ 5643 \\ + \ \ \ 287 \\ \hline 5975 \end{array}$$

11.
$$\begin{array}{r} \overset{3}{\cancel{4}}\ \overset{9}{\cancel{0}}{,}{}^{1}3\ \overset{0}{\cancel{1}}{}^{1}2 \\ - \ 1\ 4{,}9\ 0\ 8 \\ \hline 2\ 5{,}4\ 0\ 4 \end{array}$$

34

12.
$$\begin{array}{r} 609 \\ 12\overline{)7308} \\ \underline{72} \\ 10 \\ \underline{00} \\ 108 \\ \underline{108} \\ 0 \end{array}$$

13.
$$\begin{array}{r} 53 \ \textbf{R 67} \\ 100\overline{)5367} \\ \underline{500} \\ 367 \\ \underline{300} \\ 67 \end{array}$$

14.
$$\underbrace{(5 + 11)}\ \div\ 2$$
$$16 \div 2$$
$$\textbf{8}$$

15.
$$\begin{array}{r} \$2.50 \\ 2\overline{)\$5.00} \\ \underline{4} \\ 1\ 0 \\ \underline{1\ 0} \\ 00 \\ \underline{00} \\ 0 \end{array}$$

16.
$$\begin{array}{r} \$1.25 \\ 4\overline{)\$5.00} \\ \underline{4} \\ 1\ 0 \\ \underline{8} \\ 20 \\ \underline{20} \\ 0 \end{array}$$

17.
$$\begin{array}{r} \$0.25 \\ \times\ \ \ \ 10 \\ \hline \$2.50 \end{array}$$

18.
$$325\underbrace{(324\ -\ 323)}$$
$$325(1)$$
$$\textbf{325}$$

19.
$$1 + \underbrace{(2 + 3)} \ \overset{=}{} \ \underbrace{(1 + 2)} + 3$$
$$1 + 5 \qquad\qquad 3 + 3$$
$$6 \qquad\qquad\quad 6$$

20. **It felt colder at 3 p.m. The wind chill at 3 p.m.
was −10°F, and the wind chill at 11 p.m. was
−3°F. It felt colder at 3 p.m. because
−10 < −3.**

21. $60 \cdot 72 = t$
$$\begin{array}{r} 72 \\ \times\ \ \ 60 \\ \hline \textbf{4320}\ \text{times} \end{array}$$

22. $90\ \text{ft} + 90\ \text{ft} + 90\ \text{ft} + 90\ \text{ft} = \textbf{360 ft}$

23.
$$\begin{array}{r} 80 \\ -\ 30 \\ \hline \textbf{50}\ \text{more pounds} \end{array}$$

24.
$$\begin{array}{r} 30 \\ 60 \\ +\ 80 \\ \hline \textbf{170}\ \text{pounds} \end{array}$$

25.
$$\begin{array}{r} 60 \\ \times\ \ 7 \\ \hline \textbf{420}\ \text{pounds} \end{array}$$

26. **Answers will vary. Sample answer: How many
more pounds of peanuts does the mother
elephant eat each day than the baby elephant?**

27.
$$\begin{array}{r} 11 \\ 6\overline{)66} \\ \underline{6} \\ 06 \\ \underline{6} \\ 0 \end{array}$$
check:
$$\begin{array}{r} 11 \\ \times\ \ 6 \\ \hline 66 \end{array}$$
$w = \textbf{11}$

28.
$$\begin{array}{r} 60 \\ +\ 37 \\ \hline 97 \end{array}$$
check:
$$\begin{array}{r} 97 \\ -\ 60 \\ \hline 37 \end{array}$$
$m = \textbf{97}$

29.
$$\begin{array}{r} {}^{5}\cancel{6}{}^{1}0 \\ -\ 3\ 7 \\ \hline 2\ 3 \end{array}$$
check:
$$\begin{array}{r} {}^{5}\cancel{6}{}^{1}0 \\ -\ 2\ 3 \\ \hline 3\ 7 \end{array}$$
$n = \textbf{23}$

30.
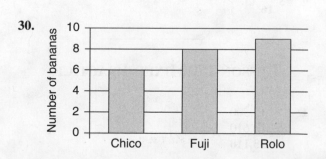

LESSON 17, WARM-UP

a. 170

b. 256

c. 192

d. 73

e. 1900

f. $2.50

g. 5

Problem Solving

Opposite faces of dot cubes always contain a total of **7 dots**.

LESSON 17, LESSON PRACTICE

a. $\frac{9}{16}, \frac{5}{8}, \frac{11}{16}, \frac{3}{4}, \frac{13}{16}, \frac{7}{8}, \frac{15}{16}, 1, 1\frac{1}{16}, 1\frac{1}{8}, 1\frac{3}{16}, 1\frac{1}{4}, 1\frac{5}{16}, 1\frac{3}{8},$ $1\frac{7}{16}, 1\frac{1}{2}$

b. $-2\frac{1}{2}$

c. $3\frac{1}{2}$

d. $1\frac{5}{6}$

e. $\frac{13}{16}$ in.

f. $2\frac{4}{16}$ or $2\frac{1}{4}$ in.

g. $3\frac{3}{16}$ in.

LESSON 17, MIXED PRACTICE

1.
$$\begin{array}{r} \overset{1}{12,500} \\ +\ 10,610 \\ \hline 23,110 \end{array}$$

2. $1969 - 1903 = d$
$$\begin{array}{r} 1969 \\ -\ 1903 \\ \hline 66 \text{ years} \end{array}$$

3. $12 \cdot 6 = t$
$$\begin{array}{r} \overset{1}{12} \\ \times\ \ 6 \\ \hline 72 \text{ yards} \end{array}$$

4. $24 \cdot 1000 = t$
$$\begin{array}{r} 24 \\ \times\ 1000 \\ \hline \$\ 24{,}000 \end{array}$$

5.
$$\begin{array}{r} 5000 \\ +\ 2000 \\ \hline 7000 \end{array}$$

6.
$$\begin{array}{r} 160 \\ 3\overline{)480} \\ \underline{3} \\ 18 \\ \underline{18} \\ 00 \\ \underline{00} \\ 0 \end{array}$$

7. $\dfrac{6 - 6}{3} = \dfrac{0}{3} = 0$

8. $b + a = c$
   $c - a = b$
   $c - b = a$

9. $\dfrac{2}{3}$

10. A square has four sides of equal length. So to find the perimeter, we add (10 cm + 10 cm + 10 cm + 10 cm), or we multiply (4 × 10 cm).

11. $2\frac{10}{16}$ or $2\frac{5}{8}$ in.

12.
$$\begin{array}{r} \$\overset{2}{\cancel{3}}.\overset{9}{\cancel{0}}{}^1 0 \\ -\ \$1.\ 7\ 5 \\ \hline \$1.\ 2\ 5 \end{array} \qquad \text{check:} \qquad \begin{array}{r} \$\overset{2}{\cancel{3}}.\overset{9}{\cancel{0}}{}^1 0 \\ -\ \$1.\ 2\ 5 \\ \hline \$1.\ 7\ 5 \end{array}$$
$y = \$1.25$

13.
$$\begin{array}{r} 20 \\ +\ 30 \\ \hline 50 \end{array} \qquad \text{check:} \qquad \begin{array}{r} 50 \\ -\ 20 \\ \hline 30 \end{array}$$
$m = 50$

14. $12\overline{)0}$ check: 12
    $\dfrac{0}{0}$ $\times\ \ 0$
                      $\overline{\ \ 0}$
  n = 0

15. $16 + 14 = 14 + w$
      $30 = 14 + w$

  $\overset{2}{\cancel{3}}{}^{1}0$ check: $\overset{1}{1}4$
  $\underline{-\ 1\ 4}$ $\underline{+\ 16}$
  $\quad 1\ 6$ $\quad 30$
  w = 16

16. $19 \times 21 \;\text{\textcircled{<}}\; 20 \times 20$
    $399 \qquad\qquad 400$

17. $100 - \underbrace{(50 - 25)}$
    $\quad 100 - 25$
    $\qquad\quad$ **75**

18. $44\overline{)5280}$ with quotient **120**
    $\underline{44}$
    $\ \ 88$
    $\ \ \underline{88}$
    $\ \ \ 00$
    $\ \ \ \ \underline{00}$
    $\ \ \ \ \ 0$

19.
    $\overset{2\ 1}{\underset{1}{}}365$
    $4\ 576$
    $\underline{+\ 50{,}287}$
    $\ \ 55{,}228$

20. Add 5 to a term to find the next term.
  5, 10, 15, 20, 25, . . .

21. **9**

22. $100\overline{)250{,}000}$ quotient **2,500**
    $\underline{200}$
    $\ \ 50\ 0$
    $\ \ \underline{50\ 0}$
    $\ \ \ \ 0\ 0$
    $\ \ \ \ \underline{0\ 0}$
    $\ \ \ \ \ \ 00$
    $\ \ \ \ \ \ \underline{00}$
    $\ \ \ \ \ \ \ \ 00$
    $\ \ \ \ \ \ \ \ \underline{0}$
    $\ \ \ \ \ \ \ \ \ 0$

23. $\$3.75$
    $\underline{\times\ \ \ 10}$
    $\$37.50$

24. $2 \cdot r = 38$
    19 rabbits
    $2\overline{)38}$
    $\underline{2}$
    18
    $\underline{18}$
    $\ 0$

25. $1 + 3 + 5 + 7 + 9 + 11 = \mathbf{36}$

26. One way to find $\frac{1}{4}$ of 52 is to divide 52 by 4.

27. (a) **4 quarters**
      $25¢\overline{)100¢}$
      $\underline{100}$
      $\ \ \ 0$

  (b) **12 quarters**
      $25¢\overline{)300¢}$
      $\underline{25}$
      $\ 5$
      $\underline{50}$
      $\ 0$

28. $\dfrac{3}{8}$ **-inch mark**

29. $4\dfrac{1}{6}$

30. $2\overline{)16}$ **8 sixteenths of an inch**
    $\underline{16}$
    $\ 0$ with quotient **8**

LESSON 18, WARM-UP

a. 92

b. 128

c. 126

d. 72

e. 1

Problem Solving

LESSON 18, LESSON PRACTICE

a.
$$\begin{array}{r} \overset{1}{26}\text{ books} \\ 36\text{ books} \\ +\ 43\text{ books} \\ \hline 105\text{ books} \end{array}$$

$$\begin{array}{r} \mathbf{35}\text{ books} \\ 3\overline{)105} \\ \underline{9} \\ 15 \end{array}$$

b. $96 + 44 + 68 + 100 = 308$
    $308 \div 4 = \mathbf{77}$

c. $\dfrac{28 + 82}{2} = \dfrac{110}{2} = \mathbf{55}$

d. $\dfrac{86 + 102}{2} = \dfrac{188}{2} = \mathbf{94}$

e. $3 + 6 + 9 + 12 + 15 = 45$
    $45 \div 5 = \mathbf{9}$

f. **12 in.**

g. **Between her thirteenth and her fourteenth birthdays**

h. **No**

LESSON 18, MIXED PRACTICE

1.
$$\begin{array}{r} \overset{1\ 1}{2068} \\ +\ 3940 \\ \hline \mathbf{6008}\text{ peanuts} \end{array}$$
Addition pattern

2. $(11 + 12) + x = 32$
    $23 + x = 32$
    $32 - 23 = 9$
    $x = \mathbf{9}$ **teeth**
    Addition pattern

3.
$$\begin{array}{r} 53¢ \\ \times\ 12 \\ \hline 1\ 06 \\ 5\ 30 \\ \hline \mathbf{\$6.36} \end{array}$$
Multiplication pattern

4.
$$\begin{array}{r} 5000 \\ -\ 2000 \\ \hline \mathbf{3000} \end{array}$$

5. $9 + 7 + 8 = 24$
    $24 \div 3 = \mathbf{8}$

6. $\dfrac{59 + 81}{2} = \dfrac{140}{2} = \mathbf{70}$

7.

$\mathbf{-4}$

8.
$$\begin{array}{r} \$0.35 \\ \times\ \ \ \ \ 100 \\ \hline \mathbf{\$35.00} \end{array}$$

9.
$$\begin{array}{r} \mathbf{1,001} \\ 10\overline{)10,010} \\ \underline{10} \\ 0\ 0 \\ \underline{0\ 0} \\ 01 \\ \underline{00} \\ 10 \\ \underline{10} \\ 0 \end{array}$$

10.
$$\begin{array}{r} \mathbf{2010\ \ R\ 10} \\ 17\overline{)34180} \\ \underline{34} \\ 01 \\ \underline{00} \\ 18 \\ \underline{17} \\ 10 \\ \underline{00} \\ 10 \end{array}$$

11.
$$\begin{array}{r} \overset{1\ 1}{\$3.64} \\ \$94.28 \\ +\ \ \$0.87 \\ \hline \mathbf{\$98.79} \end{array}$$

12.
$$\begin{array}{r} \overset{3}{\cancel{4}}\,\overset{10}{\cancel{1}},\overset{12}{\cancel{3}}\,\overset{16}{\cancel{7}}\,5 \\ -\ 1\,3,5\,7\,6 \\ \hline \mathbf{2\,7,7\,9\,9} \end{array}$$

13.
$$\begin{array}{r} 125 \\ \times\quad 16 \\ \hline 750 \\ 1250 \\ \hline \mathbf{2000} \end{array}$$

14. $\underbrace{4 \cdot 3} \cdot 2 \cdot 1 \cdot 0$

$\quad\quad \underbrace{12 \cdot 2} \cdot 1 \cdot 0$

$\quad\quad\quad \underbrace{24 \cdot 1} \cdot 0$

$\quad\quad\quad\quad 24 \cdot 0$

$\quad\quad\quad\quad\quad \mathbf{0}$

15.
$$\begin{array}{r} \overset{1}{8}4 \\ +\ 48 \\ \hline 132 \end{array} \quad \text{check:} \quad \begin{array}{r} \overset{0}{\cancel{1}}\,\overset{12}{\cancel{3}}\,2 \\ -\ 8\,4 \\ \hline 4\,8 \end{array}$$

$w\ =\ \mathbf{132}$

16.
$$\begin{array}{r} 39 \\ 6)\overline{234} \\ \underline{18} \\ 54 \\ \underline{54} \\ 0 \end{array} \quad \text{check:} \quad \begin{array}{r} 6 \\ 39)\overline{234} \\ \underline{234} \\ 0 \end{array}$$

$n\ =\ \mathbf{39}$

17. $(1 + 2) \times 3 = (1 \times 2) + m$

$\quad\quad\quad 3 \times 3 = 2 + m$

$\quad\quad\quad\quad 9 = 2 + m$

$\quad\quad\quad 9 - 2 = 7$

$\quad\quad\quad\quad\quad \boldsymbol{m\ =\ 7}$

check:

$(1 + 2) \times 3 = (1 \times 2) + 7$

$\quad\quad\quad 3 \times 3 = 2 + 7$

$\quad\quad\quad\quad 9 = 9$

18.

```
              5 cm
  ┌─────────────────────────┐
  │                         │
  │                         │ 3 cm
  │                         │
  └─────────────────────────┘
```

$5\,\text{cm} + 3\,\text{cm} + 5\,\text{cm} + 3\,\text{cm} = \mathbf{16\,cm}$

19. $2 + 4 + 6 + 8 + 10 + 12 = \mathbf{42}$

20. Multiply the value of a term by 2 to find the next term; 8

21. $500 \times 1 \enspace \overset{\bigcirc}{=} \enspace 500 \div 1$
$\quad\quad 500 \quad\quad\quad\quad 500$

22.
$$\begin{array}{r} 555 \\ 2)\overline{1110} \\ \underline{10} \\ 11 \\ \underline{10} \\ 10 \\ \underline{10} \\ 0 \end{array}$$

23. Millions

24.
$$\begin{array}{r} \overset{1}{\cancel{2}}\overset{}{0}0 \\ -\quad 8\,0 \\ \hline \mathbf{1\ 2\ 0}\ \text{heartbeats per minute} \end{array}$$

25.
$$\begin{array}{r} 200 \\ \times\quad 10 \\ \hline \mathbf{2000}\ \text{times} \end{array}$$

26. Answers may vary. Sample answer: Walking increases a resting person's heart rate by about how many heartbeats per minute?

$$\begin{array}{r} \overset{0}{\cancel{1}}\overset{1}{4}\,0 \\ -\quad 8\,0 \\ \hline \mathbf{6\ 0}\ \text{heartbeats per minute} \end{array}$$

27.
$$\begin{array}{r} \overset{1}{2}4\ \text{customers} \\ 27\ \text{customers} \\ +\ 33\ \text{customers} \\ \hline 84\ \text{customers} \end{array} \quad \begin{array}{r} 28\ \text{customers} \\ 3)\overline{84} \\ \underline{6} \\ 24 \\ \underline{24} \\ 0 \end{array}$$

28. (a)
$$\begin{array}{r} 10\ \text{dimes} \\ .10)\overline{\$1.00} \\ \underline{1\,0} \\ 00 \\ \underline{00} \\ 0 \end{array}$$

(b)
$$\begin{array}{r} 30\ \text{dimes} \\ .10)\overline{\$3.00} \\ \underline{3\,0} \\ 00 \\ \underline{00} \\ 0 \end{array}$$

29.

$2\frac{1}{4}$ in.

$1\frac{3}{4}$ in.

30. Combining, equal groups

_____

LESSON 19, WARM-UP

a. 192

b. 138

c. 138

d. 83

e. 83

f. $7.50

g. 10

Problem Solving
    NQD, DNQ, DQN, QND

_____

LESSON 19, ACTIVITY

1̸	②	③	4̸	⑤	6̸	⑦	8̸	9̸	1̸0̸
⑪	1̸2̸	⑬	14	15	16	⑰	18	⑲	20
21	22	㉓	24	25	26	27	28	㉙	30
㉛	32	33	34	35	36	㊲	38	39	40
㊶	42	㊸	44	45	46	㊼	48	49	50
51	52	㊿	54	55	56	57	58	㊾	60
㊶	62	63	64	65	66	㊷	68	69	70
㊸	72	㊻	74	75	76	77	78	㊿	80
81	82	㉝	84	85	86	87	88	㉙	90
91	92	93	94	95	96	㊲	98	99	100

LESSON 19, LESSON PRACTICE

a. 1, 2, 7, 14

$$1\overline{)14} \qquad 2\overline{)14} \qquad 7\overline{)14} \qquad 14\overline{)14}$$

$$\begin{array}{c} 14 \\ \underline{1} \\ 04 \\ \underline{4} \\ 0 \end{array} \qquad \begin{array}{c} 7 \\ \underline{14} \\ 0 \end{array} \qquad \begin{array}{c} 2 \\ \underline{14} \\ 0 \end{array} \qquad \begin{array}{c} 1 \\ \underline{14} \\ 0 \end{array}$$

b. 1, 3, 5, 15

$$1\overline{)15} \qquad 3\overline{)15} \qquad 5\overline{)15} \qquad 15\overline{)15}$$

$$\begin{array}{c} 15 \\ \underline{1} \\ 05 \\ \underline{5} \\ 0 \end{array} \qquad \begin{array}{c} 5 \\ \underline{15} \\ 0 \end{array} \qquad \begin{array}{c} 3 \\ \underline{15} \\ 0 \end{array} \qquad \begin{array}{c} 1 \\ \underline{15} \\ 0 \end{array}$$

c. 1, 2, 4, 8, 16

$$1\overline{)16} \quad 2\overline{)16} \quad 4\overline{)16} \quad 8\overline{)16} \quad 16\overline{)16}$$

$$\begin{array}{c} 16 \\ \underline{1} \\ 06 \\ \underline{6} \\ 0 \end{array} \quad \begin{array}{c} 8 \\ \underline{16} \\ 0 \end{array} \quad \begin{array}{c} 4 \\ \underline{16} \\ 0 \end{array} \quad \begin{array}{c} 2 \\ \underline{16} \\ 0 \end{array} \quad \begin{array}{c} 1 \\ \underline{16} \\ 0 \end{array}$$

d. 1, 17

$$1\overline{)17} \qquad 17\overline{)17}$$

$$\begin{array}{c} 17 \\ \underline{1} \\ 07 \\ \underline{7} \\ 0 \end{array} \qquad \begin{array}{c} 1 \\ \underline{17} \\ 0 \end{array}$$

e. 23

f. 31

g. 43

h. 42

i. 51

j. 33

k. $2 \cdot 2 \cdot 2 \cdot 2 = 16$

l. $2 \cdot 3 \cdot 3 = 18$

LESSON 19, MIXED PRACTICE

1.
$$\begin{array}{r} 42 \\ 6\overline{)252} \\ \underline{24} \\ 12 \\ \underline{12} \\ 0 \end{array}$$

2.
$$\begin{array}{r} 20 \\ \times\ \ 4 \\ \hline 80 \end{array} \qquad \begin{array}{r} 80 \\ +\ 7 \\ \hline \textbf{87 years} \end{array}$$

3.

$$\text{number line from } -3 \text{ to } 4$$

7°C

4. $7 \cdot g = 203$
$$\begin{array}{r} \textbf{29 turnips} \\ 7\overline{)203} \\ \underline{14} \\ 63 \\ \underline{63} \\ 0 \end{array}$$

5. $1 + 2 + 4 + 9 = 16$
$$\begin{array}{r} 4 \\ 4\overline{)16} \\ \underline{16} \\ 0 \end{array}$$

6. 36

7.
$$\begin{array}{r} \overset{3}{2}5\ \text{mm} \\ \times\ \ \ 6 \\ \hline \textbf{150 mm} \end{array}$$

8. 30 mm

9. **1, 2, 4, 5, 10, 20**
$$\begin{array}{cccccc}
\begin{array}{r}20\\1\overline{)20}\\ \underline{2}\\00\\ \underline{00}\\0\end{array} &
\begin{array}{r}10\\2\overline{)20}\\ \underline{2}\\00\\ \underline{00}\\0\end{array} &
\begin{array}{r}5\\4\overline{)20}\\ \underline{20}\\0\end{array} &
\begin{array}{r}4\\5\overline{)20}\\ \underline{20}\\0\end{array} &
\begin{array}{r}2\\10\overline{)20}\\ \underline{20}\\0\end{array} &
\begin{array}{r}1\\20\overline{)20}\\ \underline{20}\\0\end{array}
\end{array}$$

10. **4**
1, 3, 5, 15
$$\begin{array}{cccc}
\begin{array}{r}15\\1\overline{)15}\\ \underline{1}\\05\\ \underline{5}\\0\end{array} &
\begin{array}{r}5\\3\overline{)15}\\ \underline{15}\\0\end{array} &
\begin{array}{r}3\\5\overline{)15}\\ \underline{15}\\0\end{array} &
\begin{array}{r}1\\15\overline{)15}\\ \underline{15}\\0\end{array}
\end{array}$$

11. C. 29

12.
$$\begin{array}{r} 2,500 \\ 100\overline{)250,000} \\ \underline{200} \\ 50\ 0 \\ \underline{50\ 0} \\ 00 \\ \underline{00} \\ 00 \\ \underline{00} \\ 0 \end{array}$$

13.
$$\begin{array}{r} 20\ \text{R } 34 \\ 60\overline{)1234} \\ \underline{120} \\ 34 \\ \underline{00} \\ 34 \end{array}$$

14. $\dfrac{6 + 18 + 9}{3} = \dfrac{33}{3} = \textbf{11}$

15.
$$\begin{array}{r} \$3.45 \\ \times\ \ \ \ \ 10 \\ \hline \textbf{\$34.50} \end{array}$$

16.
$$\begin{array}{r} \$\overset{0}{\cancel{1}}\overset{9}{0}.\overset{9}{\cancel{0}}\overset{1}{\cancel{0}}0 \\ -\ \ \$1.93 \\ \hline \$8.07 \end{array} \qquad \text{check:} \qquad \begin{array}{r} \$\overset{0}{\cancel{1}}\overset{9}{0}.\overset{9}{\cancel{0}}\overset{1}{\cancel{0}}0 \\ -\ \ \$8.07 \\ \hline \$1.93 \end{array}$$
$w = \textbf{\$8.07}$

17.
$$\begin{array}{r} 4 \\ \times\ 3 \\ \hline 12 \end{array} \qquad \text{check:} \qquad \begin{array}{r} 4 \\ 3\overline{)12} \\ \underline{12} \\ 0 \end{array}$$
$w = \textbf{12}$

18. $ba = c$
$c \div a = b$
$c \div b = a$

19. $-2, 0, \dfrac{1}{2}, 1, 3$

20. $123 \div 1 \ \bigcirc\!\!> \ 123 - 1$
$\qquad 123 \qquad\qquad 122$

21. 9

22. 123,000,000

23.
$$\begin{array}{r} \$5.50 \\ 2)\overline{\$11.00} \\ \underline{10} \\ 1\ 0 \\ \underline{1\ 0} \\ 00 \\ \underline{00} \\ 0 \end{array}$$

24.
$$\begin{array}{r} 12 \text{ inches} \\ 4)\overline{48} \\ \underline{4} \\ 08 \\ \underline{8} \\ 0 \end{array}$$

25. $(51 + 49) \cdot (51 - 49)$
     $100 \quad \cdot \quad 2$
     200

26. A. **2**

27. $2 \cdot 2 \cdot 5 = 20$

28. $12 + 12 + 6 = 30$
$$\begin{array}{r} 10 \text{ dictionaries} \\ 3)\overline{30} \\ \underline{3} \\ 00 \\ \underline{00} \\ 0 \end{array}$$

29.
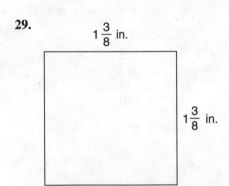

$1\frac{3}{8}$ in.

$1\frac{3}{8}$ in.

30. **If the number is even, it is divisible by 2. All
    even numbers are divisible by 2. Odd numbers
    are not divisible by 2.**

LESSON 20, WARM-UP

a. **138**

b. **192**

c. **1840**

d. **92**

e. **92**

f. **$25.00**

g. **12**

Problem Solving

There are two possible solutions. Remember the
pattern that results from subtraction by 9: the
ones digit of the difference is one greater than the
ones digit of the minuend (except for minuends
that end in 9). Fill the ones column with digits
from the set 5, 6, 7, and 8 that satisfy this pattern.
Use the digits 5, 6, 7, and 8 once each.

$$\begin{array}{r} 67 \\ -\ 9 \\ \hline 58 \end{array} \qquad \begin{array}{r} 85 \\ -\ 9 \\ \hline 76 \end{array}$$

LESSON 20, LESSON PRACTICE

a. The factors of 10 are 1, 2, 5, 10.
    The factors of 15 are 1, 3, 5, 15.
    GCF is **5.**

b. The factors of 18 are 1, 2, 3, 6, 9, 18.
    The factors of 27 are 1, 3, 9, 27.
    GCF is **9.**

c. The factors of 18 are 1, 2, 3, 6, 9, 18.
    The factors of 24 are 1, 2, 3, 4, 6, 8, 12, 24.
    GCF is **6.**

d. The factors of 12 are 1, 2, 3, 4, 6, 12.
    The factors of 18 are 1, 2, 3, 6, 9, 18.
    The factors of 24 are 1, 2, 3, 4, 6, 8, 12, 24.
    GCF is **6.**

e. The factors of 15 are 1, 3, 5, 15.
    The factors of 25 are 1, 5, 25.
    GCF is **5.**

f. The factors of 20 are 1, 2, 4, 5, 10, 20.
    The factors of 30 are 1, 2, 3, 5, 6, 10, 15, 30.
    The factors of 40 are 1, 2, 4, 5, 8, 10, 20, 40.
    GCF is **10.**

g. The factors of 12 are 1, 2, 3, 4, 6, 12.
The factors of 15 are 1, 3, 5, 15.
GCF is **3.**

h. The factors of 20 are 1, 2, 4, 5, 10, 20.
The factors of 40 are 1, 2, 4, 5, 8, 10, 20, 40.
The factors of 60 are 1, 2, 3, 4, 5, 6, 10, 12, 15, 20, 30, 60.
GCF is **20.**

LESSON 20, MIXED PRACTICE

1. $(12 \times 8) - (12 + 8)$
  96   –   20
  76

2. **1,427,000,000 km**

3. **9**

4. **$428,000**

5.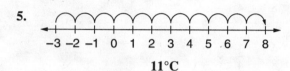
  11°C

6. $31 + 52 + 40 = 123$

$$\begin{array}{r} \textbf{41} \textbf{ points per game} \\ 3\overline{)123} \\ \underline{12} \\ 03 \\ \underline{3} \\ 0 \end{array}$$

7. The factors of 12 are 1, 2, 3, 4, 6, 12.
The factors of 20 are 1, 2, 4, 5, 10, 20.
GCF is **4.**

8. The factors of 9 are 1, 3, 9.
The factors of 15 are 1, 3, 5, 15.
The factors of 21 are 1, 3, 7, 21.
GCF is **3.**

9.
$$\begin{array}{r} \textbf{\$0.81} \\ 4\overline{)\$3.24} \\ \underline{3\,2} \\ 04 \\ \underline{4} \\ 0 \end{array}$$

10.
$$\begin{array}{r} \textbf{543} \textbf{ R 2} \\ 10\overline{)5432} \\ \underline{50} \\ 43 \\ \underline{40} \\ 32 \\ \underline{30} \\ 2 \end{array}$$

11. $\dfrac{28 + 42}{14} = \dfrac{70}{14} = 5$

12.
$$\begin{array}{r} {}^{1\;1\;\;1} \\ 56,042 \\ + \;49,985 \\ \hline \textbf{106,027} \end{array}$$

13.
$$\begin{array}{r} \textbf{3,090} \\ 12\overline{)37,080} \\ \underline{36} \\ 1\,0 \\ \underline{0\,0} \\ 1\,08 \\ \underline{1\,08} \\ 00 \\ \underline{00} \\ 0 \end{array}$$

14.
$$\begin{array}{r} \$6.47 \\ \times \quad 10 \\ \hline \textbf{\$64.70} \end{array}$$

15. $\underline{5 \times 4} \times 3 \times 2 \times 1$
  $\underline{20 \times 3} \times 2 \times 1$
  $\underline{60 \times 2} \times 1$
  120×1
  120

16.
$$\begin{array}{r} {}^{1\,1} \\ 528 \\ + \;76 \\ \hline 604 \end{array} \qquad \text{check:} \qquad \begin{array}{r} {}^{5\;\;9} \\ \cancel{6}\,\cancel{0}^{1}4 \\ - \;76 \\ \hline 5\,2\,8 \end{array}$$
  $w = \textbf{604}$

17.
$$\begin{array}{r} {}^{0\;\;{}^{1}3\;\;9} \\ \cancel{1}\,\cancel{4},\cancel{0}^{1}0\,9 \\ - \;9\,6\,7\,0 \\ \hline 4\,3\,3\,9 \end{array} \qquad \text{check:} \qquad \begin{array}{r} {}^{0\;\;{}^{1}3\;\;9} \\ \cancel{1}\,\cancel{4},\cancel{0}^{1}0\,9 \\ - \;4\,3\,3\,9 \\ \hline 9\,6\,7\,0 \end{array}$$
  $w = \textbf{4339}$

18.
$$\begin{array}{r} 15 \\ 6\overline{)90} \\ \underline{6} \\ 30 \\ \underline{30} \\ 0 \end{array} \qquad \text{check:} \qquad \begin{array}{r} {}^{3} \\ 15 \\ \times \;6 \\ \hline 90 \end{array}$$
  $w = \textbf{15}$

19.
$$\overset{1\ 1}{365}$$
$$+\ 365$$
$$\overline{730}$$
$$q = 730$$

check:
$$\overset{6\ \ ^{1}2}{7\,\cancel{3}^{1}0}$$
$$-\ 3\,6\,5$$
$$\overline{3\,6\,5}$$

20.
$$365$$
$$-\ 365$$
$$\overline{0}$$
$$p = 0$$

check:
$$365$$
$$-\ 0$$
$$\overline{365}$$

21. **4**

22. $50 - 1 \ \overset{<}{\bigcirc}\ 49 + 1$
$\quad\ \ 49 \qquad\quad 50$

23. 1, 3, 5, 7, 9, 11, 13, 15, 17, 19
19

24. Since the four sides of a square have equal lengths, we divide the perimeter, 100 cm, by 4 to find the length of each side.

25. Student estimates may vary; $2\frac{4}{16}$ or $2\frac{1}{4}$ inches

26. (a) **8 bits**

(b)
$$8$$
$$\times\ 3$$
$$\overline{\textbf{24 bits}}$$

27. $12 + 24 + 36 + 48 = 120$

$$\begin{array}{r} \textbf{30 golf balls} \\ 4\overline{)120} \\ \underline{12} \\ 00 \\ \underline{0} \\ \overline{0} \end{array}$$

28. A. **5**

29. **1, 2, 3, 4, 6, 8, 12, 24**

30. **990,000,000,000**

INVESTIGATION 2

1. **50%**

2. $\frac{1}{4}$

3. $\frac{1}{8}$

4. $25\% + 25\% + 25\% = \textbf{75\%}$

5. $\frac{1}{8} + \frac{1}{8} + \frac{1}{8} + \frac{1}{8} = \frac{4}{8}$ or $\frac{1}{2}$
$\frac{1}{2} = \textbf{50\%}$

6. $\frac{1}{6} + \frac{1}{6} + \frac{1}{6} = \frac{3}{6}$ or $\frac{1}{2}$
$\frac{1}{2} = \textbf{50\%}$

7.

8. $\frac{2}{8} = \frac{1}{4}$

9. $\frac{6}{8} = \frac{3}{4}$
3

10. $\frac{2}{6} = \frac{1}{3}$

11. $\frac{4}{6} = \frac{2}{3}$
2

12. $\frac{3}{8} + \frac{2}{8} = \frac{5}{8}$

13. $\frac{6}{6} - \frac{1}{6} = \frac{5}{6}$

14. $\frac{3}{3} - \frac{1}{3} = \frac{2}{3}$

15. $\frac{4}{4} - \frac{1}{4} = \frac{3}{4}$

16. $\frac{8}{8} - \frac{3}{8} = \frac{5}{8}$

17. $\frac{1}{3} = 33\frac{1}{3}\%$

18. $\frac{1}{6} = 16\frac{2}{3}\%$

19. ; $\frac{2}{3} < \frac{3}{4}$

20. ; $\frac{2}{3} > \frac{3}{8}$

21. ; $\frac{1}{3}$

22. one possibility:

    >

23. =

24. $\frac{7}{4} = 1\frac{3}{4}$

25. $\frac{3}{2} = 1\frac{1}{2}$

26. $1\frac{1}{2} = \frac{3}{2}$

    $\frac{3}{2} = \frac{6}{4}$

    6

27. $2 = \frac{6}{3}$

    6

28. $\frac{4}{3} = 1\frac{1}{3}$

29. $\frac{11}{6} = 1\frac{5}{6}$

30. $\frac{1}{2} = \frac{6}{12}$

    $\frac{1}{4} = \frac{3}{12}$

    $\frac{1}{3} = \frac{4}{12}$

    $\frac{1}{6} = \frac{2}{12}$

LESSON 21, WARM-UP

a. 168

b. 228

c. 83

d. 487

e. $3.50

f. 12

g. 4

Problem Solving

$$
\begin{array}{rl}
3Q &= 75¢ \\
2D &= 20¢ \\
+\ 1N &= 5¢ \\
\hline
& 100¢ \text{ or } \$1.00
\end{array}
$$

$$
\begin{array}{rl}
1HD &= 50¢ \\
1Q &= 25¢ \\
1D &= 10¢ \\
+\ 3N &= 15¢ \\
\hline
& 100¢ \text{ or } \$1.00
\end{array}
$$

$$
\begin{array}{rl}
1HD &= 50¢ \\
+\ 5D &= 50¢ \\
\hline
& 100¢ \text{ or } \$1.00
\end{array}
$$

3Q, 2D, 1N; 1HD, 1Q, 1D, 3N; 1HD, 5D

LESSON 21, LESSON PRACTICE

a. 123; No, the last digit is not even.
   234; Yes, the last digit is even.
   345; No, the last digit is not even.

b. $1 + 2 + 3 + 4 = 10$
   1234; No, the sum of the digits is not divisible by 3.
   $2 + 3 + 4 + 5 = 14$
   2345; No, the sum of the digits is not divisible by 3.
   $3 + 4 + 5 + 6 = 18$
   3456; Yes, the sum of the digits is divisible by 3.

c. 2, 3, 5, 10

d. 2, 3

SOLUTIONS

LESSON 21, MIXED PRACTICE

1. $\underbrace{(8 + 5)}_{13} \times \underbrace{(8 - 5)}_{3}$
$$13 \times 3$$
$$\mathbf{39}$$

2. $1959 - 1787 = d$

$$1\,\overset{8}{9}\,{}^{1}5\,9$$
$$-\ 1\,7\,8\,7$$
$$\overline{\ \ 1\,7\,2\ } \text{ years}$$

3. $16 \cdot w = 240$

$$\begin{array}{r} \textbf{15 bowling balls} \\ 16\overline{)240} \\ \underline{16} \\ 80 \\ \underline{80} \\ 0 \end{array}$$

4. $\dfrac{3}{4}$

5. $\dfrac{17}{30}$

6. $\dfrac{3}{100}$

7.
$$\begin{array}{r} \textbf{\$1.17} \\ 2\overline{)\$2.34} \\ \underline{2} \\ 0\,3 \\ \underline{2} \\ 14 \\ \underline{14} \\ 0 \end{array}$$

8. **Millions**

9. **256; Multiply the value of a term by 4 to find the next term.**

10. $\underbrace{64 \times 1}_{64} \;\; \textcircled{<} \;\; \underbrace{64 + 1}_{65}$

11. A. $3 + 6 + 5 = 14$
 The sum of the digits is not divisible by 9.
 B. $1 + 1 + 7 + 9 = 18$
 The sum of digits is divisible by 9.
 C. $1 + 5 + 5 + 6 = 17$
 The sum of the digits is not divisible by 9.
 B. 1179

12.
$$\begin{array}{r} 400 \\ 200 \\ +\ 200 \\ \hline \mathbf{800} \end{array}$$

13. The factors of 12 are 1, 2, 3, 4, 6, 12.
 The factors of 16 are 1, 2, 4, 8, 16.
 GCF is **4.**

14.
$$\begin{array}{r} \mathbf{40\ R\ 30} \\ 100\overline{)4030} \\ \underline{400} \\ 30 \\ \underline{00} \\ 30 \end{array}$$

15.
$$\begin{array}{r} \mathbf{2,035} \\ 24\overline{)48,840} \\ \underline{48} \\ 0\,8 \\ \underline{0\,0} \\ 84 \\ \underline{72} \\ 120 \\ \underline{120} \\ 0 \end{array}$$

16.
$$\begin{array}{r} \mathbf{113} \\ 6\overline{)678} \\ \underline{6} \\ 07 \\ \underline{6} \\ 18 \\ \underline{18} \\ 0 \end{array}$$

17.
$$\begin{array}{r} \$4.75 \\ \times\ \ \ \ 10 \\ \hline \mathbf{\$47.50} \end{array}$$

18.
$$\overset{0\ \ 9\ \ 9}{\$\cancel{1}\,\cancel{0}.\,\cancel{0}^{1}0} \qquad \text{check:} \qquad \overset{0\ \ 9\ \ 9}{\$\cancel{1}\,\cancel{0}.\,\cancel{0}^{1}0}$$
$$-\quad \$0.\,8\,7 \qquad\qquad\qquad\qquad\qquad -\quad \$9.\,1\,3$$
$$\overline{\quad\ \ \$9.\,1\,3} \qquad\qquad\qquad\qquad\qquad \overline{\quad\ \ \$0.\,8\,7}$$
$w = \mathbf{\$9.13}$

19. $\underbrace{463 + 27}_{} + m = 500$

$$490 + m = 500$$
$$500 - 490 = 10$$
$$\boldsymbol{m = 10}$$
check: $\qquad 463 + 27 + 10 = 500$

20. $-2,\ 0,\ \dfrac{1}{4},\ \dfrac{1}{2},\ 1$

21.

$$\begin{array}{r} \overset{1}{1}2 \\ 16 \\ + \ 23 \\ \hline 51 \end{array}$$

$$\begin{array}{r} 17 \\ 3\overline{)51} \\ \underline{3} \\ 21 \\ \underline{21} \\ 0 \end{array}$$

22. **1, 2, 4, 7, 14, 28**

23.

$$\begin{array}{r} \overset{6}{1}8 \ cm \\ \times \ \ \ 8 \\ \hline \textbf{144 cm} \end{array}$$

24.

10 cm

25. $(12 \times 12) - (11 \times 13)$
$\qquad 144 - 143$
$\qquad\qquad \textbf{1}$

26. **4 and 8**

27. ▨▨☐ < ▨▨☐

28. (a) **8 bits**

(b) **4 bits**

29. The factors of 20 are ①, ②, 4, ⑤, ⑩, 20.
The factors of 30 are ①, ②, 3, ⑤, 6, ⑩, 15, 30.
1, 2, 5, 10

30. Draw a circle ◯. Through the center of the circle, draw a plus sign ⊕. Then draw a times sign through the center of the circle ✳.

LESSON 22, WARM-UP

a. 216

b. 168

c. 65

d. 317

e. $6.50

f. 24

g. 25

Problem Solving

The total number of dots on opposite sides of a dot cube is 7. So the two dot cubes contain a total of 7×2, or 14 dots, on the top and bottom faces. Since 6 dots appear on the top faces, there must be $14 - 6$, or **8 dots,** on the bottom faces.

LESSON 22, LESSON PRACTICE

a.

12 musicians

$\frac{3}{4}$ could play piano. { 3 musicians / 3 musicians / 3 musicians }

$\frac{1}{4}$ could not play piano. { 3 musicians }

9 musicians

b.

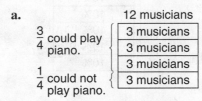

$$\begin{array}{r} \$1.50 \\ 3\overline{)\$4.50} \\ \underline{3} \\ 1\ 5 \\ \underline{1\ 5} \\ 00 \\ \underline{00} \\ 0 \end{array}$$

$\frac{2}{3}$ of $4.50 { $1.50 / $1.50 }

$\frac{1}{3}$ of $4.50 { $1.50 }

$$\begin{array}{r} \$1.50 \\ \times \ \ \ \ 2 \\ \hline \$3.00 \end{array}$$

c.

$$5)\overline{60} \quad \begin{array}{c} 12 \\ \underline{5} \\ 10 \\ \underline{10} \\ 0 \end{array}$$

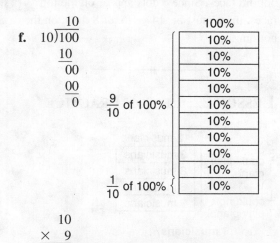

$\frac{4}{5}$ of 60

$\frac{1}{5}$ of 60

$$\begin{array}{r} 12 \\ \times \quad 4 \\ \hline \mathbf{48} \end{array}$$

d.

$$10)\overline{80} \quad \begin{array}{c} 8 \\ \underline{80} \\ 0 \end{array}$$

$\frac{3}{10}$ of 80

$\frac{7}{10}$ of 80

$$\begin{array}{r} 8 \\ \times \quad 3 \\ \hline \mathbf{24} \end{array}$$

e.

$$6)\overline{24} \quad \begin{array}{c} 4 \\ \underline{24} \\ 0 \end{array}$$

$\frac{5}{6}$ of 24

$\frac{1}{6}$ of 24

$$\begin{array}{r} 4 \\ \times \quad 5 \\ \hline \mathbf{20} \end{array}$$

f.

$$10)\overline{100} \quad \begin{array}{c} 10 \\ \underline{10} \\ 00 \\ \underline{00} \\ 0 \end{array}$$

$\frac{9}{10}$ of 100%

$\frac{1}{10}$ of 100%

$$\begin{array}{r} 10 \\ \times \quad 9 \\ \hline \mathbf{90\%} \end{array}$$

LESSON 22, MIXED PRACTICE

1. $(15 \times 12) - (15 + 12)$
$180 - 27$
153

2. $\frac{13}{50}$

3.

$$\begin{array}{r} 1760 \\ \times \quad 26 \\ \hline 10560 \\ 35200 \\ \hline 45,760 \text{ yards} \end{array} \qquad \begin{array}{r} {}^{1\ 1}45,760 \\ + \quad 385 \\ \hline \mathbf{46,145 \text{ yards}} \end{array}$$

4.

$\frac{2}{3}$ were eaten.

$\frac{1}{3}$ were not eaten.

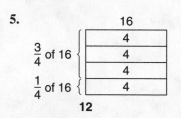

8 jelly beans

5.

$\frac{3}{4}$ of 16

$\frac{1}{4}$ of 16

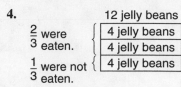

12

6.

$$10)\overline{\$3.50} \quad \begin{array}{c} \$0.35 \\ \underline{3\ 0} \\ 50 \\ \underline{50} \\ 0 \end{array}$$

$\frac{3}{10}$ of $3.50

$\frac{7}{10}$ of $3.50

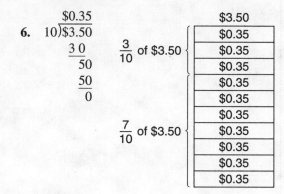

$$\begin{array}{r} \$0.35 \\ \times \qquad 3 \\ \hline \mathbf{\$1.05} \end{array}$$

7.

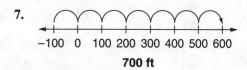

700 ft

8.

$$\begin{array}{r} {}^{1}15 \\ + \quad 8 \\ \hline 23 \end{array} \qquad \text{check:} \qquad \begin{array}{r} {}^{1}2^{1}3 \\ - \quad 1\ 5 \\ \hline 8 \end{array}$$

$w = \mathbf{23}$

9.

$$\begin{array}{r} 345 \\ \times \quad 15 \\ \hline 1725 \\ 3450 \\ \hline 5175 \end{array} \qquad \text{check:} \qquad 15)\overline{5175} \quad \begin{array}{c} 345 \\ \underline{45} \\ 67 \\ \underline{60} \\ 75 \\ \underline{75} \\ 0 \end{array}$$

$w = \mathbf{5175}$

10.
$$\overset{1\;2}{\;}$$
$0.36
$4.78
+ $34.09
$39.23

11.
$4.15
3)$12.45
12
0 4
3
15
15
0

12.
28 R 20
35)1000
70
300
280
20

13. $\dfrac{7 + 9 + 14}{3} = \dfrac{30}{3} = \mathbf{10}$

14.
124   **4500**
× 36
744
3720
4464

15. **2**

16. Factors of 12
1, 2, 3, 4, 6, 12
Factors of 15
1, 3, 5, 15
GCF is **3**.

17. **1, 2, 3, 5, 6, 10, 15, 30**

18. 2; Yes, the last digit is 0.
3; No, the sum of the digits is not divisible by 3.
5; Yes, the last digit is 0.
9; No, the sum of the digits is not divisible by 9.
10; Yes, the last digit is 0.
2, 5, 10

19.
20
5)100
10
00
00
0
20%
× 4
80%

	100%
$\frac{4}{5}$ of 100%	20%
	20%
	20%
	20%
$\frac{1}{5}$ of 100%	20%

20. $\dfrac{1}{3}$ $\lessgtr$ $\dfrac{1}{2}$

21. **C. 39**

22. $(3 + 3) - (3 \times 3)$
6 — 9
−3

23.
$$\overset{1}{43}$$
+ 27
70

35
2)70
6
10
10
0

24. 15 cm + 10 cm + 15 cm + 10 cm = **50 cm**

25. $2\dfrac{1}{4}$ **in.**

26.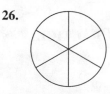

Apple pie

Cherry pie

A slice of cherry pie

27.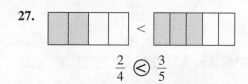
$\dfrac{2}{4}$ $\lessgtr$ $\dfrac{3}{5}$

28.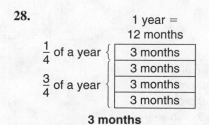

	1 year = 12 months
$\frac{1}{4}$ of a year	3 months
$\frac{3}{4}$ of a year	3 months
	3 months
	3 months

3 months

29. (a) **8 bits**
(b) **2 bits**

30. $c - t = p$
$p + t = c$
$t + p = c$

LESSON 23, WARM-UP

a. 310

b. 180

c. 96

d. 1550

e. $4.50

f. 42

g. 2

Problem Solving

9 cubes per layer $\times$ 3 layers = **27 cubes**

LESSON 23, LESSON PRACTICE

a. $\dfrac{\text{number of dogs}}{\text{number of cats}} = \dfrac{12}{19}$

b. $\dfrac{\text{number of girls}}{\text{number of boys}} = \dfrac{13}{17}$

$$\begin{array}{r} \overset{2}{\cancel{3}}{}^{1}0 \\ -\ 1\ 7 \\ \hline 1\ 3\ \text{girls} \end{array}$$

c. $\dfrac{\text{number of trucks}}{\text{number of cars}} = \dfrac{2}{7}$

LESSON 23, MIXED PRACTICE

1.
$$\begin{array}{r} 30\ \text{cm} \\ \times\ \ 10\ \text{cm} \\ \hline \textbf{300 cm} \end{array}$$

2.

$\dfrac{2}{3}$ are finished. { 9 problems / 9 problems

$\dfrac{1}{3}$ are not finished. { 9 problems

18 problems

3. $100 \cdot 36 = t$

$$\begin{array}{r} 36 \\ \times\ \ 100 \\ \hline \textbf{3600 inches} \end{array}$$

4.
$$\begin{array}{r} \overset{2}{\cancel{3}}{}^{1}1 \\ -\ 2\ 5 \\ \hline 6\ \text{days} \end{array}$$

$\dfrac{6}{31}$

5.

$\dfrac{3}{5}$ of 25

$\dfrac{2}{5}$ of 25

6.
$$\begin{array}{r} 3.60 \\ 10\overline{)\$36.00} \\ \underline{30} \\ 6\,0 \\ \underline{6\,0} \\ 00 \\ \underline{00} \\ 0 \end{array}$$

$\dfrac{7}{10}$ of $36.00

$\dfrac{3}{10}$ of $36.00

$$\begin{array}{r} \$3.60 \\ \times\ \ \ \ 7 \\ \hline \mathbf{\$25.20} \end{array}$$

7. $\dfrac{7}{8}$

8. $1\dfrac{1}{8}$

9. **25%**

10.
$$\begin{array}{r} \$3.75 \\ \times\ \ \ \ 16 \\ \hline 22\ 50 \\ 37\ 50 \\ \hline \mathbf{\$60.00} \end{array}$$

11.
$$\begin{array}{r} \mathbf{\$0.15} \\ 25\overline{)\$3.75} \\ \underline{2\ 5} \\ 1\ 25 \\ \underline{1\ 25} \\ 0 \end{array}$$

12. **Millions**

13. One way to find $\frac{2}{3}$ of a number is to first divide the number by 3; then multiply the answer by 2.

14.
$$\begin{array}{r} 1\ 00 \\ \$0.35)\overline{\$35.00} \\ 35 \\ \hline 0\ 0 \\ 0\ 0 \\ \hline 00 \\ 00 \\ \hline 0 \end{array}$$
check
$$\begin{array}{r} \$0.35 \\ \times\ \ \ \ 100 \\ \hline \$35.00 \end{array}$$

$n = 100$

15.
$$\begin{array}{r} {}^0\cancel{1}\,{}^9\cancel{0}.\,{}^1\cancel{2}\,{}^1 0 \\ -\ \ \ \$3.\,4\,6 \\ \hline \$6.\,7\,4 \end{array}$$
check:
$$\begin{array}{r} {}^0\cancel{1}\,{}^9\cancel{0}.\,{}^1\cancel{2}\,{}^1 0 \\ -\ \ \ \$6.\,7\,4 \\ \hline \$3.\,4\,6 \end{array}$$

$m = \$6.74$

16. $\frac{3}{4}$ $<$ 1

17.

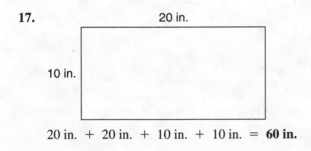

20 in. + 20 in. + 10 in. + 10 in. = **60 in.**

18. 2, 4, 8, 16, 32, 64 . . . .
64

19.
$$\begin{array}{r} 3000 \\ +\ 5000 \\ \hline 8000 \end{array}$$

20. $\underbrace{12 \div 6}\ -\ 2$ $<$ $12 \div \underbrace{(6-2)}$
$\qquad\quad 2 - 2 \qquad\qquad\ \ 12 \div 4$
$\qquad\qquad 0 \qquad\qquad\qquad\quad 3$

21. Factors of 24
1, 2, 3, 4, 6, 8, 12, 24
Factors of 32
1, 2, 4, 8, 16, 32
GCF is **8.**

22. $1 + 3 + 5 + 7 + 9 + 11 + 13 = $ **49**

23. (a) **4**
    (b) **2**

24. $12\frac{1}{2}\%$

25. $\frac{4}{8}$

26. $\dfrac{\text{number of boys}}{\text{number of girls}} = \dfrac{10}{13}$

$$\begin{array}{r} 23 \\ -\ 10 \\ \hline 13\ \text{girls} \end{array}$$

27. 23, 29

28. **B.** ⟷

29. 252; No, last digit is not a 5 or 0.
525; No, last digit is not even.
250; Yes, last digit is 0 and even.
C. 250

30. $\dfrac{\text{win}}{\text{loss}} = \dfrac{5}{9}$

LESSON 24, WARM-UP

a. 144

b. 300

c. 86

d. 1250

e. $5.50

f. 34

g. 5

Problem Solving
XYZ, XZY, YXZ, YZX, ZXY, ZYX; There are 4 possible orders of finish in which Xavier is not first.

SOLUTIONS

LESSON 24, LESSON PRACTICE

a. $\dfrac{3}{8} + \dfrac{4}{8} = \dfrac{7}{8}$

b. $\dfrac{3}{4} + \dfrac{1}{4} = \dfrac{4}{4} = 1$

c. $\dfrac{1}{8} + \dfrac{1}{8} + \dfrac{1}{8} = \dfrac{3}{8}$

d. $\dfrac{4}{8} - \dfrac{1}{8} = \dfrac{3}{8}$

e. $\dfrac{3}{4} - \dfrac{2}{4} = \dfrac{1}{4}$

f. $\dfrac{1}{4} - \dfrac{1}{4} = \dfrac{0}{4} = 0$

LESSON 24, MIXED PRACTICE

1.
$$\begin{array}{r} \$6.00 \\ \times \quad 5 \\ \hline \$30.00 \end{array} \qquad \begin{array}{r} \$30.00 \\ + \ \$5.00 \\ \hline \mathbf{\$35.00} \end{array}$$
Multiplication pattern (equal groups); addition pattern (combining)

2.

1 dozen cookies

$\dfrac{3}{4}$ eaten $\left\{\begin{array}{|c|}\hline \text{3 cookies} \\\hline \text{3 cookies} \\\hline \text{3 cookies} \\\hline\end{array}\right.$

$\dfrac{1}{4}$ not eaten $\left\{\begin{array}{|c|}\hline \text{3 cookies} \\\hline\end{array}\right.$

9 cookies

3.
$$\begin{array}{r} 220 \\ 8\overline{)1760} \\ \underline{16} \\ 16 \\ \underline{16} \\ 00 \\ \underline{00} \\ 0 \end{array}$$

1 mile = 1760 yards

$\dfrac{1}{8}$ of a mile $\left\{\begin{array}{|c|}\hline \text{220 yards} \\\hline\end{array}\right.$

$\dfrac{7}{8}$ of a mile $\left\{\begin{array}{|c|}\hline \text{220 yards} \\\hline \text{220 yards} \\\hline \text{220 yards} \\\hline \text{220 yards} \\\hline \text{220 yards} \\\hline \text{220 yards} \\\hline \text{220 yards} \\\hline\end{array}\right.$

220 yards

4. $\dfrac{1}{4} + \dfrac{2}{4} = \dfrac{3}{4}$

5. $\dfrac{7}{8} - \dfrac{4}{8} = \dfrac{3}{8}$

6. $\dfrac{1}{2} + \dfrac{1}{2} = \dfrac{2}{2} = 1$

7. $\dfrac{1}{2} - \dfrac{1}{2} = \dfrac{0}{2} = 0$

8.

75%

9. $\dfrac{\text{fiction books}}{\text{nonfiction books}} = \dfrac{41}{23}$

10. Add 123 and 321 and divide by 2.

11.
$$\begin{array}{r} \overset{2}{25} \text{ miles} \\ \times \quad 4 \\ \hline \mathbf{100} \text{ miles} \end{array}$$

12. A. 21

13.
$$\begin{array}{r} \mathbf{111 \ R \ 1} \\ 9\overline{)1000} \\ \underline{9} \\ 10 \\ \underline{9} \\ 10 \\ \underline{9} \\ 1 \end{array}$$

14.
$$\begin{array}{r} \mathbf{700 \ R \ 22} \\ 32\overline{)22{,}422} \\ \underline{22\ 4} \\ 02 \\ \underline{00} \\ 22 \\ \underline{00} \\ 22 \end{array}$$

15.
$$\begin{array}{r} \mathbf{\$3.50} \\ 100\overline{)\$350.00} \\ \underline{300} \\ 50\ 0 \\ \underline{50\ 0} \\ 00 \\ \underline{00} \\ 0 \end{array}$$

16. $\frac{1}{2} \; \enclose{circle}{>} \; \frac{1}{4}$

17. No; an estimate; I rounded 172 to 200 and 636 to 600; 200 + 600 = 800; 800 < 900, so Mr. Johnson will not be charged extra for the van.

18. 44°F

19. $\underbrace{(35 \times 35)} \; - \; \underbrace{(5 \times 5)}$

      1225   –   25

             1200

20. 33,000,000

21. Factors of 21
1, 3, 7, 21
Factors of 28
1, 2, 4, 7, 14, 28
GCF is **7.**

22. 123; No, the sum is not divisible by 9.
234; Yes, the sum is divisible by 9.
345; No the sum is not divisible by 9.
B. 234

23. $\frac{4}{4}$

24.
```
        8          check:      20
     × 20                   8)160
      160                     16
   w = 160                    00
                              00
                               0
```

25.
```
      12
   7)84          check:      12
     7                      × 7
     14                      84
     14
      0
   x = 12
```

26.
```
     7
   4 8̶ ¹1         check:        ¹
   − 3 7 6                    376
     1 0 5                  + 105
   w = 105                    481
```

27.
```
     ¹
    286          check:       ⁷
   + 592                    8̶ ¹7 8
    878                    − 2 8 6
   m = 878                   5 9 2
```

28. Madison

29.
```
    14000
  − 10000
    4000   About 4000 more people
```

30.

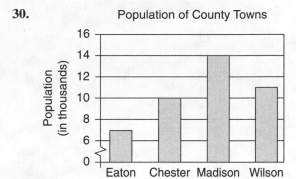

Population of County Towns

LESSON 25, WARM-UP

a. 258

b. 225

c. 86

d. 2500

e. $3.75

f. 15

g. 4

Problem Solving

The 3 possible coin combinations were outlined in the Lesson 21 Problem Solving exercise: 3Q, 2D, 1N; 1HD, 1Q, 1D, 3N; 1HD, 5D. At least one **dime** is contained in each combination.

LESSON 25, LESSON PRACTICE

a. $3\frac{4}{8}$ or $3\frac{1}{2}$ in.
```
      3 4/8
   8)28
     24
      4
```

b. $14\frac{2}{7}\%$
```
   7)100
     7
     30
     28
      2
```

c. $10\overline{)467}$ $\quad46\frac{7}{10}$

$\quad\quad\frac{40}{67}$

$\quad\quad\frac{60}{7}$

d. $12 \times 1 = \mathbf{12}$
$12 \times 2 = \mathbf{24}$
$12 \times 3 = \mathbf{36}$
$12 \times 4 = \mathbf{48}$

e. $8 \times 1 = \mathbf{8}$
$8 \times 2 = \mathbf{16}$
$8 \times 3 = \mathbf{24}$
$8 \times 4 = \mathbf{32}$
$8 \times 5 = \mathbf{40}$
$8 \times 6 = \mathbf{48}$

f. $8 \times 3 = 24$
$12 \times 2 = 24$
24

g. $6\overline{)35}$ $\quad5\frac{5}{6}$

$\quad\quad\frac{30}{5}$

h. $10\overline{)49}$ $\quad4\frac{9}{10}$

$\quad\quad\frac{40}{9}$

i. $12\overline{)65}$ $\quad5\frac{5}{12}$

$\quad\quad\frac{60}{5}$

LESSON 25, MIXED PRACTICE

1. $\left(\dfrac{1}{2} + \dfrac{1}{2}\right) - \left(\dfrac{1}{3} + \dfrac{1}{3}\right)$

$\dfrac{2}{2} - \dfrac{2}{3}$

$1 - \dfrac{2}{3}$

$\dfrac{3}{3} - \dfrac{2}{3}$

$\dfrac{1}{3}$

2. **Carlos can find the average distance of the three punts by adding 35 yards, 30 yards, and 37 yards and then dividing by 3.**

3. **149,600,000 kilometers**

4. $\dfrac{3}{8}$ in. $+ \dfrac{1}{8}$ in. $+ \dfrac{3}{8}$ in. $+ \dfrac{1}{8}$ in. $= \dfrac{8}{8}$ in. or **1 in.**

5. $4 \cdot p = 30$

$4\overline{)30}$ $\quad7\frac{2}{4}$ or $7\frac{1}{2}$ **inches**

$\quad\quad\frac{28}{2}$

6. $\dfrac{3}{3} - \dfrac{2}{3} = \dfrac{1}{3}$

7. $\dfrac{1}{2}$ of 12 $\;\bigcirc\!\!\!\!>\;$ $\dfrac{1}{3}$ of 12

$2\overline{)12}$ $\quad6$ $\quad\quad3\overline{)12}$ $\quad4$

$\quad\frac{12}{0}$ $\quad\quad\quad\frac{12}{0}$

8.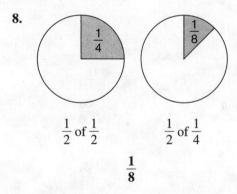

$\dfrac{1}{2}$ of $\dfrac{1}{2}$ $\quad\quad\quad$ $\dfrac{1}{2}$ of $\dfrac{1}{4}$

$\dfrac{1}{8}$

9. $9\overline{)100\%}$ $\quad11\frac{1}{9}\%$

$\quad\quad\frac{9}{10}$

$\quad\quad\frac{9}{1}$

10. (a) **6**

(b) **3**

11. $\dfrac{1}{3}$

12. $7\overline{)365}$ → $52\frac{1}{7}$

$$\begin{array}{r} 52\frac{1}{7} \\ 7\overline{)365} \\ 35 \\ \hline 15 \\ 14 \\ \hline 1 \end{array}$$

13. $\frac{2}{3} + \frac{2}{3} + \frac{2}{3} = \frac{6}{3} = \mathbf{2}$

14. $\frac{6}{6} - \frac{5}{6} = \mathbf{\frac{1}{6}}$

15. $\underbrace{30 \times 40}_{} \div 60$

$1200 \div 60$

20

16. $\frac{5}{12} - \frac{5}{12} = \frac{0}{12} = \mathbf{0}$

17.
$$\begin{array}{r} \overset{1}{\cancel{2}}^{1}0 \\ -7 \\ \hline 1\;3 \text{ losses} \end{array}$$

$\dfrac{\text{win}}{\text{loss}} = \mathbf{\dfrac{7}{13}}$

18. $10 \cdot 25 = t$

$$\begin{array}{r} \$0.25 \\ \times 10 \\ \hline \mathbf{\$2.50} \end{array}$$

19. Factors of 24
1, 2, 3, 4, 6, 8, 12, 24
Factors of 30
1, 2, 3, 5, 6, 10, 15, 30
GCF is **6.**

20.
$$\begin{array}{r} 1 \\ 100\overline{)100} \\ 100 \\ \hline 0 \end{array}$$

21. $\frac{8}{8} - \frac{5}{8} = \frac{3}{8}$

$m = \mathbf{\dfrac{3}{8}}$

check: $\quad \frac{5}{8} + \frac{3}{8} = \frac{8}{8} = 1$

22.
$$\begin{array}{r} 12 \\ 12\overline{)144} \\ 12 \\ \hline 24 \\ 24 \\ \hline 0 \end{array}$$
check:
$$\begin{array}{r} 12 \\ 12\overline{)144} \\ 12 \\ \hline 24 \\ 24 \\ \hline 0 \end{array}$$

$n = \mathbf{12}$

23.
$$\begin{array}{r} 3000 \\ 6000 \\ +5000 \\ \hline \mathbf{14,000} \end{array}$$

24.
$$\begin{array}{r} 20 \\ 3\overline{)60} \\ 6 \\ \hline 00 \\ 00 \\ \hline 0 \end{array}$$

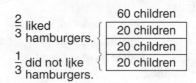

$\frac{2}{3}$ liked hamburgers.

$\frac{1}{3}$ did not like hamburgers.

40 children

25. **About 2 inches;**
$1\frac{14}{16}$ **or** $1\frac{7}{8}$ **inches**

26. **One possibility: Jan could draw segments from the center of the circle to the places where 12, 4, and 8 would be on a clock face.**

27.
$$\begin{array}{r} 3\frac{3}{4} \\ 4\overline{)15} \\ 12 \\ \hline 3 \end{array}$$

28.

$\frac{3}{4}$ $<$ $\frac{4}{5}$

29. $25 \times 1 = \mathbf{25}$
$25 \times 2 = \mathbf{50}$
$25 \times 3 = \mathbf{75}$
$25 \times 4 = \mathbf{100}$

30. 910; No, sum is not divisible by 9.
8910; Yes, sum is divisible by 9 and last digit is 0.
78,910; No, sum is not divisible by 9.
B. 8910

LESSON 26, WARM-UP

a. 238

b. 224

c. 93

d. 600

e. $3.25

f. 16

g. 2

Problem Solving

The total number of dots on a dot cube is 21. The total number of dots on opposite sides of a dot cube is 7. So Tad must have counted $21 - 7$, or **14 dots.**

LESSON 26, LESSON PRACTICE

a. $\dfrac{1}{4}$

b. $\dfrac{3}{4}$

c.
$$12\tfrac{1}{2}\%$$
$$+\ 12\tfrac{1}{2}\%$$
$$\overline{24\tfrac{2}{2}}$$
$$\tfrac{2}{2} = 1$$
$$24 + 1 = \mathbf{25\%}$$

d.
$$16\tfrac{2}{3}\%$$
$$+\ 66\tfrac{2}{3}\%$$
$$\overline{82\tfrac{4}{3}}$$
$$\tfrac{4}{3} = 1\tfrac{1}{3}$$
$$82 + 1\tfrac{1}{3} = \mathbf{83\tfrac{1}{3}\%}$$

e.
$$3\tfrac{3}{4}$$
$$+\ 2\tfrac{3}{4}$$
$$\overline{5\tfrac{6}{4}}$$
$$\tfrac{6}{4} = 1\tfrac{2}{4} = 1\tfrac{1}{2}$$
$$5 + 1\tfrac{1}{2} = \mathbf{6\tfrac{1}{2}}$$

f.
$$1\tfrac{1}{8}$$
$$+\ 2\tfrac{7}{8}$$
$$\overline{3\tfrac{8}{8}}$$
$$\tfrac{8}{8} = 1$$
$$3 + 1 = \mathbf{4}$$

g.
$$3$$
$$+\ 2\tfrac{2}{3}$$
$$\overline{\mathbf{5\tfrac{2}{3}}}$$

h.
$$\tfrac{3}{4}$$
$$+\ 4$$
$$\overline{\mathbf{4\tfrac{3}{4}}}$$

LESSON 26, MIXED PRACTICE

1.
$$3\tfrac{3}{4}\ \text{mi}$$
$$+\ 3\tfrac{3}{4}\ \text{mi}$$
$$\overline{6\tfrac{6}{4}\ \text{mi}}$$
$$\tfrac{6}{4} = 1\tfrac{2}{4} = 1\tfrac{1}{2}$$
$$6 + 1\tfrac{1}{2} = \mathbf{7\tfrac{1}{2}\ miles}$$

2.
$$\begin{array}{r} \mathbf{3}\ \textbf{years old} \\ 12\overline{)36} \\ \underline{36} \\ 0 \end{array}$$

3. 1 dozen $= 12$
$$\tfrac{1}{2} \text{ dozen} = 6$$
$$12 + 12 + 6 = \mathbf{30\ children}$$

4.   1 m = 100 cm
1 km = 1000 m

$$\begin{array}{r} 1000 \\ \times \quad 100 \\ \hline \textbf{100,000 centimeters} \end{array}$$

5. $\frac{2}{3}$ in. + $\frac{2}{3}$ in. + $\frac{2}{3}$ in. = $\frac{6}{3}$ in. = $\frac{2}{1}$ in.

   = 2 in.

6. $\frac{1}{2}$ + $\frac{1}{2}$ $\bigcirc\!\!\!>$ $\frac{1}{2}$ × $\frac{1}{2}$

   $\frac{2}{2}$ $\qquad$ $\frac{1}{4}$

   1

7. $\begin{array}{r} 5\frac{7}{8} \\ + \quad 7\frac{5}{8} \\ \hline 12\frac{12}{8} \end{array}$

   $\frac{12}{8} = \frac{3}{2} = 1\frac{1}{2}$

   $12 + 1\frac{1}{2} = \mathbf{13\frac{1}{2}}$

8. $12\frac{1}{2}\%$

   $12\frac{1}{2}\%$

   $\underline{+\ 12\frac{1}{2}\%}$

   $36\frac{3}{2}\%$

   $\frac{3}{2} = 1\frac{1}{2}$

   $36 + 1\frac{1}{2} = \mathbf{37\frac{1}{2}\%}$

9. $\mathbf{\frac{12}{12}}$

10.   Factors of 15
1, 3, 5, 15
Factors of 25
1, 5, 25
GCF is **5.**

11.   **Add 8 to the value of a term to find the next term; 56**

12. $\begin{array}{r} 2\frac{4}{5} \\ 5{\overline{\smash{\big)}\,14}} \\ \underline{10} \\ 4 \end{array}$

13. $\frac{2}{5}$ + $\frac{4}{5}$ = $\frac{6}{5}$ = $\mathbf{1\frac{1}{5}}$

14. $1 - \frac{2}{3}$ $\qquad$ check: $\qquad$ $\frac{2}{3} + \frac{1}{3} = \frac{3}{3} = 1$

   $\frac{3}{3} - \frac{2}{3} = \frac{1}{3}$

   $n = \mathbf{\frac{1}{3}}$

15.   Factors of 12
1, 2, 3, 4, 6, 12
Factors of 18
1, 2, 3, 6, 9, 18
GCF is **6.**

16. $1 - \frac{3}{4}$

   $\frac{4}{4} - \frac{3}{4} = \mathbf{\frac{1}{4}}$

17. $\begin{array}{r} 3\frac{3}{4} \\ + \quad 3 \\ \hline \mathbf{6\frac{3}{4}} \end{array}$

18. $2\frac{1}{2} - 2\frac{1}{2} = \mathbf{0}$

19.   4671; No, the last digit is not even.
3858; Yes; the last digit is even and the sum of the digits is divisible by 3.
6494; No, the sum of the digits is not divisible by 3.
B. 3858

20.   **31, 37**

21. $\begin{array}{r} 5000 \\ - \quad 4000 \\ \hline \mathbf{1000} \end{array}$

22. $\mathbf{\frac{3}{4}}$

23. $\begin{array}{r} \overset{1}{}85 \\ 85 \\ \overset{2}{}90 \\ + \quad 100 \\ \hline 360 \end{array}$ $\qquad$ $\begin{array}{r} 90 \\ 4{\overline{\smash{\big)}\,360}} \\ \underline{36} \\ 00 \\ \underline{00} \\ 0 \end{array}$

24.

$\frac{3}{5}$ of $30 $\left\{\begin{array}{l}\end{array}\right.$

$\frac{2}{5}$ of $30 $\left\{\begin{array}{l}\end{array}\right.$

$30
$6
$6
$6
$6
$6

$18

25. (a) **25 millimeters**

(b) **1 inch**

26. $-1,\ 0,\ \frac{1}{2},\ 1$

27.
$$\begin{array}{r} 55 \text{ inches} \\ - \ 25 \text{ inches} \\ \hline \end{array}$$
About 30 inches

28.
$$\begin{array}{r} \overset{1}{2}5 \text{ in.} \\ 55 \text{ in.} \\ + \ 40 \text{ in.} \\ \hline 120 \text{ in.} \end{array} \qquad \begin{array}{r} 40 \text{ inches} \\ 3\overline{)120} \\ \underline{12} \\ 00 \\ \underline{00} \\ 0 \end{array}$$

29.
$$\begin{array}{r} \overset{3}{\cancel{4}}{}^{1}0 \text{ inches} \\ - \ 2\ 5 \text{ inches} \\ \hline \end{array}$$
About 1 5 inches

30. **Answers may vary. Sample answer: What was the approximate total rainfall during the first three years?**
$$\begin{array}{r} \overset{1}{2}5 \text{ in.} \\ 55 \text{ in.} \\ + \ 40 \text{ in.} \\ \hline \end{array}$$
About 120 inches

LESSON 27, WARM-UP

a. **364**

b. **198**

c. **82**

d. **306**

e. **$2.75**

f. **43**

g. **10**

Problem Solving

The missing digit in the hundreds column can be only 7 or 8. If 8 is used in the hundreds column, then only 9 and 7 can be used in the ones column. Then 1 and 6 must be used in the tens column.
$$\begin{array}{r} 819 \\ - \ 452 \\ \hline 367 \end{array}$$

LESSON 27, LESSON PRACTICE

a. **Diameter**

b. **Circumference**

c. **Radius**

d. $\begin{array}{r} 5 \text{ in.} \\ 2\overline{)10} \\ \underline{10} \\ 0 \end{array}$

LESSON 27, MIXED PRACTICE

1. $\underbrace{(55 + 45)}_{100} \times \underbrace{(55 - 45)}_{10}$

1000

2.

$\frac{3}{4}$ is water. $\left\{\begin{array}{l}\end{array}\right.$

$\frac{1}{4}$ is not water. $\left\{\begin{array}{l}\end{array}\right.$

20 pounds
5 pounds
5 pounds
5 pounds
5 pounds

15 pounds

3. $306 - 249 = d$
or
$306 - d = 249$
$$\begin{array}{r} \overset{2}{\cancel{3}}\overset{9}{\cancel{0}}{}^{1}6 \\ - \ 2\ 4\ 9 \\ \hline 5\ 7 \text{ fleas} \end{array}$$

4. $\begin{array}{r} 2\frac{1}{2} \text{ in.} \\ 2\overline{)5} \\ \underline{4} \\ 1 \end{array}$

5. 122; No, the sum is not divisible by 3.
123; No, the last digit is not even.
132; Yes, the sum is divisible by 3 and the last digit is even.
C. 132

6. **1,230,000**

7. $10p = \$4.90$

$$\begin{array}{r} \$0.49 \\ 10\overline{)\$4.90} \\ \underline{4\,0} \\ 90 \\ \underline{90} \\ 0 \end{array}$$

49¢ per pound

8. **24**

9.

39

10.

$10

11. $\dfrac{1}{6} + \dfrac{2}{6} + \dfrac{3}{6} = \dfrac{6}{6} = \mathbf{1}$

12. $\dfrac{7}{8} - \dfrac{3}{8} = \dfrac{4}{8} = \mathbf{\dfrac{1}{2}}$

13. $\dfrac{6}{6} - \dfrac{5}{6} = \mathbf{\dfrac{1}{6}}$

14. $\dfrac{2}{8} + \dfrac{5}{8} = \mathbf{\dfrac{7}{8}}$

15. (a) **8**

(b) **4**

16. $\mathbf{\dfrac{2}{3}}$

17. $\dfrac{1}{8}$

18. $\dfrac{1}{2}$

19.

$$\begin{array}{r} \mathbf{40\,R\,20} \\ 52\overline{)2100} \\ \underline{208} \\ 20 \\ \underline{00} \\ 20 \end{array}$$

20.

$$\begin{array}{r} \mathbf{9\ inches} \\ 4\overline{)36} \\ \underline{36} \\ 0 \end{array}$$

21.

$$\begin{array}{r} \mathbf{1\tfrac{1}{6}} \\ 6\overline{)7} \\ \underline{6} \\ 1 \end{array}$$

22.

$$\begin{array}{r} \mathbf{24} \\ 18\overline{)432} \\ \underline{36} \\ 72 \\ \underline{72} \\ 0 \end{array}$$

23. $\underbrace{(55 + 45)}_{100} \div \underbrace{(55 - 45)}_{10}$

$$100 \div 10$$

10

24. 502; No, the last digit is not 0.
205; No, the last digit is not 0.
250; Yes, the last digit is 0.
C. 250

25. **One method is to add the digits of the number. If the sum of the digits is divisible by 9, the number is also divisible by 9.**

26. **2**

27. **Circumference**

28. $\mathbf{\dfrac{4}{5}}$

29.

$$37\frac{1}{2}\%$$
$$-\ 12\frac{1}{2}\%$$
$$\overline{\quad 25\frac{0}{2}\%\quad}$$

25%

30.

$$33\frac{1}{3}\%$$
$$+\ 16\frac{2}{3}\%$$
$$\overline{\quad 49\frac{3}{3}\%\quad}$$

$$\frac{3}{3} = 1$$
$$49 + 1 = \mathbf{50\%}$$

LESSON 28, WARM-UP

a. **336**

b. **255**

c. **85**

d. **1250**

e. **$1.75**

f. **18**

g. **5**

Problem Solving

> **Three of the six faces of each small block were painted.** (The other three faces of each block were in the interior of the large cube.)

LESSON 28, LESSON PRACTICE

a. **Observe student.**

b. **Obtuse**

c. **Right**

d. **Acute**

e. 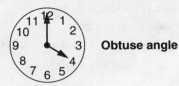 **Obtuse angle**

f.

Right angle

g. $\angle 1$ and $\angle 3$

h. ═══════

i.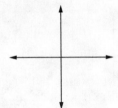

j. $\angle F$ (or $\angle HFG$ or $\angle GFH$)

k. $\angle G$ (or $\angle FGH$ or $\angle HGF$)

LESSON 28, MIXED PRACTICE

1. $\dfrac{1}{3} + \dfrac{2}{3} + \dfrac{3}{3} = \dfrac{6}{3} = \mathbf{2}$

2.

	140 men
$\dfrac{2}{5}$ rode with Little John.	28 men
	28 men
	28 men
$\dfrac{3}{5}$ did not ride with Little John.	28 men
	28 men

56 men

3. $32g = 768$

```
       24 peanuts
32)768
   64
   ---
   128
   128
   ---
     0
```

4. $1776 - 1492 = d$

$$
\begin{array}{r}
1\,\overset{6}{\cancel{7}}{}^{1}7\,6 \\
-\ 1\,4\,9\,2 \\
\hline
2\,8\,4 \ \text{years}
\end{array}
$$

5.
$$
\begin{array}{r}
7\frac{2}{3} \\
3\overline{)23} \\
\underline{21} \\
2
\end{array}
$$

6.
$$
\begin{array}{r}
1\frac{2}{3} \\
+\ 1\frac{2}{3} \\
\hline
2\frac{4}{3}
\end{array}
$$

$$\frac{4}{3} = 1\frac{1}{3}$$

$$2 + 1\frac{1}{3} = 3\frac{1}{3}$$

7.
$$
\begin{array}{r}
3 \\
+\ 4\frac{2}{3} \\
\hline
7\frac{2}{3}
\end{array}
$$

8.
$$
\begin{array}{r}
3\frac{5}{6} \\
-\ 1\frac{4}{6} \\
\hline
2\frac{1}{6}
\end{array}
$$

9. $\dfrac{1}{2}$

10.

	$24.00
$\frac{2}{3}$ of $24.00	$8.00
	$8.00
$\frac{1}{3}$ of $24.00	$8.00

$16.00

11. (a)
$$
\begin{array}{r}
10\% \\
10\overline{)100\%} \\
\underline{10} \\
00 \\
\underline{00} \\
0
\end{array}
$$

$$
\begin{array}{r}
10\% \\
\times\ \ 3 \\
\hline
30\%
\end{array}
$$

12. $\dfrac{1}{4}$

13.
$$
\begin{array}{r}
250 \\
-\ 240 \\
\hline
10
\end{array}
$$

14. $1 - \dfrac{1}{4}$

$$\frac{4}{4} - \frac{1}{4} = \frac{3}{4}$$

$$m = \frac{3}{4}$$

check: $\quad \dfrac{1}{4} + \dfrac{3}{4} = \dfrac{4}{4} = 1$

15.
$$
\begin{array}{r}
\overset{3}{\cancel{4}}\,\overset{1}{\cancel{2}}{}^{1}3 \\
-\ 2\,9\,7 \\
\hline
1\,2\,6
\end{array}
\qquad
\text{check:}
\qquad
\begin{array}{r}
\overset{3}{\cancel{4}}\,\overset{1}{\cancel{2}}{}^{1}3 \\
-\ 1\,2\,6 \\
\hline
2\,9\,7
\end{array}
$$

$$w = 126$$

16. (a) $\angle PSQ$ or $\angle QSP$

(b) $\angle QSR$ or $\angle RSQ$

17.

22 correct answers

$$
\begin{array}{r}
22 \\
20 \\
23 \\
+\ 23 \\
\hline
88
\end{array}
\qquad
\begin{array}{r}
22 \\
4\overline{)88} \\
\underline{8} \\
08 \\
\underline{8} \\
0
\end{array}
$$

18.

12 inches

$$
\begin{array}{r}
3\overline{)36} \\
\underline{3} \\
06 \\
\underline{6} \\
0
\end{array}
$$

19. Factors of 24
1, 2, 3, 4, 6, 8, 12, 24
Factors of 36
1, 2, 3, 4, 6, 9, 12, 18, 36
Factors of 60
1, 2, 3, 4, 5, 6, 10, 12, 15, 20, 30, 60
GCF is **12.**

20.
$$
\begin{array}{r}
\overset{0}{\cancel{1}}\overset{9}{\cancel{0}},\overset{1}{0}\,\overset{0}{\cancel{1}}\overset{0}{\cancel{1}}0 \\
-\ \ \ 9\,9\,0\,9 \\
\hline
1\,0\,1
\end{array}
$$

21. $\underline{(100 \times 100)} - \underline{(100 \times 99)}$

$10,000 - 9900$

$\mathbf{100}$

22.

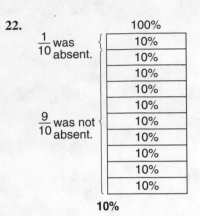

$\frac{1}{10}$ was absent. / $\frac{9}{10}$ was not absent.

100% / 10% (×10) / 10%

23.

$$10\overline{)5097} \quad 509\frac{7}{10}$$

$$\underline{50}$$
$$09$$
$$\underline{0}$$
$$97$$
$$\underline{90}$$
$$7$$

24.

24 eggs

$\frac{3}{4}$ of 24 { 6 eggs / 6 eggs / 6 eggs

$\frac{1}{4}$ of 24 { 6 eggs

18 eggs

25. (a) $3\frac{3}{16}$ inches

(b) **8 centimeters**

26.
$6 \times 1 = 6 \qquad 8 \times 1 = 8$
$6 \times 2 = 12 \qquad 8 \times 2 = 16$
$6 \times 3 = 18 \qquad 8 \times 3 = 24$
$6 \times 4 = 24 \qquad 8 \times 4 = 32$
$6 \times 5 = 30 \qquad 8 \times 5 = 40$

6, 12, 18, (24), 30

8, 16, (24), 32, 40

27. $\frac{1}{6}$

28.
$\begin{array}{r} 13 \\ - 7 \\ \hline \end{array}$
6 white stripes

$\frac{\text{red stripes}}{\text{white stripes}} = \frac{7}{6}$

29. $3 \cdot 3 \cdot 3 = 27$

30. 234; Yes, sum of digits is divisible by 9.
345; No, sum of digits is not divisible by 9.
567; Yes, sum of digits is divisible by 9.
B. 345

LESSON 29, WARM-UP

a. **301**

b. **256**

c. **92**

d. **375**

e. **$2.75**

f. **35**

g. **50**

Problem Solving
HHH, HHT, HTH, HTT,
THH, THT, TTH, TTT

LESSON 29, LESSON PRACTICE

a. $\frac{1}{2} \times \frac{4}{5} = \frac{4}{10}$

$\frac{4 \div 2}{10 \div 2} = \frac{2}{5}$

b. $\frac{1}{4} \times \frac{2}{3} = \frac{2}{12}$

$\frac{2 \div 2}{12 \div 2} = \frac{1}{6}$

c. $\frac{2}{3} \times \frac{3}{4} = \frac{6}{12}$

$\frac{6 \div 6}{12 \div 6} = \frac{1}{2}$

d. $\frac{5}{6} \times \frac{6}{5} = \frac{30}{30}$

$$30\overline{)30} \quad 1$$
$$\underline{30}$$
$$0$$

e. $\dfrac{5}{1} \times \dfrac{2}{3} = \dfrac{10}{3}$

$$3\overline{)10}\;\;\mathbf{3\tfrac{1}{3}}$$
$$\dfrac{9}{1}$$

f. $\dfrac{2}{1} \times \dfrac{4}{3} = \dfrac{8}{3}$

$$3\overline{)8}\;\;\mathbf{2\tfrac{2}{3}}$$
$$\dfrac{6}{2}$$

g. $\dfrac{9 \div 3}{12 \div 3} = \mathbf{\dfrac{3}{4}}$

h. $\dfrac{6 \div 2}{10 \div 2} = \mathbf{\dfrac{3}{5}}$

i. $\dfrac{18 \div 6}{24 \div 6} = \mathbf{\dfrac{3}{4}}$

j.
$$\begin{array}{r} 30 \\ -\ 20 \\ \hline 10 \text{ boys} \end{array}$$

$\dfrac{\text{number of boys}}{\text{number of girls}} = \dfrac{10}{20} = \mathbf{\dfrac{1}{2}}$

LESSON 29, MIXED PRACTICE

1.
$$\begin{array}{r} 2000 \\ \times\ \ \ \ \ 8 \\ \hline \mathbf{16{,}000 \text{ pounds}} \end{array}$$

2.
$$\begin{array}{r} 16 \\ \times\ 16 \\ \hline 96 \\ 160 \\ \hline \mathbf{256 \text{ jelly beans}} \end{array}$$

3. $\left(\dfrac{1}{2} + \dfrac{1}{2}\right) - \left(\dfrac{1}{2} \times \dfrac{1}{2}\right)$

$$\dfrac{2}{2} \quad - \quad \dfrac{1}{4}$$
$$1 \quad - \quad \dfrac{1}{4}$$
$$\dfrac{4}{4} \quad - \quad \dfrac{1}{4}$$
$$\mathbf{\dfrac{3}{4}}$$

4. $\dfrac{\text{win}}{\text{loss}} = \dfrac{6}{8} = \mathbf{\dfrac{3}{4}}$

5. $\dfrac{16 \div 8}{24 \div 8} = \mathbf{\dfrac{2}{3}}$

6. $\dfrac{1}{8} + \dfrac{3}{8} = \dfrac{4}{8}$

$\dfrac{4 \div 4}{8 \div 4} = \mathbf{\dfrac{1}{2}}$

7. $\dfrac{1}{2} \times \dfrac{2}{3} = \dfrac{2}{6}$

$\dfrac{2 \div 2}{6 \div 2} = \mathbf{\dfrac{1}{3}}$

8. $\dfrac{7}{12} - \dfrac{3}{12} = \dfrac{4}{12}$

$\dfrac{4 \div 4}{12 \div 4} = \mathbf{\dfrac{1}{3}}$

9.

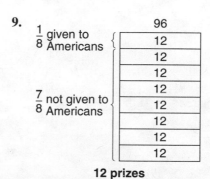

$\dfrac{1}{8}$ given to Americans

$\dfrac{7}{8}$ not given to Americans

12 prizes

10. ..., **13**, **16**, **19**, ...

Add 3 to a term to find the next term.

11.
$$\begin{array}{r} 12 \\ -\ \ 5 \\ \hline 7 \end{array} \qquad \mathbf{\dfrac{7}{12}}$$

12.
$$\begin{array}{r} \$3.60 \\ \times\ \ \ \ \ 100 \\ \hline \mathbf{\$360.00} \end{array}$$

13.
$$100\overline{)50{,}000}\;\;\mathbf{500}$$
$$\dfrac{50\ 0}{00}$$
$$\dfrac{00}{00}$$
$$\dfrac{00}{0}$$

14. $4\overline{)18}$ $\dfrac{4\frac{2}{4} = 4\frac{1}{2}}{}$

$\dfrac{16}{2}$

15.

$$\begin{array}{c}\text{-8 -6 -4 -2 0 2 4 6 8 10 12 14}\end{array}$$

23°F

16. $m + \underbrace{496 + 2684} = 3217$

$\quad\quad m + 3180 \;\;= 3217$

$\begin{array}{r} 3\,\overset{1}{2}{\overset{1}{1}}7 \\ -\ 3\,1\,8\,0 \\ \hline 3\,7 \end{array}$ check: $\begin{array}{r} \overset{1}{2}37 \\ \overset{1}{}496 \\ +\ 2684 \\ \hline 3217 \end{array}$

$m = 37$

17. $\begin{array}{r} \overset{0}{1}\overset{9}{\cancel{0}}\overset{9}{\cancel{0}}{}^{1}0 \\ -\ 8\,5\,7 \\ \hline 1\,4\,3 \end{array}$ check: $\begin{array}{r} \overset{0}{1}\overset{9}{\cancel{0}}\overset{9}{\cancel{0}}{}^{1}0 \\ -\ 1\,4\,3 \\ \hline 8\,5\,7 \end{array}$

$n = 143$

18. $24\overline{)480}$ check: $\begin{array}{r} 24 \\ \times\ \ 20 \\ \hline 480 \end{array}$

$\dfrac{48}{00}$

$\dfrac{00}{0}$

$x = 20$

19. $\underbrace{7 \cdot 11} \cdot 13$

$\quad 77 \cdot 13$

$\quad\;\; \mathbf{1001}$

20. To estimate 4963 ÷ 39, first round 4963 to 5000 and round 39 to 40. Then divide 5000 by 40.

21. $\dfrac{2}{3} \times \dfrac{3}{2} \;\textcircled{=}\; 1$

$\dfrac{6}{6} = 1$

22. $10\,\text{mm} + 10\,\text{mm} = 20\,\text{mm}$
$60\,\text{mm} - 20\,\text{mm} = 40\,\text{mm}$
$40\,\text{mm} \div 2 = \mathbf{20\,mm}$

23. $\begin{array}{r} 12 \\ -\ 40 \\ \hline -\ 28 \end{array}$

24. $\left(\dfrac{1}{2} \times \dfrac{1}{2}\right) - \dfrac{1}{4}$

$\quad \dfrac{1}{4} - \dfrac{1}{4}$

$\quad\quad \mathbf{0}$

25. (a) $\angle A$ and $\angle C$

(b) $\angle B$ and $\angle D$

26. $\dfrac{2}{3} \times \dfrac{3}{5} = \dfrac{6}{15}$

$\dfrac{6 \div 3}{15 \div 3} = \dfrac{\mathbf{2}}{\mathbf{5}}$

27. $\dfrac{3}{4} \times \dfrac{4}{3} = \dfrac{12}{12}$

$\dfrac{12 \div 12}{12 \div 12} = \mathbf{1}$

28. $2\overline{)24}$ $\dfrac{\text{radius}}{\text{diameter}} = \dfrac{12}{24} = \dfrac{\mathbf{1}}{\mathbf{2}}$

$\dfrac{2}{04}$

$\dfrac{4}{0}$

29.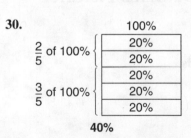

Acute angle

30.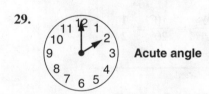

$\dfrac{2}{5}$ of 100%

$\dfrac{3}{5}$ of 100%

40%

LESSON 30, WARM-UP

a. 288

b. 210

c. 94

d. 535

e. $7.25

f. 36

g. 49

Problem Solving

The single-digit factor must be 7 (6 is too small and 8 is too large to multiply by a number in the 230's and get a product in the 1600's). Since $8 \times 7 = 56$, 8 must be placed in the three-digit factor and 6 must be placed in the product.

$$\begin{array}{r} 238 \\ \times \quad 7 \\ \hline 1666 \end{array}$$

LESSON 30, LESSON PRACTICE

a. Multiples of 6: 6, 12, 18, (24), 30, 36, . . .
Multiples of 8: 8, 16, (24), 32, 40, 48, . . .
LCM is **24.**

b. Multiples of 3: 3, 6, 9, 12, (15), 18, 21, . . .
Multiples of 5: 5, 10, (15), 20, 25, 30, . . .
LCM is **15.**

c. Multiples of 5: 5, (10), 15, 20, 25, . . .
Multiples of 10: (10), 20, 30, 40, 50, . . .
LCM is **10.**

d. $\dfrac{1}{6}$

e. $\dfrac{3}{2}$

f. $\dfrac{5}{8}$

g. $\dfrac{3}{1}$

h. $\dfrac{3}{8} \times \dfrac{8}{3} = 1$

i. $4 \times \dfrac{1}{4} = 1$

j. $\dfrac{6}{1} \times \dfrac{1}{6} = 1$

k. $\dfrac{8}{7} \times \dfrac{7}{8} = 1$

l. $\dfrac{5}{2}$

m. $\dfrac{12}{5}$

LESSON 30, MIXED PRACTICE

1. 3, 6, 9, (12)
4, 8, (12)
$12 - 12 = \textbf{0}$

2.

117 pounds

$\dfrac{2}{3}$ is water. $\left\{\begin{array}{|c|}\hline \text{39 pounds} \\ \hline \text{39 pounds} \\ \hline\end{array}\right.$

$\dfrac{1}{3}$ is not water. $\left\{\begin{array}{|c|}\hline \text{39 pounds} \\ \hline\end{array}\right.$

About 78 pounds

3. $15\overline{)120}$
$\underline{120}$
0

$42 \cdot 8 = p$

$$\begin{array}{r} \overset{1}{42} \\ \times \quad 8 \\ \hline \end{array}$$
336 pieces of popcorn

4. $12 \times 1 = \textbf{12}$
$12 \times 2 = \textbf{24}$
$12 \times 3 = \textbf{36}$
$12 \times 4 = \textbf{48}$

5. Multiples of 4: 4, 8, (12), 16, 20, 24, 28, . . .
Multiples of 6: 6, (12), 18, 24, 30, 36, . . .
LCM is **12.**

6. $\begin{array}{r}\overset{5}{\cancel{6}}{}^{1}0 \\ -\ 1\ 2 \\ \hline 4\ 8\end{array}$ $\dfrac{\text{commercial time}}{\text{noncommercial time}} = \dfrac{12}{48} = \dfrac{1}{4}$

$\dfrac{12 \div 12}{48 \div 12} = \dfrac{1}{4}$

7. $\dfrac{2}{5} + \dfrac{2}{5} + \dfrac{2}{5} = \dfrac{6}{5} = 1\dfrac{1}{5}$

8. $\dfrac{10}{10} - \dfrac{1}{10} = \dfrac{\mathbf{9}}{\mathbf{10}}$

9. $\dfrac{11}{12} - \dfrac{1}{12} = \dfrac{10}{12}$

$\dfrac{10 \div 2}{12 \div 2} = \dfrac{\mathbf{5}}{\mathbf{6}}$

10. $\dfrac{3}{4} \times \dfrac{4}{3} = \dfrac{12}{12}$

$\dfrac{12 \div 12}{12 \div 12} = \dfrac{1}{1} = \mathbf{1}$

11. $\dfrac{5}{1} \times \dfrac{3}{4} = \dfrac{15}{4}$

$\begin{array}{r} 3\tfrac{3}{4} \\ 4\overline{)15} \\ \underline{12} \\ 3 \end{array}$

12. $\dfrac{5}{2} \times \dfrac{5}{3} = \dfrac{25}{6}$

$\begin{array}{r} 4\tfrac{1}{6} \\ 6\overline{)25} \\ \underline{24} \\ 1 \end{array}$

13. Factors of 24: 1, 2, 3, 4, 6, 8, 12, 24

14.
$\begin{array}{r} \overset{1}{}\$3.00 \\ \$24.00 \\ \underline{+\ \ \$6.50} \\ \mathbf{\$33.50} \end{array}$

15.
$\begin{array}{r} \overset{4}{\$\cancel{5}}.\overset{1}{0}\,0 \\ \underline{-\ \$1.\,5\,0} \\ \mathbf{\$3.\,5\,0} \end{array}$

16.
$\begin{array}{r} 600 \\ \underline{\times\ \ \ \ 400} \\ \mathbf{240{,}000} \end{array}$

17. $\angle C$

18. $\dfrac{2}{3} \times \dfrac{2}{3} \;\textcircled{<}\; \dfrac{2}{3} \times 1$

$\dfrac{4}{9} \qquad\qquad \dfrac{2}{3} \times \dfrac{3}{3} = \dfrac{6}{9}$

19.
$\begin{array}{r} \mathbf{5{,}000} \\ 100\overline{)500{,}000} \\ \underline{500}\phantom{{,}000} \\ -0\,0 \\ \underline{0\,0} \\ 0\,0 \\ \underline{0\,0} \\ 0\,0 \\ \underline{0\,0} \\ 0 \end{array}$

20.
$\begin{array}{r} \mathbf{244} \\ 35\overline{)8540} \\ \underline{70} \\ 154 \\ \underline{140} \\ 140 \\ \underline{140} \\ 0 \end{array}$

21.
$\begin{array}{r} \mathbf{14\tfrac{2}{7}\%} \\ 7\overline{)100\%} \\ \underline{7} \\ 30 \\ \underline{28} \\ 2 \end{array}$

22. $\dfrac{4 \div 4}{12 \div 4} = \dfrac{\mathbf{1}}{\mathbf{3}}$

23.
$\begin{array}{r} \overset{1}{3}75 \\ 632 \\ \underline{+\ 571} \\ 1578 \end{array}$
$\qquad$
$\begin{array}{r} \mathbf{526} \\ 3\overline{)1578} \\ \underline{15} \\ 07 \\ \underline{6} \\ 18 \\ \underline{18} \\ 0 \end{array}$

24.
$\begin{array}{r} \mathbf{6\ inches} \\ 6\overline{)36} \\ \underline{36} \\ 0 \end{array}$

25. 1

26. $\dfrac{5}{2}$

27. $\dfrac{8}{3}$

28. $\dfrac{1}{2}$

29.
$$10\overline{)45} \quad \frac{4\frac{5}{10} = 4\frac{1}{2}}{\underset{5}{\underline{40}}}$$

30. $\frac{3}{4} \times \frac{4}{1} = \frac{12}{4} = \frac{3}{1} = $ **3 inches**

INVESTIGATION 3

1. m∠AOC = **15°**

2. m∠AOE = **45°**

3. m∠AOF = **90°**

4. m∠AOH = **142°**

5. m∠IOH = **38°**

6. m∠IOE = **135°**

INVESTIGATION 3, ACTIVITY

1. **45°**

2. **90°**

3. **30°**

4. **150°**

5. **135°**

6. **165°**

7. **60°**

8. **165°**

9. **90°**

10. **180°**

11. **105°**

12. **150°**

7.

8.

9.

10.

11.

12.

13. See student work. All sides should have the same length, and all angles should have the same measure.

14. *AB* = 6 in.; *AC* = 6 in.

15. **60°**

16. See student work. Sides *ST* and *SU* should have the same length, and ∠S should measure 90°.

17. m∠T = 45°; m∠U = 45°

18. **14 cm**

LESSON 31, WARM-UP

a. **100**

b. **222**

c. **57**

d. **$8.75**

e. **21**

f. **60**

g. **8**

Problem Solving

If all 60 passengers sat two to a seat, then passengers would fill $60 \div 2$, or 30 seats. Since there are 40 seats total, there would be $40 - 30$, or **10 seats,** left empty.

If one passenger sat in every seat, then there would be $60 - 40$, or 20 passengers, remaining. If each of the 20 remaining passengers is added to a seat, there would be $40 - 20$, or **20 seats,** with just one passenger.

LESSON 31, LESSON PRACTICE

a. 6 square units in a row
$\times$ 4 rows
24 square units

b. 7 square units in a row
$\times$ 7 rows
49 square units

c. $8 \text{ mm} \times 5 \text{ mm} =$ **40 sq. mm**

d. $12 \text{ mm} \times 12 \text{ mm} =$ **144 sq. mm**

e. $5 \times 5 = 25$
5 inches

f. $5 \text{ in.} \times 4 =$ **20 inches**

LESSON 31, MIXED PRACTICE

1. 4, 8, ⑫
3, 6, 9, ⑫

$$12\overline{)12} \quad \frac{1}{}$$
$$\frac{12}{0}$$

2. **580,000,000 miles**

3. $3\overline{)10} \quad 3\frac{1}{3}$
$\frac{9}{1}$

4. 4 square stickers in a row
$\times$ 2 rows
8 square stickers

5. 10 tiles in a row
$\times$ 10 rows
100 tiles

6. $12 \text{ in.} \times 8 \text{ in.} =$ **96 square inches**

7. $7 \times 7 =$ **49**

8.

$\frac{2}{3}$ of 24 $\left\{\begin{array}{c} 8 \\ 8 \end{array}\right.$ $\frac{1}{3}$ of 24 $\left\{\begin{array}{c} 8 \end{array}\right.$
with 24 at top

16

9. $\overset{3}{\cancel{4}}{}^{1}2$ check: 24
$- \ 2\ 4$ $+ \ 18$
$1\ 8$ 42
$f =$ **18**

10. $\frac{1}{8} + \frac{1}{8} = \frac{2}{8} = \frac{1}{4}$

11. $\frac{5}{6} - \frac{1}{6} = \frac{4}{6} = \frac{2}{3}$

12. $\frac{2}{3} \cdot \frac{1}{2} = \frac{2}{6} = \frac{1}{3}$

13. $\frac{2}{3} \times \frac{5}{1} = \frac{10}{3}$

$3\overline{)10} \quad 3\frac{1}{3}$
$\frac{9}{1}$

14. 400
$\times \quad 500$
200,000

15. $10\overline{)\$20.00} \quad \$2.00$
$\frac{20}{0\ 0}$
$\frac{0\ 0}{00}$
$\frac{00}{0}$

16.
$$\begin{array}{r} 63 \\ \times\ 47¢ \\ \hline 441 \\ 2520 \\ \hline \mathbf{2961}¢\ \text{or}\ \mathbf{\$29.61} \end{array}$$

17.
$$\begin{array}{r} \mathbf{210\ R\ 3} \\ 22\overline{)4623} \\ \underline{44} \\ 22 \\ \underline{22} \\ 03 \\ \underline{00} \\ 3 \end{array}$$

18. Smallest odd prime number is 3

$$\frac{1}{3}$$

19. One third equals $33\frac{1}{3}\%$

$$\begin{array}{r} 33\frac{1}{3}\% \\ +\ 33\frac{1}{3}\% \\ \hline 66\frac{2}{3}\% \end{array}$$

20.
$$\begin{array}{r} 100 \\ -\ 90 \\ \hline 10 \end{array} \quad \begin{array}{r} {}^{0}\cancel{1}\,{}^{9}\cancel{0}{}^{1}0 \\ -\ 8\,9 \\ \hline 1\,1 \end{array} \quad \begin{array}{r} 111 \\ -\ 100 \\ \hline 11 \end{array} \quad \begin{array}{r} 109 \\ -\ 100 \\ \hline 9 \end{array}$$

D. 109

21. **3,670,000,000 miles**

22.
$$\begin{array}{r} 7\ \text{inches} \\ 2\overline{)14} \\ \underline{14} \\ 0 \end{array}$$

$$\frac{\text{radius}}{\text{diameter}} = \frac{7\ \text{in.}}{14\ \text{in.}} = \frac{1}{2}$$

23. $\dfrac{3}{9} = \dfrac{1}{3}$

24. $2\dfrac{5}{8}$ **in.**

25. $\dfrac{3}{10} \times \dfrac{3}{10} = \dfrac{\mathbf{9}}{\mathbf{100}}$

26. $\dfrac{4}{3}$

27. $\dfrac{8}{8}$

28.

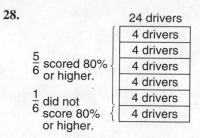

29. (a) $\angle PMQ$ or $\angle QMP$

(b) $\angle RMQ$ or $\angle QMR$

30. **If the room is rectangular, first measure the length of the room and the width of the room. Then multiply the length by the width to calculate the floor area of the room.**

LESSON 32, WARM-UP

a. 300

b. 1580

c. 73

d. $6.50

e. 120

f. 6

g. 10

Problem Solving

1, 2, and **3 dots**

The total number of dots on a dot cube is 21, so Monifa could not see $21 - 6$, or **15 dots.**

LESSON 32, LESSON PRACTICE

a. $(2 \times 100) + (5 \times 1)$

b. $(1 \times 1000) + (7 \times 100) + (6 \times 10)$

c. $(8 \times 1000) + (5 \times 10)$

d. $6000 + 400 = $ **6400**

e. $700 + 5 = $ **705**

f.
$$
\begin{array}{r}
{\scriptstyle 10\,:\,65} \\
\cancel{11}\!:\!\cancel{05}\text{ a.m.} \\
-\quad 7\!:\!15\text{ a.m.} \\
\hline
3\!:\!50
\end{array}
$$
3 hr 50 min

g.
$$
\begin{array}{r}
11\!:\!50\text{ p.m.} \\
+\quad 3\!:\!30 \\
\hline
14\!:\!80 \\
15\!:\!20
\end{array}
$$
3:20 a.m.

LESSON 32, MIXED PRACTICE

1.
$$
\underbrace{(24 + 7)}_{31} \times \underbrace{(18 - 6)}_{12}
$$
$$
372
$$

2. $1836 - 1786 = d$
$$
\begin{array}{r}
{\scriptstyle 7} \\
1\,\cancel{8}\,{}^{1}3\,6 \\
-\ 1\,7\,8\,6 \\
\hline
5\ 0\text{ years}
\end{array}
$$

3.
$$
\begin{array}{r}
\$0.14\text{ per ounce} \\
16\overline{)\$2.24} \\
\underline{1\,6} \\
64 \\
\underline{64} \\
0
\end{array}
$$

4.
$$
\begin{array}{r}
6\!:\!50\text{ a.m.} \\
+\ 3\!:\!30 \\
\hline
9\!:\!80
\end{array}
$$
10:20 a.m.

5.
$$
\begin{array}{r}
2 \\
20\overline{)40} \\
\underline{40} \\
0
\end{array}
\qquad
\begin{array}{r}
5 \\
20\overline{)100} \\
\underline{100} \\
0
\end{array}
$$
$$
\frac{2}{5}
$$

6.
$$
\begin{array}{r}
90\text{ ft} \\
\times\ 90\text{ ft} \\
\hline
\mathbf{8100}\text{ square feet}
\end{array}
$$

7.
$$
\begin{array}{r}
90\text{ ft} \\
\times\ 4 \\
\hline
\mathbf{360}\text{ ft}
\end{array}
$$

8. 1, 3, 5, 7, 9, 11, 13, 15
This is a sequence of positive odd numbers. Add 2 to the value of a term to find the next term. 15

9. $(7 \times 1000) + (5 \times 100)$

10.
$$
\begin{array}{r}
{\scriptstyle 0\ 9} \\
\cancel{1}\,\cancel{0}\,{}^{1}0\,0 \\
-\quad 9\,9\,0 \\
\hline
1\,0
\end{array}
\quad
\begin{array}{r}
{\scriptstyle 0\ 9\ 9} \\
\cancel{1}\,\cancel{0}\,\cancel{0}\,{}^{1}0 \\
-\quad 9\,0\,9 \\
\hline
9\,1
\end{array}
\quad
\begin{array}{r}
1009 \\
-\ 1000 \\
\hline
9
\end{array}
\quad
\begin{array}{r}
1090 \\
-\ 1000 \\
\hline
90
\end{array}
$$
C. 1009

11.
$$
\begin{array}{r}
{\scriptstyle 1} \\
\$623 \\
\$494 \\
+\ \$380 \\
\hline
\$1497
\end{array}
\qquad
\begin{array}{r}
\$499 \\
3\overline{)\$1497} \\
\underline{12} \\
29 \\
\underline{27} \\
27 \\
\underline{27} \\
0
\end{array}
$$

12.
$$
\begin{array}{r}
\$0.05 \\
\times\quad 100 \\
\hline
\mathbf{\$5.00}
\end{array}
$$

13. $\dfrac{5}{2}$

14.

	$24
$\frac{3}{4}$ of $24	$6
	$6
	$6
$\frac{1}{4}$ of $24	$6

$18

15. $\dfrac{3}{5} + \dfrac{3}{5} = \dfrac{6}{5} = \mathbf{1\dfrac{1}{5}}$

16. $\dfrac{3}{4} - \dfrac{1}{4} = \dfrac{2}{4} = \mathbf{\dfrac{1}{2}}$

17. $\dfrac{3}{4} \times \dfrac{1}{3} = \dfrac{3}{12} = \mathbf{\dfrac{1}{4}}$

18. $\dfrac{3}{10} \times \dfrac{7}{10} = \mathbf{\dfrac{21}{100}}$

19. $1\dfrac{2}{3} - 1\dfrac{1}{3} = \mathbf{\dfrac{1}{3}}$

20. One fourth of a circle is 25%.

$$\begin{array}{r} 25\% \\ \times \quad 3 \\ \hline \mathbf{75\%} \end{array}$$

21.
$$\begin{array}{r} 53 \\ + \ 12 \\ \hline 65 \end{array} \qquad \text{check:} \qquad \begin{array}{r} 65 \\ - \ 53 \\ \hline 12 \end{array}$$

$$w = \mathbf{65}$$

22.
$$\begin{array}{r} 30 \\ 8\overline{)240} \\ \underline{24} \\ 00 \\ \underline{00} \\ 0 \end{array} \qquad \begin{array}{r} 30 \\ \times \ 8 \\ \hline 240 \end{array}$$

$$q = \mathbf{30}$$

23.
$$\begin{array}{r} 36 \\ - \ 15 \\ \hline 21 \ \text{girls} \end{array}$$

$$\frac{\text{boys}}{\text{girls}} = \frac{15}{21} = \mathbf{\frac{5}{7}}$$

24. Multiples of 4
4, 8, ⑫, 16, 20, 24
Multiples of 6
6, ⑫, 18, 24, 30
LCM is **12.**

25.

26. $\dfrac{24}{30} = \mathbf{\dfrac{4}{5}}$

27.
$$\begin{array}{r} {}^{10\,:\,75}\!\cancel{11}\!:\!\cancel{15} \ \text{a.m.} \\ - \quad 6\!:\!45 \ \text{a.m.} \\ \hline 4\!:\!30 \end{array}$$

4 hr 30 min

28. $\underbrace{(3 \times 100) + (5 \times 1)}_{305} \ \boxed{<} \ 350$

29. $\dfrac{3}{10}$

30. To find the cost per ounce, divide the price of the box of cereal by the weight of the cereal in ounces.

LESSON 33, WARM-UP

a. 500

b. 203

c. 84

d. $7.25

e. 240

f. 12

g. 7

Problem Solving
  3 faces: 8 blocks (corner blocks)
  2 faces: 12 blocks (middle blocks along edges)
  1 face: 6 blocks (center blocks)

LESSON 33, LESSON PRACTICE

a. $\dfrac{80}{100}$

$$\frac{80 \div 20}{100 \div 20} = \mathbf{\frac{4}{5}}$$

b. $\dfrac{5}{100}$

$$\frac{5 \div 5}{100 \div 5} = \mathbf{\frac{1}{20}}$$

c. $\dfrac{25}{100}$

$$\frac{25 \div 25}{100 \div 25} = \mathbf{\frac{1}{4}}$$

d. $\dfrac{24}{100}$

$$\frac{24 \div 4}{100 \div 4} = \mathbf{\frac{6}{25}}$$

e. $\dfrac{23}{100}$

f. $\dfrac{10}{100}$

$\dfrac{10 \div 10}{100 \div 10} = \dfrac{1}{10}$

g. $\dfrac{20}{100}$

$\dfrac{20 \div 20}{100 \div 20} = \dfrac{1}{5}$

h. $\dfrac{2}{100}$

$\dfrac{2 \div 2}{100 \div 2} = \dfrac{1}{50}$

i. $\dfrac{75}{100}$

$\dfrac{75 \div 25}{100 \div 25} = \dfrac{3}{4}$

LESSON 33, MIXED PRACTICE

1. $\underbrace{(10 \times 15)}_{150} \div \underbrace{(10 + 15)}_{25}$

$ \mathbf{6}$

2. $6650 - 3766 = d$

$$
\begin{array}{r}
{}^{5}{\llap{6}}\,{}^{5}{\llap{6}}\,{}^{4}{\llap{5}}{}^{1}0 \\
-\ 3\,7\,6\,6 \\
\hline
2\,8\,8\,4 \text{ kilometers}
\end{array}
$$

3. **15,000,000,000 years**

4. $(3 \times 1000) + (4 \times 10)$

5. $600 + 2 = \mathbf{602}$

6. $\dfrac{10}{10}; \dfrac{100}{100}$

7. $\dfrac{3}{5}$

8. 12 in. + 8 in. + 12 in. + 8 in. = **40 in.**

9.
$$
\begin{array}{r}
12 \text{ tiles in a row} \\
\times\ \ 8 \text{ rows} \\
\hline
\mathbf{96 \text{ square tiles}}
\end{array}
$$

10. 56; No, the sum of the digits is not divisible by 3.
75; No, the last digit is not even.
83; No, the last digit is not even and the sum of the digits is not divisible by 3.
48; Yes, the last digit is even and the sum of the digits is divisible by 3.
D. 48

11.
$$
\begin{array}{r}
5000 \\
-\ 2000 \\
\hline
\mathbf{3000}
\end{array}
$$

12.
$$
\begin{array}{r}
\$4.30 \\
\times\ \ \ \ 100 \\
\hline
\mathbf{\$430.00}
\end{array}
$$

13.
$$
\begin{array}{r}
\mathbf{\$16.08} \\
25\overline{)\$402.00} \\
\underline{25} \\
152 \\
\underline{150} \\
2\ 0 \\
\underline{0\ 0} \\
2\ 00 \\
\underline{2\ 00} \\
0
\end{array}
$$

14. $\dfrac{3}{5} \times \dfrac{20}{1} = \dfrac{60}{5} = \dfrac{12}{1} = \mathbf{12}$

$$
\begin{array}{r}
12 \\
5\overline{)60} \\
\underline{5} \\
10 \\
\underline{10} \\
0
\end{array}
$$

15. $\dfrac{4}{5} + \dfrac{4}{5} = \dfrac{8}{5} = \mathbf{1\dfrac{3}{5}}$

$$
\begin{array}{r}
1\tfrac{3}{5} \\
5\overline{)8} \\
\underline{5} \\
3
\end{array}
$$

16. $\dfrac{5}{8} - \dfrac{1}{8} = \dfrac{4}{8}$

$\dfrac{4 \div 4}{8 \div 4} = \dfrac{1}{2}$

17. $\dfrac{5}{2} \times \dfrac{3}{2} = \dfrac{15}{4} = \mathbf{3\dfrac{3}{4}}$

$$
\begin{array}{r}
3\tfrac{3}{4} \\
4\overline{)15} \\
\underline{12} \\
3
\end{array}
$$

18. $\dfrac{3}{10} \times \dfrac{3}{100} = \dfrac{9}{1000}$

19. 2, 4, 6, 8, 10, 12, 14, 16, 18, 20

This is a sequence of positive even numbers. Add 2 to the value of a term to find the next term. 20

20.
$$\begin{array}{r} 24 \\ + 23 \\ \hline 47 \end{array} \qquad \text{check:} \qquad \begin{array}{r} 47 \\ - 24 \\ \hline 23 \end{array}$$
$Q = \mathbf{47}$

21. $\dfrac{2}{1}$ or **2**   check: $\dfrac{1}{2} \times \dfrac{2}{1} = \dfrac{2}{2} = 1$

$w = \mathbf{2}$

22. $3 \cdot 5$

23. **About 2 meters**

24. $\dfrac{5}{30}$

$\dfrac{5 \div 5}{30 \div 5} = \dfrac{1}{6}$

25. $1\dfrac{7}{10}$

26. (a) $\dfrac{70}{100}$

$\dfrac{70 \div 10}{100 \div 10} = \dfrac{7}{10}$

(b) $\dfrac{30}{100}$

$\dfrac{30 \div 10}{100 \div 10} = \dfrac{3}{10}$

27.

$\dfrac{4}{5}$ were correct.

$\dfrac{1}{5}$ were not correct.

20 answers
4 answers
4 answers
4 answers
4 answers
4 answers

16 answers

28. **If the numerator is more than half of the denominator, the fraction is greater than $\frac{1}{2}$. If the numerator is less than half of the denominator, the fraction is less than $\frac{1}{2}$.**

29.
$$\begin{array}{r} 8\!:\!45 \text{ p.m.} \\ + \ 6\!:\!30 \\ \hline 14\!:\!75 \\ = \ 15\!:\!15 \end{array} \qquad \begin{array}{r} 15\!:\!15 \\ - \ 12\!:\!00 \\ \hline \mathbf{3\!:\!15 \text{ a.m.}} \end{array}$$

30. $\dfrac{1}{16}, \dfrac{1}{8}, \dfrac{1}{4}, \dfrac{1}{2}$

LESSON 34, WARM-UP

a. **900**

b. **1130**

c. **63**

d. **$9.50**

e. **500**

f. **15**

g. **12**

Problem Solving
  3 red;
  3 black;
  2 black, 1 red;
  2 red, 1 black

LESSON 34, LESSON PRACTICE

a. **Thousandths**

b. **4**

c. **2**

d. **Ones**

e. **Cent**

f. **Mill**

SOLUTIONS

LESSON 34, MIXED PRACTICE

1.

24 members

$\frac{3}{8}$ were tenors.
- 3 members
- 3 members
- 3 members

$\frac{5}{8}$ were not tenors.
- 3 members
- 3 members
- 3 members
- 3 members
- 3 members

9 tenors

2.
$$\begin{array}{r} 8 \text{ ounces} \\ \times\ 3 \\ \hline \textbf{24 ounces} \end{array}$$

3.
$$\begin{array}{r} \overset{1}{8}:47 \text{ a.m.} \\ +\ \ 1:15 \\ \hline 9:62 \\ \textbf{10:02 a.m.} \end{array}$$

4. (a) $\dfrac{60}{100}$

$\dfrac{60 \div 20}{100 \div 20} = \dfrac{3}{5}$

(b) $\dfrac{40}{100}$

$\dfrac{40 \div 20}{100 \div 20} = \dfrac{2}{5}$

5. $\dfrac{100}{100} \enspace \textcircled{=} \enspace \dfrac{10}{10}$

  $\quad 1 \qquad\qquad 1$

6. $600 + 5 = \textbf{605}$

7. **6**

8. (a)
$$\begin{array}{r} \textbf{6 inches} \\ 4\overline{)24} \text{ inches} \\ \underline{24} \\ 0 \end{array}$$

(b) 6 in. $\times$ 6 in. = **36 square inches**

9.

A

$1\frac{1}{2}$ in. $2\frac{1}{2}$ in.

C 2 in. B

10. Multiples of 6
6,12,18,⟨24⟩, 30
Multiples of 8
8,16,⟨24⟩, 32,40
LCM is **24.**

11.
$$\begin{array}{r} \textbf{\$0.56} \\ 10\overline{)\$5.60} \\ \underline{5\,0} \\ 60 \\ \underline{60} \\ 0 \end{array}$$

12. $\dfrac{9}{10} \cdot \dfrac{9}{10} = \dfrac{\textbf{81}}{\textbf{100}}$

13.
$$\begin{array}{r} \textbf{30} \\ 30\overline{)900} \\ \underline{90} \\ 00 \\ \underline{00} \\ 0 \end{array}$$

14. **36,800**

15.
$$\begin{array}{r} 24 \\ 6\overline{)144} \\ \underline{12} \\ 24 \\ \underline{24} \\ 0 \end{array}$$
check:
$$\begin{array}{r} 24 \\ \times\ 6 \\ \hline 144 \end{array}$$

$d = \textbf{24}$

16.
$$\begin{array}{r} 144 \\ \times\ \ 6 \\ \hline 864 \end{array}$$
check:
$$\begin{array}{r} 144 \\ 6\overline{)864} \\ \underline{6} \\ 26 \\ \underline{24} \\ 24 \\ \underline{24} \\ 0 \end{array}$$

$d = \textbf{864}$

17. $\dfrac{5}{2} + \dfrac{5}{2} \enspace \textcircled{=} \enspace \dfrac{2}{1} \times \dfrac{5}{2}$

$\quad \dfrac{10}{2} \qquad\qquad \dfrac{10}{2}$

$\quad\ 5 \qquad\qquad\ 5$

18. $\dfrac{3}{8} + \dfrac{3}{8} = \dfrac{6}{8}$

$\dfrac{6 \div 2}{8 \div 2} = \dfrac{\textbf{3}}{\textbf{4}}$

19. $\dfrac{11}{12} - \dfrac{1}{12} = \dfrac{10}{12}$

$\dfrac{10 \div 2}{12 \div 2} = \mathbf{\dfrac{5}{6}}$

20. $\dfrac{5}{4} \times \dfrac{3}{2} = \dfrac{15}{8}$

$\begin{array}{r} \mathbf{1\frac{7}{8}} \\ 8\overline{)15} \\ 8 \\ \hline 7 \end{array}$

21. $\dfrac{6}{30}$

$\dfrac{6 \div 6}{30 \div 6} = \mathbf{\dfrac{1}{5}}$

22. $86 \cdot \$5.75 = t$

$\begin{array}{r} \$5.75 \\ \times\quad 86 \\ \hline 3450 \\ 46000 \\ \hline \mathbf{\$494.50} \end{array}$

23. -6

24. $(80 \div 40) - (8 \div 4)$

$\qquad 2 - 2$

$\qquad\qquad \mathbf{0}$

25. 8

26.
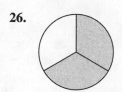

27. $\begin{array}{r} \mathbf{435\frac{5}{12}} \\ 12\overline{)5225} \\ 48 \\ \hline 42 \\ 36 \\ \hline 65 \\ 60 \\ \hline 5 \end{array}$

28. $\begin{array}{r} {}^{1} \\ 12 \text{ ounces} \\ 11 \text{ ounces} \\ +\ 7 \text{ ounces} \\ \hline 30 \end{array}$ $\qquad \begin{array}{r} \mathbf{10 \text{ ounces}} \\ 3\overline{)30 \text{ ounces}} \\ 3 \\ \hline 00 \\ 00 \\ \hline 0 \end{array}$

29. $tr = d$

$d \div t = r$

$d \div r = t$

30. $\begin{array}{r} \mathbf{15} \\ 5\overline{)75} \\ 5 \\ \hline 25 \\ 25 \\ \hline 0 \end{array}$ $\qquad \begin{array}{r} \mathbf{15} \\ 10\overline{)150} \\ 10 \\ \hline 50 \\ 50 \\ \hline 0 \end{array}$

LESSON 35, WARM-UP

a. 1300

b. 344

c. 76

d. $15.00

e. 120

f. 360

g. 4

Problem Solving

Think, "What three-digit number plus 1 equals a four-digit number?"

$\begin{array}{r} \mathbf{999} \\ +\quad 1 \\ \hline \mathbf{1000} \end{array}$

LESSON 35, LESSON PRACTICE

a. $\dfrac{1}{10}$

b. $\dfrac{21}{100}$

c. $\dfrac{321}{1000}$

d. 0.3

e. 0.17

f. 0.123

g. Five hundredths

h. Fifteen thousandths

i. One and two tenths

j. $\frac{7}{10}$; 0.7

k. $\frac{31}{100}$; 0.31

l. $\frac{731}{1000}$; 0.731

m. 5.6

n. 11.12

o. 0.125

_____

LESSON 35, MIXED PRACTICE

1. $\frac{3}{4} \times \frac{3}{5} = \frac{9}{20}$

2.
360 seeds

$\frac{3}{4}$ sprouted. $\left\{ \begin{array}{l} \boxed{\text{90 seeds}} \\ \boxed{\text{90 seeds}} \\ \boxed{\text{90 seeds}} \end{array} \right.$

$\frac{1}{4}$ did not sprout. $\left\{ \boxed{\text{90 seeds}} \right.$

270 carrot seeds

3.
$\begin{array}{l} \overset{1}{}11\!:\!45 \text{ a.m.} \\ +\ \ 2\!:\!15 \\ \hline 13\!:\!60 \\ =\ 14\!:\!00 \end{array}$ $\begin{array}{l} 14\!:\!00 \\ -\ 12\!:\!00 \\ \hline \mathbf{2\!:\!00} \textbf{ p.m.} \end{array}$

4. (a) $\frac{23}{100}$

    (b) **0.23**

5. Ten and one hundredth

6. 10.5

7. (a) $\frac{25}{100}$

    $\frac{25 \div 25}{100 \div 25} = \frac{1}{4}$

    (b) $\frac{75}{100}$

    $\frac{75 \div 25}{100 \div 25} = \frac{3}{4}$

8. $5000 + 600 + 40 = \mathbf{5640}$

9. 3

10.
$\begin{array}{r} 20 \text{ mm} \\ \times\ 10 \text{ mm} \\ \hline \mathbf{200} \textbf{ sq. mm} \end{array}$

11. $20 \text{ mm} + 10 \text{ mm} + 20 \text{ mm} + 10 \text{ mm}$
     $= \mathbf{60} \textbf{ mm}$

12.
$\begin{array}{r} 100 \text{ centimeters} \\ \times\ \ \ 10 \\ \hline \mathbf{1000} \textbf{ centimeters} \end{array}$

13. $-1, 0, 0.001, 0.01, 0.1, 1.0$

14. **About 1 meter**

15. $\frac{3}{5} + \frac{2}{5} = \frac{5}{5}$

     $\frac{5 \div 5}{5 \div 5} = \frac{1}{1} = \mathbf{1}$

16. $\frac{5}{8} - \frac{5}{8} = \frac{0}{8} = \mathbf{0}$

17. $\frac{2}{3} \times \frac{3}{4} = \frac{6}{12}$

     $\frac{6 \div 6}{12 \div 6} = \frac{1}{2}$

18. (a) $\frac{5}{2}$

     (b) $\frac{5}{2} \times \frac{2}{1} = \frac{10}{2} = \mathbf{5}$

19.
$$\begin{array}{r} 3\frac{2}{6} = 3\frac{1}{3} \\ 6\overline{)20} \\ \underline{18} \\ 2 \end{array}$$

20.
$$\begin{array}{r} 16\frac{4}{6}\% = 16\frac{2}{3}\% \\ 6\overline{)100\%} \\ \underline{6} \\ 40 \\ \underline{36} \\ 4 \end{array}$$

21. $3\frac{4}{4} - 1\frac{1}{4} = 2\dfrac{3}{4}$

22. $5 \;\bigcirc\!\!\!= \; 4\frac{4}{4}$
$\qquad\qquad 5$

23. $16\dfrac{2}{3}\%$

24. $3 \times 18 \div 6 \;\bigcirc\!\!\!= \; 3 \times (18 \div 6)$
$\quad\;\; 54 \div 6 \qquad\qquad 3 \times 3$
$\qquad\;\; 9 \qquad\qquad\qquad\;\; 9$

25. -14

26.
$$\begin{array}{r} 3 \\ 2\overline{)6} \\ \underline{6} \\ 0 \end{array} \quad \begin{array}{r} 3 \\ 20\overline{)60} \\ \underline{60} \\ 0 \end{array} \quad \begin{array}{r} 3 \\ 4\overline{)12} \\ \underline{12} \\ 0 \end{array} \quad \begin{array}{r} 3\frac{1}{8} \\ 8\overline{)25} \\ \underline{24} \\ 1 \end{array}$$
$\qquad$ **D.** $\dfrac{25}{8}$

27. $\dfrac{3}{10} \times \dfrac{7}{10} = \dfrac{21}{100}$

28. $\dfrac{21}{100}$; **0.21**

29.
$$\begin{array}{r} 8 \\ 50\overline{)400} \\ \underline{400} \\ 0 \end{array} \quad \begin{array}{r} 8 \\ 100\overline{)800} \\ \underline{800} \\ 0 \end{array}$$

30. $4 \cdot r = 50$
$$\begin{array}{r} 12\frac{2}{4} = 12\frac{1}{2} \text{ inches} \\ 4\overline{)50} \\ \underline{4} \\ 10 \\ \underline{8} \\ 2 \end{array}$$

LESSON 36, WARM-UP

a. 1700

b. 2875

c. 47

d. $12.50

e. 160

f. 48

g. 8

Problem Solving

$$\begin{array}{r} 1HD = 50¢ \\ 4D = 40¢ \\ +\quad 2N = 10¢ \\ \hline 100¢ \text{ or } \$1.00 \end{array}$$

$$\begin{array}{r} 3Q = 75¢ \\ 1D = 10¢ \\ +\; 3N = 15¢ \\ \hline 100¢ \text{ or } \$1.00 \end{array}$$

$$\begin{array}{r} 2Q = 50¢ \\ +\; 5D = 50¢ \\ \hline 100¢ \text{ or } \$1.00 \end{array}$$

$$\begin{array}{r} 1HD = 50¢ \\ 1Q = 25¢ \\ +\quad 5N = 25¢ \\ \hline 100¢ \text{ or } \$1.00 \end{array}$$

Possible answers: 1HD, 4D, 2N; 3Q, 1D, 3N; 2Q, 5D; 1HD, 1Q, 5N

LESSON 36, LESSON PRACTICE

a.

$$3 \xrightarrow{2 + \frac{2}{2}} 2\frac{2}{2}$$
$$\underline{-\ 2\frac{1}{2}} \qquad \underline{-\ 2\frac{1}{2}}$$
$$\frac{1}{2}$$

b.

$$2 \xrightarrow{1 + \frac{4}{4}} 1\frac{4}{4}$$
$$\underline{-\ \frac{1}{4}} \qquad \underline{-\ \frac{1}{4}}$$
$$1\frac{3}{4}$$

c.

$$4 \xrightarrow{3 + \frac{4}{4}} 3\frac{4}{4}$$
$$\underline{-\ 2\frac{1}{4}} \qquad \underline{-\ 2\frac{1}{4}}$$
$$1\frac{3}{4}$$

d.

$$3 \xrightarrow{2 + \frac{12}{12}} 2\frac{12}{12}$$
$$\underline{-\ \frac{5}{12}} \qquad \underline{-\ \frac{5}{12}}$$
$$2\frac{7}{12}$$

e.

$$10 \xrightarrow{9 + \frac{2}{2}} 9\frac{2}{2}$$
$$\underline{-\ 2\frac{1}{2}} \qquad \underline{-\ 2\frac{1}{2}}$$
$$7\frac{1}{2}$$

f.

$$6 \xrightarrow{5 + \frac{10}{10}} 5\frac{10}{10}$$
$$\underline{-\ 1\frac{3}{10}} \qquad \underline{-\ 1\frac{3}{10}}$$
$$4\frac{7}{10}$$

g.

$$4 \xrightarrow{3 + \frac{6}{6}} 3\frac{6}{6}$$
$$\underline{-\ 1\frac{5}{6}} \qquad \underline{-\ 1\frac{5}{6}}$$
$$2\frac{1}{6}\ \text{pies}$$

h. Answers may vary. Sample answer: There are two pies on the shelf. The little boy ate $\frac{2}{3}$ of a pie. How many pies are left on the shelf?

$$2 \xrightarrow{1 + \frac{3}{3}} 1\frac{3}{3}$$
$$\underline{-\ \frac{2}{3}} \qquad \underline{-\ \frac{2}{3}}$$
$$1\frac{1}{3}\ \text{pies}$$

LESSON 36, MIXED PRACTICE

1. $\dfrac{25}{100}$

$$\frac{25 \div 25}{100 \div 25} = \frac{1}{4}$$

2.

$$1 \xrightarrow{\frac{4}{4}} \frac{4}{4}$$
$$\underline{-\ \frac{3}{4}} \qquad \underline{-\ \frac{3}{4}}$$
$$\frac{1}{4}$$

3.

$$\begin{array}{r} \textbf{1760} \text{ yards} \\ 3\overline{)5280} \\ \underline{3} \\ 22 \\ \underline{21} \\ 18 \\ \underline{18} \\ 00 \\ \underline{00} \\ 0 \end{array}$$

4. 3

5. One and three tenths

6. 0.05

7. (a) $\dfrac{31}{100}$

    (b) **0.31**

8. $400 + 3 =$ **403**

9. **3**

10. (a) **3 inches**

    (b) $\begin{array}{r} 3 \text{ inches} \\ \times\ \ 4 \\ \hline \textbf{12 inches} \end{array}$

11. $\angle AMB$ (or $\angle BMA$) and $\angle DMC$
    (or $\angle CMD$)

12. $3\dfrac{1}{4} + 2\dfrac{1}{4} = 5\dfrac{2}{4} = \mathbf{5\dfrac{1}{2}}$

13. $\begin{array}{c} 3 \\ -\ 1\dfrac{1}{4} \end{array} \xrightarrow{\ 2+\frac{4}{4}\ } \begin{array}{c} 2\dfrac{4}{4} \\ -\ 1\dfrac{1}{4} \\ \hline 1\dfrac{3}{4} \end{array}$

14. $3\dfrac{1}{3} + 2\dfrac{2}{3} = 5\dfrac{3}{3} = \mathbf{6}$

15. $\dfrac{3}{4} \times \dfrac{28}{1} = \dfrac{84}{4}$

    $\begin{array}{r} 21 \\ 4\overline{)84} \\ 8 \\ \hline 04 \\ 4 \\ \hline 0 \end{array}$

16. $\dfrac{3}{4} \times \dfrac{4}{6} = \dfrac{12}{24}$

    $\dfrac{12 \div 12}{24 \div 12} = \dfrac{\mathbf{1}}{\mathbf{2}}$

17.

    $\begin{array}{l} \\ \text{spent } \frac{5}{6} \\ \\ \text{did not} \\ \text{spend } \frac{1}{6} \end{array} \left\{ \begin{array}{|c|} \hline \$24 \\ \hline \$4 \\ \hline \$4 \\ \hline \$4 \\ \hline \$4 \\ \hline \$4 \\ \hline \$4 \\ \hline \end{array} \right.$

    $20

18. $\begin{array}{r} \overset{1}{4}2 \\ 57 \\ +\ 63 \\ \hline 162 \end{array}$ $\begin{array}{r} 54 \\ 3\overline{)162} \\ 15 \\ \hline 12 \\ 12 \\ \hline 0 \end{array}$

19. **1, 2, 4, 5, 10, 20**

20. (a) Multiples of 9
    9, ⑱, 27, 36, 45
    Multiples of 6
    6, 12, ⑱, 24, 30
    LCM is **18.**

    (b) Factors of 9
    1, ③, 9
    Factors of 6
    1, 2, ③, 6
    GCF is **3.**

21. $\begin{array}{r} 12 \\ \times\ \ 6 \\ \hline 72 \end{array}$ check: $\begin{array}{r} 6 \\ 12\overline{)72} \\ 72 \\ \hline 0 \end{array}$

    $m = \mathbf{72}$

22. $\begin{array}{r} 2 \\ 6\overline{)12} \\ 12 \\ \hline 0 \end{array}$ check: $\begin{array}{r} 6 \\ 2\overline{)12} \\ 12 \\ \hline 0 \end{array}$

    $n = \mathbf{2}$

23. **59,000,000**

24. $\begin{array}{r} 800 \\ \times\ \ 700 \\ \hline \mathbf{560,000} \end{array}$

25. **50 mm**

26. $\dfrac{1}{6} + \dfrac{1}{3} = \dfrac{\mathbf{1}}{\mathbf{2}}$

    $\dfrac{1}{2} - \dfrac{1}{3} = \dfrac{\mathbf{1}}{\mathbf{6}}$

    $\dfrac{1}{2} - \dfrac{1}{6} = \dfrac{\mathbf{1}}{\mathbf{3}}$

27. $\dfrac{9}{10} \times \dfrac{9}{100} = \dfrac{\mathbf{81}}{\mathbf{1000}}$

28. **0.081**

29. (a) $\dfrac{4}{3}$

(b) $\dfrac{4}{3} + \dfrac{4}{3} + \dfrac{4}{3} = \dfrac{12}{3} = \mathbf{4}$

30.   radius $= 12$ ft

diameter $= 12$ ft $\times 2 = 24$ ft

$\dfrac{\text{radius}}{\text{diameter}} = \dfrac{12 \div 12}{24 \div 12} = \dfrac{1}{2}$

LESSON 37, WARM-UP

a. 2100

b. 387

c. 47

d. $20.00

e. 260

f. 250

g. 1

Problem Solving

Opposite sides cannot be seen at the same time, so any combination containing 1 and 6 dots, 2 and 5 dots, or 3 and 4 dots is excluded. Only one combination is left that totals ten dots.

1 dot, 4 dots, 5 dots

LESSON 37, LESSON PRACTICE

a.
$$\begin{array}{r} 3.46 \\ +\ 0.2 \\ \hline \mathbf{3.66} \end{array}$$

b.
$$\begin{array}{r} 8.28 \\ -\ 6.1 \\ \hline \mathbf{2.18} \end{array}$$

c.
$$\begin{array}{r} 0.735 \\ +\ 0.21 \\ \hline \mathbf{0.945} \end{array}$$

d.
$$\begin{array}{r} 0.543 \\ -\ 0.21 \\ \hline \mathbf{0.333} \end{array}$$

e.
$$\begin{array}{r} 0.43 \\ 0.1 \\ +\ 0.413 \\ \hline \mathbf{0.943} \end{array}$$

f.
$$\begin{array}{r} 0.\overset{2}{\cancel{3}}{}^{1}0 \\ -\ 0.2\,7 \\ \hline \mathbf{0.\,0\,3} \end{array}$$

g.
$$\begin{array}{r} \overset{1}{0.6} \\ +\ 0.7 \\ \hline \mathbf{1.3} \end{array}$$

h.
$$\begin{array}{r} \overset{0}{\cancel{1}}.\overset{9}{\cancel{0}}{}^{1}0 \\ -\ 0.2\,4 \\ \hline \mathbf{0.\,7\,6} \end{array}$$

i.
$$\begin{array}{r} \overset{1}{0.9} \\ +\ 0.12 \\ \hline \mathbf{1.02} \end{array}$$

j.
$$\begin{array}{r} \overset{0}{\cancel{1}}.{}^{1}2\,3 \\ -\ 0.4 \\ \hline \mathbf{0.\,8\,3} \end{array}$$

LESSON 37, MIXED PRACTICE

1. $\dfrac{60}{100}$

$\dfrac{60 \div 20}{100 \div 20} = \dfrac{\mathbf{3}}{\mathbf{5}}$

2. $8 + 4 = \mathbf{12\ pencils}$

3. $375 + n = 1000$

$$\begin{array}{r} \overset{0}{\cancel{1}}\,\overset{9}{\cancel{0}}\,\overset{9}{\cancel{0}}{}^{1}0 \\ -\quad 3\,7\,5 \\ \hline \mathbf{6\,2\,5} \end{array}$$

4.
$$\begin{array}{r} \overset{1}{3.4} \\ 0.62 \\ +\ 0.3 \\ \hline \mathbf{4.32} \end{array}$$

5.
$$\begin{array}{r} 4.56 \\ -\ 3.2 \\ \hline \textbf{1.36} \end{array}$$

6.
$$\begin{array}{r} \overset{1\ 1}{\$0.37} \\ \$0.23 \\ +\ \$0.48 \\ \hline \textbf{\$1.08} \end{array}$$

7.
$$\begin{array}{r} \$\overset{4}{\cancel{5}}.\ \overset{9}{\cancel{0}}{}^{1}0 \\ -\ \$0.\ 0\ 5 \\ \hline \textbf{\$4.\ 9\ 5} \end{array}$$

8. **10,000**

9. (a) **10 feet**

(b)
$$\begin{array}{r} 10\ \text{feet} \\ \times\ \ \ \ 4 \\ \hline \textbf{40 feet} \end{array}$$

10. **3**

11. **D. 0.01**

12.
$$\begin{array}{r} \underline{30\ \times\ 40}\ \ \ \times\ \ 40 \\ 1200\ \ \ \ \times\ \ 40 \\ \hline \textbf{48,000} \end{array}$$

13.
$$\begin{array}{r} \textbf{1070} \\ 3\overline{)3210} \\ \underline{3} \\ 02 \\ \underline{00} \\ 21 \\ \underline{21} \\ 00 \\ \underline{00} \\ 0 \end{array}$$

14.
$$\begin{array}{r} \textbf{1,070} \\ 30\overline{)32,100} \\ \underline{30} \\ 2\ 1 \\ \underline{0\ 0} \\ 2\ 10 \\ \underline{2\ 10} \\ 00 \\ \underline{00} \\ 0 \end{array}$$

15.
$$\begin{array}{r} \overset{0\ \ 9\ \ 9\ \ 9}{\$\cancel{1}\ \cancel{0},\ \cancel{0}\ \cancel{0}{}^{1}0} \\ -\ \ \ \ \$3\ 4\ 5 \\ \hline \textbf{\$9\ 6\ 5\ 5} \end{array}$$

16.
$$\frac{3}{4} + \frac{3}{4} = \frac{6}{4}$$
$$\begin{array}{r} 1\frac{2}{4} = \mathbf{1\frac{1}{2}} \\ 4\overline{)6} \\ \underline{4} \\ 2 \end{array}$$

17.
$$3 \quad \xrightarrow{\ 2\ +\ \frac{5}{5}\ } \quad 2\frac{5}{5}$$
$$-\ 1\frac{3}{5} \qquad\qquad\quad -\ 1\frac{3}{5}$$
$$\qquad\qquad\qquad\qquad \mathbf{1\frac{2}{5}}$$

18.
$$\frac{3}{3} - \frac{2}{2}$$
$$1\ -\ 1\ =\ \mathbf{0}$$

19. $1\frac{1}{3} + 2\frac{1}{3} + 3\frac{1}{3} = 6\frac{3}{3} = \mathbf{7}$

20.
$$\frac{1}{4} + \frac{3}{4} \enspace \textcircled{>} \enspace \frac{1}{4} \times \frac{3}{4}$$
$$\frac{4}{4} \qquad\qquad \frac{3}{16}$$
$$1$$

21.
$$\begin{array}{r} \mathbf{14\frac{2}{7}} \\ 7\overline{)100} \\ \underline{7} \\ 30 \\ \underline{28} \\ 2 \end{array}$$

22.
$$\begin{array}{rr} & \mathbf{92\,lb} \\ 90\ \text{lb} & 3\overline{)276}\ \text{lb} \\ {}_{1}84\ \text{lb} & \underline{27} \\ +\ 102\ \text{lb} & 06 \\ \hline 276\ \text{lb} & \underline{06} \\ & 0 \end{array}$$

23. Multiples of 4
4, 8, 12, 16, ⃝20, 24, 28
Multiples of 5
5, 10, 15, ⃝20, 25, 30
LCM is **20.**

24.
$$\begin{array}{r} \$38.50 \\ - \ \$34.00 \\ \hline 4.50 \end{array}$$
−4.50

25. $10\dfrac{1}{10}$

26. $\dfrac{3}{10} \times \dfrac{9}{10} = \dfrac{27}{100} = \mathbf{0.27}$

27. Fractions may vary but must include such numbers as $\frac{2}{2}, \frac{3}{3}$, and $\frac{4}{4}$. If the numerator and denominator of a fraction are equal (but not zero), the fraction equals 1.

28. If we multiply the dividend and divisor by the same number, the resulting problem has the same quotient as the original problem. Here the dividend and divisor were both doubled (that is, multiplied by 2), so the quotients of the two problems are the same.

29.
$$\begin{array}{r} \overset{3:83}{4:2\!\!\!/3} \ \text{p.m.} \\ - \ 2:50 \ \text{p.m.} \\ \hline 1:33 \end{array}$$
1 hr 33 min

30.

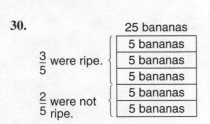

15 bananas

LESSON 38, WARM-UP

a. 2500

b. 750

c. 37

d. $3.75

e. 75

f. $4.00

g. 8

Problem Solving

3 different colors; Teresa could paint each pair of opposite faces (top and bottom, for example) the same color.

LESSON 38, LESSON PRACTICE

a.
$$\begin{array}{r} 4 \\ + \ 2.1 \\ \hline 6.1 \end{array}$$

b.
$$\begin{array}{r} 4.3 \\ - \ 2 \\ \hline 2.3 \end{array}$$

c.
$$\begin{array}{r} 3 \\ + \ 0.4 \\ \hline 3.4 \end{array}$$

d.
$$\begin{array}{r} \overset{3}{4}{}^{1}3.2 \\ - \ 5 \\ \hline 38.2 \end{array}$$

e.
$$\begin{array}{r} 0.23 \\ 4 \\ + \ 3.7 \\ \hline 7.93 \end{array}$$

f.
$$\begin{array}{r} 6.3 \\ - \ 6 \\ \hline 0.3 \end{array}$$

g.
$$\begin{array}{r} 12.5 \\ + \ 10 \\ \hline 22.5 \end{array}$$

h.
$$\begin{array}{r} 75.25 \\ - \ 25 \\ \hline 50.25 \end{array}$$

i. $9 \times 9 = \mathbf{81}$

j. 9

k. $6^2 + 8^2 = 36 + 64 = \mathbf{100}$

l. $\sqrt{100} - \sqrt{49} = 10 - 7 = \mathbf{3}$

m. $15 \times 15 = \mathbf{225}$

n. **12**

o. $(6 \cdot 6)(\text{ft} \cdot \text{ft}) = \mathbf{36\ ft^2}$

p. **8 m**

q. $5 \times 5 = 25$
$6 \times 6 = 36$
$7 \times 7 = 49$
$8 \times 8 = 64$
25, 36, 49, 64

LESSON 38, MIXED PRACTICE

1. Factors of 54
1, 2, 3, 6, ⑨, 18, 27, 54
Factors of 45
1, 3, 5, ⑨, 15, 45
GCF is **9.**

2. $3 \cdot w = \mathbf{126}$
    42 weeks
$3\overline{)126}$
 $\underline{12}$
  06
  $\underline{06}$
   0

3. $1948 - 1869 = d$

$1 \overset{8}{\cancel{9}} \overset{13}{\cancel{4}} 8$
$- 1\,8\,6\,9$
   7 9 years old

4. 3
   $\underline{+ 1.2}$
     4.2

5. 3.6
   $\underline{+ 4}$
     7.6

6. 5.63
   $\underline{- 1.2}$
     4.43

7. $\overset{1}{5}.376$
   $\underline{+ 0.24}$
     5.616

8. 4.75
   $\underline{- 0.6}$
     4.15

9. $4 - 3 = \mathbf{1}$

10. (a) $\dfrac{47}{100}$

  (b) **0.47**

11. **9043**

12. **5**

13. (a) $\sqrt{81} = \mathbf{9\ inches}$

  (b) 9
    $\underline{\times\ 4}$
    36 inches

14. Multiples of 2
2, 4, 6, 8, 10, ⑫, 14, 16
Multiples of 3
3, 6, 9, ⑫, 15, 18, 21
Multiples of 4
4, 8, ⑫, 16, 20, 24, 28
LCM is **12.**

15. $1\dfrac{2}{3} + 2\dfrac{2}{3} = 3\dfrac{4}{3} = \mathbf{4\dfrac{1}{3}}$

16.
$\qquad\qquad\quad 8 + \dfrac{4}{4}$
$\quad 9 \qquad\longrightarrow\qquad 8\dfrac{4}{4}$
$\underline{- 1\dfrac{1}{4}}\qquad\qquad\quad \underline{- 1\dfrac{1}{4}}$
$\qquad\qquad\qquad\qquad\quad \mathbf{7\dfrac{3}{4}}$

17. $\dfrac{3}{4} \times \dfrac{4}{5} = \dfrac{12}{20}$

$\dfrac{12 \div 4}{20 \div 4} = \mathbf{\dfrac{3}{5}}$

18. $\dfrac{7}{10} \times \dfrac{11}{10} = \mathbf{\dfrac{77}{100}}$

19. (a) $\mathbf{\dfrac{3}{2}}$

  (b) $\dfrac{3}{2} \times \dfrac{2}{1} = \dfrac{6}{2} = \mathbf{3}$

20. $\dfrac{6 \div 3}{9 \div 3} = \dfrac{2}{3}$

21. 1, 2, 3, 5, 6, 10, 15, 30

22. (a) $\dfrac{35 \div 5}{100 \div 5} = \dfrac{7}{20}$

    (b) $\dfrac{65 \div 5}{100 \div 5} = \dfrac{13}{20}$

23. 186,000

24. $\dfrac{3}{1} = 3$

    $m = 3$

25. $\dfrac{22 + 23 + 24}{3} = \dfrac{69}{3}$

```
     23
  3)69
     6
     09
     09
      0
```

26. $24 \div 8 \enspace \boxed{=} \enspace 240 \div 80$
        3   3

27. $\dfrac{7}{10} \times \dfrac{21}{100} = \dfrac{147}{1000} = \mathbf{0.147}$

28.
```
       $0.08
  10)$0.80
      0 0
      ___
       80
       80
       __
        0
```

29. D. $\dfrac{4}{5}$

30. First divide the perimeter by 4 to find the length of each side. Then square the length of a side (multiply the length by itself) to find the area.

LESSON 39, WARM-UP

a. 2900

b. 270

c. 38

d. $9.50

e. 175

f. 25

g. $\dfrac{1}{2}$

Problem Solving

Megan needs to pull two more socks from the drawer. This would give her a total of four socks, and since there are only three sock colors, Megan would be certain to have at least one matching pair.

LESSON 39, LESSON PRACTICE

a.
```
      15
  ×  0.3
  ____
     4.5
```

b.
```
     1.5
  ×   3
  ____
     4.5
```

c.
```
     1.5
  ×  0.3
  ____
    0.45
```

d.
```
    0.15
  ×    3
  _____
    0.45
```

e.
```
     1.5
  ×  1.5
  ____
      75
     150
  _____
    2.25
```

f.
```
    0.15
  ×   10
  _____
    1.50  or  1.5
```

g.
```
    0.25
  ×  0.5
  _____
   0.125
```

h.
$$\begin{array}{r} 0.025 \\ \times\ \ \ 100 \\ \hline \textbf{2.500} \text{ or } \textbf{2.5} \end{array}$$

i.
$$\begin{array}{r} 0.8 \\ \times\ 0.8 \\ \hline \textbf{0.64} \end{array}$$

j.
$$\begin{array}{r} 1.2 \\ \times\ 1.2 \\ \hline 24 \\ 120 \\ \hline \textbf{1.44} \end{array}$$

LESSON 39, MIXED PRACTICE

1. **29,035 feet**

2.
$$\begin{array}{r} 9{,}678\frac{1}{3} \text{ yards} \\ 3\overline{)29{,}035} \\ \underline{27} \\ 2\,0 \\ \underline{1\,8} \\ 23 \\ \underline{21} \\ 25 \\ \underline{24} \\ 1 \end{array}$$

3.
$$\begin{array}{r} 12\frac{4}{8} = 12\frac{1}{2}\text{¢ in 1 bit} \\ 8\overline{)100} \\ \underline{8} \\ 20 \\ \underline{16} \\ 4 \end{array}$$

$$\begin{array}{r} {}^{1\,3}12.5 \\ \times\ \ \ \ 6 \\ \hline \textbf{75.0¢} \text{ or } \textbf{75¢} \end{array}$$

4.
$$\begin{array}{r} {}^{2}0.25 \\ \times\ \ 0.5 \\ \hline \textbf{0.125} \end{array}$$

5.
$$\begin{array}{r} \$1.80 \\ \times\ \ \ \ 10 \\ \hline \textbf{\$18.00} \end{array}$$

6.
$$\begin{array}{r} {}^{2}63 \\ \times\ 0.7 \\ \hline \textbf{44.1} \end{array}$$

7.
$$\begin{array}{r} 1.23 \\ 4 \\ +\ 0.5 \\ \hline \textbf{5.73} \end{array}$$

8.
$$\begin{array}{r} {}^{0}\ {}^{1}1 \\ \cancel{1}\,\cancel{2}.{}^{1}3\,4 \\ -\ \ \ 5.\ 6 \\ \hline \textbf{6.\ 7\ 4} \end{array}$$

9.
$$\begin{array}{r} 1.1 \\ \times\ 1.1 \\ \hline 11 \\ 110 \\ \hline \textbf{1.21} \end{array}$$

10. (a) **10.3**

(b) $10\frac{3}{10}$

11. **Answers may vary, but the product will be the smallest number of the three (so it will be listed first).**

12. **0.123**

13. $600 + 40 = \textbf{640}$

14.
$$\begin{array}{r} 10 \\ 4\overline{)40} \end{array}$$

$$\begin{array}{r} 10 \text{ tiles in a row} \\ \times\ 10 \text{ rows} \\ \hline \textbf{100 tiles} \end{array}$$

15. Multiples of 2
2, 4, ⑥, 8, 10, 12
Multiples of 3
3, ⑥, 9, 12, 15, 18
Multiples of 6
⑥, 12, 18, 24, 30, 36
LCM is **6.**

16.
$$\begin{array}{r} 2\frac{4}{8} = \textbf{2}\frac{1}{2} \\ 8\overline{)20} \\ \underline{16} \\ 4 \end{array}$$

17. $\left(\frac{1}{3} + \frac{2}{3}\right) - 1$

$\left(\frac{3}{3}\right) - 1$

$\frac{3}{3} - \frac{3}{3} = \frac{0}{3} = \textbf{0}$

18. $\dfrac{3}{5} \times \dfrac{2}{3} = \dfrac{6}{15}$

$\dfrac{6 \div 3}{15 \div 3} = \dfrac{2}{5}$

19. $\dfrac{8}{9} \times \dfrac{9}{8} = \dfrac{72}{72} = 1$

20. $\dfrac{4 \div 2}{6 \div 2} = \dfrac{2}{3}$

21.
$$\begin{array}{r} \overset{12\,:\,60}{\cancel{1}:\cancel{00}} \text{ a.m.} \\ -\ 2:30 \\ \hline \textbf{10:30 p.m.} \end{array}$$

22.
$$\begin{array}{r} \overset{2}{2}6 \\ 29 \\ 28 \\ +\ 25 \\ \hline 108 \end{array} \qquad \begin{array}{r} \textbf{27 correct answers} \\ 4\overline{)108} \\ \underline{8} \\ 28 \\ \underline{28} \\ 0 \end{array}$$

23.
$$\begin{array}{r} \textbf{200} \\ 40\overline{)8000} \\ \underline{80} \\ 00 \\ \underline{00} \\ 00 \\ \underline{00} \\ 0 \end{array}$$

24. $\underset{1}{365 - 364} \;\textcircled{>}\; \underset{-1}{364 - 365}$

25. 3

26. $2\dfrac{3}{16}$ inches

27. (a) $\dfrac{5}{3}$

(b) $\dfrac{5}{3} \times \dfrac{2}{1} = \dfrac{10}{3} = 3\dfrac{1}{3}$

$$\begin{array}{r} 3\dfrac{1}{3} \\ 3\overline{)10} \\ \underline{9} \\ 1 \end{array}$$

28.
$$\begin{array}{r} \textbf{26} \\ 15\overline{)390} \\ \underline{30} \\ 90 \\ \underline{90} \\ 0 \end{array} \qquad \begin{array}{r} \textbf{26} \\ 5\overline{)130} \\ \underline{10} \\ 30 \\ \underline{30} \\ 0 \end{array} \qquad \begin{array}{r} \textbf{26} \\ 10\overline{)260} \\ \underline{20} \\ 60 \\ \underline{60} \\ 0 \end{array}$$

29.
$$\begin{array}{r} 0.5 \text{ m} \\ \times\ 0.3 \text{ m} \\ \hline \textbf{0.15 m}^2 \end{array}$$

30. $\dfrac{60 \div 60}{120 \div 60} = \dfrac{1}{2}$

LESSON 40, WARM-UP

a. **3300**

b. **2250**

c. **35**

d. **$6.75**

e. **480**

f. **48**

g. **7**

Problem Solving

The hundreds place in the top factor must be 1. (A larger number would result in a four-digit product.) So the hundreds place in the product must be 9. Think, "What number times 9 equals a number with 2 in the ones place?" (8). The tens place in the top factor must be 0 to prevent regrouping in the hundreds place.

$$\begin{array}{r} 108 \\ \times\quad 9 \\ \hline 972 \end{array}$$

LESSON 40, LESSON PRACTICE

a.
$$\begin{array}{r} 0.2 \\ \times\ 0.3 \\ \hline \textbf{0.06} \end{array}$$

b.
$$\begin{array}{r} 4.\overset{5}{\cancel{6}}{}^{1}0 \\ -\ 0.4\,6 \\ \hline \textbf{4.1 4} \end{array}$$

c.
$$\begin{array}{r} 0.1 \\ \times\ 0.01 \\ \hline \textbf{0.001} \end{array}$$

d.
$$\begin{array}{r} 0.\overset{3}{\cancel{4}}{}^{1}0 \\ -\ 0.3\,2 \\ \hline \mathbf{0.0\,8} \end{array}$$

e.
$$\begin{array}{r} 0.12 \\ \times\ \ 0.4 \\ \hline \mathbf{0.048} \end{array}$$

f.
$$\begin{array}{r} \overset{0}{\cancel{1}}.\overset{9}{\cancel{0}}{}^{1}0 \\ -\ 0.9\,8 \\ \hline \mathbf{0.0\,2} \end{array}$$

g.
$$\begin{array}{r} 0.3 \\ \times\ 0.3 \\ \hline \mathbf{0.09} \end{array}$$

h.
$$\begin{array}{r} 0.12 \\ \times\ 0.12 \\ \hline 24 \\ 120\ \ \\ \hline \mathbf{0.0144} \end{array}$$

i. **10.011**

j. $\dfrac{8 \div 8}{32 \div 8} = \dfrac{1}{4} = \mathbf{25\%}$

LESSON 40, MIXED PRACTICE

1. $\dfrac{4 \div 4}{32 \div 4} = \dfrac{1}{8} = \mathbf{12\dfrac{1}{2}\%}$

2. $1920 - 1788 = d$
$$\begin{array}{r} 1\,\overset{8}{\cancel{9}}\,\overset{11}{\cancel{2}}{}^{1}0 \\ -\ 1\,7\,8\,8 \\ \hline \mathbf{1\,3\,2\ years} \end{array}$$

3.
$$\begin{array}{r} \overset{12:120}{\cancel{2:00}}\ \text{p.m.} \\ -\ 3:30\ \ \ \ \ \\ \hline 9:90\ \text{a.m.} \\ \mathbf{10:30\ a.m.} \end{array}$$

4.
$$\begin{array}{r} \overset{2}{\cancel{3}}.{}^{1}0 \\ -\ 0.3 \\ \hline \mathbf{2.7} \end{array}$$

5.
$$\begin{array}{r} 1.\overset{1}{\cancel{2}}{}^{1}0 \\ -\ 0.1\,2 \\ \hline \mathbf{1.0\,8} \end{array}$$

6.
$$\begin{array}{r} 1.0 \\ -\ 0.1 \\ \hline \mathbf{0.9} \end{array}$$

7.
$$\begin{array}{r} 0.12 \\ \times\ 0.2 \\ \hline \mathbf{0.024} \end{array}$$

8.
$$\begin{array}{r} 0.1 \\ \times\ 0.1 \\ \hline \mathbf{0.01} \end{array}$$

9.
$$\begin{array}{r} 4.8 \\ \times\ 0.23 \\ \hline 14\ \ \\ 960\ \ \\ \hline \mathbf{1.104} \end{array}$$

10. **1.02**

11. $60{,}000 + 800 = \mathbf{60{,}800}$

12.
$$\begin{array}{r} 8\ \ \\ 4\overline{)32} \\ 32\ \\ \hline 0\ \end{array}$$

8 tiles in a row
$$\begin{array}{r} \times\ 8\ \text{rows} \\ \hline \mathbf{64\ tiles} \end{array}$$

13. Multiples of 2
2, 4, 6, ⑧, 10, 12
Multiples of 4
4, ⑧, 12, 16, 20, 24
Multiples of 8
⑧, 16, 24, 32, 40
LCM is **8.**

14. $6\dfrac{2}{3} + 4\dfrac{2}{3} = 10\dfrac{4}{3} = \mathbf{11\dfrac{1}{3}}$

15.
$$5 \xrightarrow{\ 4 + \frac{8}{8}\ } 4\dfrac{8}{8}$$
$$\begin{array}{r} -\ 3\dfrac{3}{8} \end{array} \qquad \begin{array}{r} -\ 3\dfrac{3}{8} \\ \hline \mathbf{1\dfrac{5}{8}} \end{array}$$

16. $\dfrac{5}{8} \times \dfrac{2}{3} = \dfrac{10}{24}$

$\dfrac{10 \div 2}{24 \div 2} = \dfrac{5}{12}$

17. $2\dfrac{5}{6} + 5\dfrac{2}{6} = 7\dfrac{7}{6} = 8\dfrac{1}{6}$

18. $\dfrac{1}{2} \times \dfrac{2}{2} \;\textcircled{=}\; \dfrac{1}{2} \times \dfrac{3}{3}$

$\quad \dfrac{2}{4} \qquad\qquad \dfrac{3}{6}$

$\quad \dfrac{1}{2} \qquad\qquad \dfrac{1}{2}$

19.
$$\begin{array}{r} {\scriptstyle 0\ 9\ 9} \\ \cancel{1}\ \cancel{0}\ \cancel{0}{}^{1}0 \\ -\ \ \ 5\ 6\ 7 \\ \hline 4\ 3\ 3 \end{array}$$

$w = \mathbf{433}$

20. 2, 4, 5, 10, 20, 25, 50

21. $81 + 3 = \mathbf{84}$

22. **\$4200**

23. 15 assignments + 10 assignments
$\qquad\qquad\qquad\; +\;$ 5 assignments
$\qquad\qquad\qquad\; =\;$ **30 assignments**

24. $\dfrac{5 \div 5}{30 \div 5} = \dfrac{1}{6}$

25. $\dfrac{15 \div 15}{30 \div 15} = \dfrac{1}{2} = \mathbf{50\%}$

26. See student work.

27. $0.3 \times 0.2 = 0.06$
$0.06 \div 0.3 = 0.2$
$0.06 \div 0.2 = 0.3$

28.
$$\begin{array}{r} 16 \\ 15\overline{)240} \\ 15 \\ \hline 90 \\ 90 \\ \hline 0 \end{array} \qquad \begin{array}{r} 16 \\ 5\overline{)80} \\ 5 \\ \hline 30 \\ 30 \\ \hline 0 \end{array} \qquad \begin{array}{r} 16 \\ 10\overline{)160} \\ 10 \\ \hline 60 \\ 60 \\ \hline 0 \end{array}$$

29. $\dfrac{40 \div 20}{100 \div 20} = \dfrac{2}{5}$

$\dfrac{2}{5} \times \dfrac{25}{1} = \dfrac{50}{5} =$ 10 red balloons

$\begin{array}{r} 25 \\ -\ 10 \\ \hline 15 \end{array}$ $\qquad \dfrac{\text{blue balloons}}{\text{red balloons}} = \dfrac{15 \div 5}{10 \div 5} = \dfrac{3}{2}$

30. $5\dfrac{9}{10}$

INVESTIGATION 4

1. quantitative

2. qualitative

3. quantitative

4. qualitative

5. qualitative

6. See student work. The closed-option question should provide a limited number of options. (For example, "How many hours of television do you watch per week: less than 4 hours, 4–8 hours, or more than 8 hours?") The open-option question should not limit options. (For example, "What is your favorite television show?") The combination question should provide options but include an opportunity to provide a different answer. (For example, "What type of television show is your favorite: comedy, drama, cartoon, or other? If choice is other, please specify.")

7. Pet store shoppers are more likely to own pets than people from the general population. Support for a leash law might be lower among those surveyed than among the general population.

8. Orchestra members are more likely to have a high interest in music than students in general. Thus, movies preferred by orchestra members might be more musically oriented than movies preferred by students in general.

9. People generally prefer to be considered "sensible" over being "not sensible." So a "yes" answer would be more likely due to the bias.

10. Politeness toward the host might make children less likely to give a "yes" answer.

Extensions

a. See student work.

b. See student work.

c. See student work.

d. See student work. All times should be rounded to the nearest hour.

e. See student work.

LESSON 41, WARM-UP

a. 1000

b. 675

c. 55

d. $5.25

e. 350

f. 60

g. 12

Problem Solving

(a) 78 feet + 27 feet + 78 feet + 27 feet = **210 feet**

(b) 27 feet + 9 feet = **36 feet**

(c) 78 feet + 36 feet + 78 feet + 36 feet = **228 feet**

LESSON 41, LESSON PRACTICE

a. $\dfrac{50}{100} = \dfrac{1}{2}$

b. $\dfrac{10}{100} = \dfrac{1}{10}$

c. $\dfrac{25}{100} = \dfrac{1}{4}$

d. $\dfrac{75}{100} = \dfrac{3}{4}$

e. $\dfrac{20}{100} = \dfrac{1}{5}$

f. $\dfrac{1}{100}$

g. $\dfrac{65}{100} = \mathbf{0.65}$

h. $\dfrac{7}{100} = \mathbf{0.07}$

i. $\dfrac{30}{100} = \mathbf{0.30}$ or **0.3**

j. $\dfrac{8}{100} = \mathbf{0.08}$

k. $\dfrac{60}{100} = \mathbf{0.60}$ or **0.6**

l. $\dfrac{1}{100} = \mathbf{0.01}$

m. **35**; 10% of 350 is $\frac{1}{10}$ of 350. Divide 350 by 10 to get 35.

n. **12**; 25% of 48 is $\frac{1}{4}$ of 48. Divide 48 by 4 to get 12.

o.
$$\begin{array}{r} 15.00 \\ \times\ \ 0.08 \\ \hline \$1.2000 \end{array}$$
$1.20

p. **$1**; $9\frac{1}{2}$% rounds to 10% and $9.98 rounds to $10. Find 10% $\left(\frac{1}{10}\right)$ of $10, which is $1.

q.
$$\begin{array}{r} 3\ 0 \text{ questions} \\ \times\ 0.80 \\ \hline 24.00 \end{array}$$
24 questions

LESSON 41, MIXED PRACTICE

1.
```
    2 0  questions
 ×  0.80
  16.00
```
16 questions

2.
```
    $8.50
 ×   0.08
  $0.68 00
```
$0.68

3.
```
   220  yards
 ×    4
   880 yards
```

4. Convert 20% to a fraction $\left(\frac{1}{5}\right)$ or a decimal (0.20), and multiply that number by 30.

5. $20.00 - 9.18 = b$

$$
\begin{array}{r}
\$ \overset{1}{2}\overset{9}{\cancel{0}}.\overset{9}{\cancel{0}}{}^{1}0 \\
- \ \$ \ 9.18 \\
\hline
\$ 10.82
\end{array}
$$

6. $16 \cdot C = 288$

```
      18 chairs
16)288
   16
   128
   128
     0
```

7.
```
   126        123
   102      3)369
 + 141        3
   369        06
              6
              09
               9
               0
```

8.
```
   2.5 m
 × 2 m
   5.0 m²
```
5 m²

9. $-1, 0, \dfrac{2}{3}, \dfrac{3}{2}$

10. 2, 3, 5, 7, 11, 13, 17, 19

11. D. 9; the sum of the digits is not divisible by 9.

12.
```
   6̸⁵0̈¹0
 -  1 5
   4 5  losses
```
$$\frac{\text{win}}{\text{loss}} = \frac{15}{45} = \frac{1}{3}$$

13.
```
    ¹28 inches
 ×    2
   56 inches
```

14. (a) **Ivy**

(b) **Main**

15. B. 45°

16.
```
     2.5
 ×   2.5
   125
   500
   6.25
```

17. 9

18. $\dfrac{40}{100} = \dfrac{2}{5}$

19.
```
 0.09        0.09
           ×  10
            $0.90
```

20. $\dfrac{3}{2}$

21.
```
     500
 7)3500        check:        500
   35                      ×   7
   00                       3500
   00
   00
   00
    0
```
$m = 500$

22.
```
 $ 1̸⁰0̸⁹.0̸⁹¹0      check:      $6.25
 -  $ 6.25                  + $3.75
   $ 3.75                   $10.00
```
$w = \$3.75$

23. $n = \dfrac{3}{2}$ check: $\dfrac{2}{3} \times \dfrac{3}{2} = \dfrac{6}{6} = 1$

24.
$$\begin{array}{r} \overset{1}{3}7 \\ +\ 76 \\ \hline 113 \end{array}$$
check:
$$\begin{array}{r} \overset{0}{\cancel{1}}\ \overset{10}{\cancel{1}}3 \\ -\ \ 3\ 7 \\ \hline 7\ 6 \end{array}$$
$x = 113$

25.
$$\begin{array}{r} \overset{3}{\cancel{4}}.\overset{1}{0} \\ -\ 2.5 \\ \hline 1.5 \end{array}$$
$$\begin{array}{r} 6.25 \\ +\ 1.5 \\ \hline \mathbf{7.75} \end{array}$$

26. $3\dfrac{3}{4} + 2\dfrac{3}{4} = 5\dfrac{6}{4} = 6\dfrac{2}{4} = 6\dfrac{1}{2}$

27. $1 - 1 = \mathbf{0}$

28. $\dfrac{5}{6} \cdot \dfrac{3}{5} = \dfrac{15}{30} = \dfrac{1}{2}$

29. $\dfrac{3}{4} \times \dfrac{48}{1} = \dfrac{144}{4} = \mathbf{36}$

$$\begin{array}{r} 36 \\ 4\overline{)144} \\ \underline{12} \\ 24 \\ \underline{24} \\ 0 \end{array}$$

30. **$3; Round 9% to 10% and $32.17 to $30.
Then find 10% $\left(\dfrac{1}{10}\right)$ of $30.**

LESSON 42, WARM-UP

a. **500**

b. **875**

c. **75**

d. **$9.25**

e. **500**

f. **58**

g. **3**

Problem Solving

$25 \times 2 = 50$
$50 - 19 = \mathbf{31}$

LESSON 42, LESSON PRACTICE

a. $\dfrac{1}{3} \times \dfrac{4}{4} = \dfrac{4}{12}$ **4**

b. $\dfrac{2}{3} \times \dfrac{2}{2} = \dfrac{4}{6}$ **4**

c. $\dfrac{3}{4} \times \dfrac{2}{2} = \dfrac{6}{8}$ **6**

d. $\dfrac{3}{4} \times \dfrac{3}{3} = \dfrac{9}{12}$ **9**

e. $\dfrac{2}{3} \times \dfrac{4}{4} = \dfrac{8}{12}$

$\dfrac{1}{4} \times \dfrac{3}{3} = \dfrac{3}{12}$

$\dfrac{8}{12} + \dfrac{3}{12} = \dfrac{11}{12}$

f. $\dfrac{1}{6} \times \dfrac{2}{2} = \dfrac{2}{12}$

$\dfrac{5}{12} - \dfrac{2}{12} = \dfrac{3}{12} = \dfrac{1}{4}$

LESSON 42, MIXED PRACTICE

1. $\dfrac{1}{2} \times \dfrac{3}{3} = \dfrac{3}{6}$ $\dfrac{2}{3} \times \dfrac{2}{2} = \dfrac{4}{6}$

$\dfrac{3}{6} + \dfrac{4}{6} = \dfrac{7}{6} = 1\dfrac{1}{6}$

2. **200,000,000,000 stars**

3.
$$\begin{array}{r} 120 \text{ yards} \\ \times\ \ 40 \text{ yards} \\ \hline \mathbf{4800 \text{ square yards}} \end{array}$$

4.
$$\begin{array}{r} 0.40 \\ \times\ \ 30 \\ \hline 12.00 \end{array}$$
12

5. $\dfrac{1}{2} \times \dfrac{4}{4} = \dfrac{4}{8}$ **4**

6. $\dfrac{1}{2} \times \dfrac{5}{5} = \dfrac{5}{10}$ **5**

7. $\sqrt{81} = 9$
$$\begin{array}{r} 4.32 \\ 0.6 \\ +\ 9 \\ \hline 13.92 \end{array}$$

8.
$$\begin{array}{r} \overset{5}{\cancel{6}}.\overset{12}{\cancel{3}}0 \\ -\ 0.54 \\ \hline 5.76 \end{array}$$

9.
$$\begin{array}{r} 0.15 \\ \times\ 0.15 \\ \hline 75 \\ 150 \\ \hline 0.0225 \end{array}$$

10. $\dfrac{7}{6}$

11. **4**

12. Multiples of 3
3, 6, 9, ⑫, 15, 18
Multiples of 4
4, 8, ⑫, 16, 20, 24
Multiples of 6
6, ⑫, 18, 24, 30, 36
LCM is **12.**

13. $5\dfrac{3}{5} + 4\dfrac{4}{5} = 9\dfrac{7}{5} = \mathbf{10\dfrac{2}{5}}$

14. $\sqrt{36} = 6$

$$\begin{array}{r} 6 \\ -\ 4\dfrac{2}{3} \\ \hline \end{array} \longrightarrow \begin{array}{r} 5\dfrac{3}{3} \\ -\ 4\dfrac{2}{3} \\ \hline 1\dfrac{1}{3} \end{array}$$

with $5 + \dfrac{3}{3}$

15. $\dfrac{8}{3} \times \dfrac{1}{2} = \dfrac{8}{6} = 1\dfrac{2}{6} = \mathbf{1\dfrac{1}{3}}$

16. $\dfrac{6}{5} \times \dfrac{3}{1} = \dfrac{18}{5} = \mathbf{3\dfrac{3}{5}}$

17. $1 = \dfrac{4}{4}$ $\dfrac{4}{4} - \dfrac{1}{4} = \dfrac{\mathbf{3}}{\mathbf{4}}$

18.
$$\begin{array}{r} \dfrac{10}{10} \longrightarrow 1 \\ -\ \dfrac{5}{5} \longrightarrow 1 \\ \hline 0 \end{array}$$
$1 - 1 = \mathbf{0}$

19. Answers may vary but should include fractions such as $\dfrac{2}{6}$, $\dfrac{3}{9}$, and $\dfrac{4}{12}$.

20. **2 and 17**

21. Round the scores to 12,000, 10,000, and 14,000. Add the scores and get 36,000; then divide by 3 to find the average **(12,000 points)**.

22. **200;** Round 8176 to 8000, and round 41 to 40. Then divide 8000 by 40.

23.
$\dfrac{2}{3}$ of 12 { 4 doughnuts / 4 doughnuts
$\dfrac{1}{3}$ of 12 { 4 doughnuts
12 doughnuts

8 doughnuts

24. $\dfrac{3}{4} \times \dfrac{2}{2} = \dfrac{6}{8}$

$\dfrac{7}{8} - \dfrac{6}{8} = \dfrac{\mathbf{1}}{\mathbf{8}}$

25.
$$\begin{array}{r} \overset{1}{0.4}\ \text{m} \\ 0.2\ \text{m} \\ 0.4\ \text{m} \\ +\ 0.2\ \text{m} \\ \hline 1.2\ \text{m} \end{array}$$

26.
$$\begin{array}{r} 0.4\ \text{m} \\ \times\ 0.2\ \text{m} \\ \hline 0.08\ \text{m}^2 \end{array}$$

27. $r - s = d$
$s + d = r$
$d + s = r$

28. Divisor is 4; dividend is 20; quotient is 5.

29.
$$\begin{array}{r} 11\text{:}45 \text{ a.m.} \\ +\ 2\text{:}30 \\ \hline 13\text{:}75 \\ =\ 14\text{:}15 \end{array} \qquad \begin{array}{r} 14\text{:}15 \\ -\ 12\text{:}00 \\ \hline \mathbf{2\text{:}15 \text{ p.m.}} \end{array}$$

30. (a) $\dfrac{6}{5}$

(b) $\dfrac{6}{5} \times \dfrac{3}{1} = \dfrac{18}{5} = 3\dfrac{3}{5}$

LESSON 43, WARM-UP

a. 900

b. 520

c. 65

d. $25

e. 600

f. $7.00

g. 6

Problem Solving

The lowest number that can be rolled is 1×3, or 3. The greatest number that can be rolled is 6×3, or 18. The numbers 3–18 comprise **16 different numbers.**

LESSON 43, LESSON PRACTICE

a. $15 \div 1 = 15$

b. $133 \div 7 = 19$

$$\begin{array}{r} 19 \\ 7\overline{)133} \\ \underline{7} \\ 63 \\ \underline{63} \\ 0 \end{array}$$

c.
$$\begin{array}{r} \overset{4}{\cancel{5}}.{}^{1}0 \\ -\ 3.2 \\ \hline 1.8 \end{array} \qquad \text{check:} \qquad \begin{array}{r} \overset{4}{\cancel{5}}.{}^{1}0 \\ -\ 1.8 \\ \hline 3.2 \end{array}$$
$d = \mathbf{1.8}$

d. $\dfrac{4}{5} + \dfrac{1}{5} = \dfrac{5}{5}$ check: $\dfrac{5}{5} - \dfrac{1}{5} = \dfrac{4}{5}$
$f = \mathbf{1}$

e.
$$4 \xrightarrow{\ 3 + \frac{5}{5}\ } 3\dfrac{5}{5}$$
$$-\ 1\dfrac{1}{5} \qquad\qquad -\ 1\dfrac{1}{5}$$
$$m = \mathbf{2\dfrac{4}{5}}$$

check:
$$\begin{array}{r} 2\dfrac{4}{5} \\ +\ 1\dfrac{1}{5} \\ \hline 3\dfrac{5}{5} = 4 \end{array}$$

f. $\dfrac{3}{8} \times \dfrac{8}{3} = \dfrac{24}{24} = 1$
$w = \mathbf{\dfrac{8}{3}}$

LESSON 43, MIXED PRACTICE

1.
$$\begin{array}{r} \$120 \\ \times\ 0.08 \\ \hline \$9.60 \end{array} \qquad \begin{array}{r} \$120.00 \\ +\ \$9.60 \\ \hline \mathbf{\$129.60} \end{array}$$

2. $150 - 128 = d$
$$\begin{array}{r} 1\ \overset{4}{\cancel{5}}{}^{1}0 \\ -\ 1\ 2\ 8 \\ \hline \mathbf{2\ 2} \textbf{ empty places} \end{array}$$

3.
$$\begin{array}{r} 19.75 \text{ seconds} \\ -\ 19.32 \text{ seconds} \\ \hline \mathbf{0.43} \textbf{ second} \end{array}$$

4. $\dfrac{2}{3} \times \dfrac{2}{2} = \dfrac{4}{6}$ **4**

5. $\dfrac{1}{2} \times \dfrac{3}{3} = \dfrac{3}{6}$ **3**

6. $n = \dfrac{3}{2}$ check: $\dfrac{2}{3} \times \dfrac{3}{2} = \dfrac{6}{6} = 1$

7.

$$6 \quad \xrightarrow{5 + \frac{5}{5}} \quad 5\frac{5}{5}$$

$$\underline{- 1\frac{4}{5}} \qquad \underline{- 1\frac{4}{5}}$$

$$w = 4\frac{1}{5}$$

check:

$$6 \quad \xrightarrow{5 + \frac{5}{5}} \quad 5\frac{5}{5}$$

$$\underline{- 4\frac{1}{5}} \qquad \underline{- 4\frac{1}{5}}$$

$$1\frac{4}{5}$$

8. $4\frac{1}{4} + 6\frac{3}{4} = 10\frac{4}{4} = 11$

$m = \textbf{11}$

check:

$$11 \quad \xrightarrow{10 + \frac{4}{4}} \quad 10\frac{4}{4}$$

$$\underline{- 4\frac{1}{4}} \qquad \underline{- 4\frac{1}{4}}$$

$$6\frac{3}{4}$$

9.

$$\begin{array}{r} 2.45 \\ + \ 3 \\ \hline 5.45 \end{array} \qquad \text{check:} \qquad \begin{array}{r} 5.45 \\ - \ 2.45 \\ \hline 3.00 \end{array}$$

$c = \textbf{5.45}$

10.

$$\begin{array}{r} 1\overset{1}{2}.\overset{9}{\cancel{0}}{}^{1}0 \\ - \ 1.4\ 3 \\ \hline 1\ 0.5\ 7 \end{array} \qquad \text{check:} \qquad \begin{array}{r} 1\overset{1}{2}.\overset{9}{\cancel{0}}{}^{1}0 \\ - \ 1\ 0.5\ 7 \\ \hline 1.4\ 3 \end{array}$$

$d = \textbf{10.57}$

11. $\dfrac{5}{8} \times \dfrac{1}{5} = \dfrac{5}{40} = \dfrac{\textbf{1}}{\textbf{8}}$

12. $\dfrac{3}{4} \times \dfrac{5}{1} = \dfrac{15}{4} = \textbf{3}\dfrac{\textbf{3}}{\textbf{4}}$

13. $3\dfrac{7}{8} - 1\dfrac{3}{8} = 2\dfrac{4}{8} = \textbf{2}\dfrac{\textbf{1}}{\textbf{2}}$

14. B. 33

15. $\dfrac{2}{2} \ \textcircled{=} \ \dfrac{2}{2} \times \dfrac{2}{2}$

$$1 \qquad 1 \times 1$$

$$1$$

16. $\textbf{-12}$

17. $\textbf{9.12}$

18. $\textbf{67,000,000}$

19. 0.37×100

$$\begin{array}{r} 0.37 \\ \times \ \ 100 \\ \hline \textbf{37.00} \ \text{ or } \textbf{37} \end{array}$$

20.

$$\begin{array}{r} 0.6 \\ \times \ 0.4 \\ \hline 0.24 \end{array} \qquad \begin{array}{r} 0.24 \\ \times \ 0.2 \\ \hline \textbf{0.048} \end{array}$$

21.

$$\begin{array}{r} 20 \\ 4\overline{)80} \text{ feet} \\ 8 \\ \hline 00 \\ 00 \\ \hline 0 \end{array} \qquad \begin{array}{r} 20 \\ \times \ 20 \\ \hline \textbf{400} \text{ square feet} \end{array}$$

22.

$$\begin{array}{r} 6\frac{4}{16} = \textbf{6}\frac{\textbf{1}}{\textbf{4}} \\ 16\overline{)100} \\ 96 \\ \hline 4 \end{array}$$

23. (a) $\textbf{25} \div \textbf{4} = \textbf{6}\dfrac{\textbf{1}}{\textbf{4}}$

$$\begin{array}{r} 6\frac{1}{4} \\ 4\overline{)25} \\ 24 \\ \hline 1 \end{array}$$

(b) $\textbf{9} \div \textbf{1} = \textbf{9}$

24. Multiples of 4

4, 8, 12, 16, 20, ⃝24, 28

Multiples of 6

6, 12, 18, ⃝24, 30, 36

Multiples of 8

8, 16, ⃝24, 32, 40

LCM is **24.**

25. $\dfrac{1}{2}, \dfrac{9}{16}, \dfrac{5}{8}$

26. $1\dfrac{5}{8}$ **inches**

27. $4\dfrac{7}{10}$

28. $\dfrac{1}{2} \times \dfrac{5}{5} = \dfrac{5}{10}$ $\dfrac{1}{5} \times \dfrac{2}{2} = \dfrac{2}{10}$

$\dfrac{5}{10} + \dfrac{2}{10} = \dfrac{7}{10}$

29.

$\dfrac{2}{5}$ were occupied.
$\dfrac{3}{5}$ were not occupied.

20 seats
4 seats
4 seats
4 seats
4 seats
4 seats

8 seats $\dfrac{40}{100} = \dfrac{2}{5}$

30. (a) **Right**

(b) **Acute**

(c) **Obtuse**

(d) **Right**

LESSON 44, WARM-UP

a. **1300**

b. **461**

c. **85**

d. **$11.25**

e. **700**

f. **$0.15**

g. **7**

Problem Solving

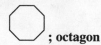 ; **octagon**

LESSON 44, LESSON PRACTICE

a. **0.05**

b. **50**

c. **1.25**

d. **4**

e. 0.2 $\bigcirc\!\!\!> $ 0.15

f. 12.5 $\bigcirc\!\!\!>$ 1.25

g. 0.012 $\bigcirc\!\!\!<$ 0.12

h. 0.31 $\bigcirc\!\!\!>$ 0.039

i. 0.4 $\bigcirc\!\!\!=$ 0.40

j. **0.015, 0.12, 0.125, 0.2**

LESSON 44, MIXED PRACTICE

1. 4, 8, ⑫

5, 10, ⑮

12 + 15 = **27**

2. $\begin{array}{r} 5280 \text{ feet} \\ \times 5 \\ \hline \mathbf{26{,}400 \text{ feet}} \end{array}$

3. $\begin{array}{r} 2\,\overset{8}{\cancel{9}},\,\overset{9}{\cancel{0}}{}^{1}3\,5 \\ -\ 1\,4,\,4\,9\,5 \\ \hline \mathbf{1\,4,\,5\,4\,0} \text{ feet} \end{array}$

4. $\begin{array}{r} 2\,\overset{8}{\cancel{9}},{}^{1}0\,3\,5 \text{ feet} \\ -\ 2\,6,\,4\,0\,0 \text{ feet} \\ \hline \mathbf{2\,6\,3\,5} \text{ feet} \end{array}$

5. $\begin{array}{r} 5\dfrac{1}{3} \\ -\ 4 \\ \hline w = \mathbf{1\dfrac{1}{3}} \end{array}$ check: $\begin{array}{r} 5\dfrac{1}{3} \\ -\ 1\dfrac{1}{3} \\ \hline 4 \end{array}$

6. $6\dfrac{4}{5} + 1\dfrac{3}{5} = 7\dfrac{7}{5} = 8\dfrac{2}{5}$

$m = \mathbf{8\dfrac{2}{5}}$ check: $\begin{array}{r} 7\dfrac{7}{5} \\ -\ 6\dfrac{4}{5} \\ \hline 1\dfrac{3}{5} \end{array}$

7.
$$\overset{1\ 1}{6.74}$$
$$+\ 0.285$$
$$\overline{7.025}$$

check:
$$\overset{1\ 1}{6.74}$$
$$0.285$$
$$+\ 4$$
$$\overline{11.025}$$

$$\overset{0}{\cancel{1}}{}^{1}1.025$$
$$-\ 7.025$$
$$\overline{4.000}$$
$$f = \mathbf{4}$$

8.
$$0.\overset{3}{\cancel{4}}{}^{1}0$$
$$-\ 0.33$$
$$\overline{0.07}$$

check:
$$0.\overset{3}{\cancel{4}}{}^{1}0$$
$$-\ 0.07$$
$$\overline{0.33}$$

$$d = \mathbf{0.07}$$

9.
$$67\frac{3}{4}$$
$$-\ 1\frac{1}{4}$$
$$\overline{66\frac{2}{4}} = \mathbf{66\frac{1}{2}} \textbf{ inches}$$

10. **$3.60;** 10% of $36 is $\frac{1}{10}$ of $36. Find $\frac{1}{10}$ of $36 by dividing $36 by 10.

11. **0.032**

12. $\frac{1}{6} \times \frac{24,042}{1} = \frac{24,042}{6}$

$$\begin{array}{r} \mathbf{4,007} \\ 6\overline{)24,042} \\ \underline{24} \\ 0\ 0 \\ \underline{0\ 0} \\ 04 \\ \underline{00} \\ 42 \\ \underline{42} \\ 0 \end{array}$$

13. (a) 0.25 $\gtrless$ 0.125

(b) 25% $\gtrless$ 12.5%

14. $600 + 4 = \mathbf{604}$

15. $17 \div 1 = \mathbf{17}$

16. (a) $\dfrac{8}{5}$

(b) $\dfrac{8}{5} \times \dfrac{3}{1} = \dfrac{24}{5} = \mathbf{4\dfrac{4}{5}}$

17. Multiples of 2
2, 4, 6, 8, 10, ⑫, 14
Multiples of 3
3, 6, 9, ⑫, 15, 18
Multiples of 4
4, 8, ⑫, 16, 20, 24
Multiples of 6
6, ⑫, 18, 24, 30, 36
LCM is **12.**

18.
$$\begin{array}{r} 1.3 \\ \times\ 1.3 \\ \hline 39 \\ 130 \\ \hline \mathbf{1.69} \end{array}$$

19. $\dfrac{3}{4} \times \dfrac{3}{3} = \dfrac{9}{12}$ **9**

20. $\dfrac{2}{3} \times \dfrac{4}{4} = \dfrac{8}{12}$ **8**

21.
$$\begin{array}{r} \overset{1}{2}6 \\ 37 \\ 42 \\ +\ 43 \\ \hline 148 \end{array}$$

$$\begin{array}{r} \mathbf{37} \\ 4\overline{)148} \\ \underline{12} \\ 28 \\ \underline{28} \\ 0 \end{array}$$

22. **365,000**

23.
$$\overset{2}{\cancel{3}}{}^{1}0$$
$$-\ 1\ 2$$
$$\overline{1\ 8} \textbf{ boys}$$

$$\frac{\text{boys}}{\text{girls}} = \frac{18}{12} = \mathbf{\frac{3}{2}}$$

24. (a) **1, 2, 4, 5, 10, 20, 25, 50, 100**

(b) **2 and 5**

25. $\dfrac{9}{100}$; **0.09**

26. $\dfrac{3}{4} \times \dfrac{3}{3} = \dfrac{9}{12}$ $\qquad \dfrac{2}{3} \times \dfrac{4}{4} = \dfrac{8}{12}$

$$\frac{9}{12} + \frac{8}{12} = \frac{17}{12} = \mathbf{1\frac{5}{12}}$$

27. **B. 40%. Nearly half of the rectangle is shaded. Since $\frac{1}{2}$ equals 50%, the shaded part is close to but less than 50%.**

28.
$$\begin{array}{r} {}^{12:\,135}\\ \cancel{2:15}\ \text{p.m.} \\ -\ 10:30\ \text{a.m.} \\ \hline 2:105 \\ =\ 3:45 \end{array}$$

3 hours 45 minutes

29. C. 1.1

30. $10\dfrac{1}{10}$

c.
$$\begin{array}{r} 1.5 \\ 3\overline{)4.5} \\ \underline{3} \\ 1\,5 \\ \underline{1\,5} \\ 0 \end{array}$$

d.
$$\begin{array}{r} 0.15 \\ 4\overline{)0.60} \\ \underline{4} \\ 20 \\ \underline{20} \\ 0 \end{array}$$

LESSON 45, WARM-UP

a. 1700

b. 875

c. 95

d. $10.75

e. 750

f. $4.00

g. 7

Problem Solving

$1 + 2 + 3 + 4 + 5 + 6 =$ **21 blocks**

e.
$$\begin{array}{r} 0.07 \\ 2\overline{)0.14} \\ \underline{0\,0} \\ 14 \\ \underline{14} \\ 0 \end{array}$$

f.
$$\begin{array}{r} 0.08 \\ 5\overline{)0.40} \\ \underline{0\,0} \\ 40 \\ \underline{40} \\ 0 \end{array}$$

g.
$$\begin{array}{r} 0.075 \\ 4\overline{)0.300} \\ \underline{0\,0} \\ 30 \\ \underline{28} \\ 20 \\ \underline{20} \\ 0 \end{array}$$

LESSON 45, LESSON PRACTICE

a.
$$\begin{array}{r} 1.8\ \text{miles} \\ 2\overline{)3.6}\ \text{miles} \\ \underline{2} \\ 1\,6 \\ \underline{1\,6} \\ 0 \end{array}$$

b.
$$\begin{array}{r} 1.6\ \text{meters} \\ 4\overline{)6.4}\ \text{meters} \\ \underline{4} \\ 2\,4 \\ \underline{2\,4} \\ 0 \end{array}$$

h.
$$\begin{array}{r} 0.002 \\ 6\overline{)0.012} \\ \underline{00} \\ 12 \\ \underline{12} \\ 0 \end{array}$$

i.
$$\begin{array}{r} 0.14 \\ 10\overline{)1.40} \\ \underline{1\,0} \\ 40 \\ \underline{40} \\ 0 \end{array}$$

j.
$$\begin{array}{r} 0.14 \\ 5\overline{)0.70} \\ \underline{5} \\ 20 \\ \underline{20} \\ 0 \end{array}$$

k.
$$\begin{array}{r} 0.025 \\ 4\overline{)0.100} \\ \underline{0} \\ 10 \\ \underline{8} \\ 20 \\ \underline{20} \\ 0 \end{array}$$

LESSON 45, MIXED PRACTICE

1. $\dfrac{3}{5}$

2.
$$\begin{array}{r} 50 \\ 20\overline{)1000} \\ \underline{100} \\ 00 \\ \underline{00} \\ 0 \end{array}$$
50 $20 bills

3. $\dfrac{2}{3} \times \dfrac{24}{1} = \dfrac{48}{3}$
$$\begin{array}{r} 16 \text{ shots made} \\ 3\overline{)48} \\ \underline{3} \\ 18 \\ \underline{18} \\ 0 \end{array}$$
$16 \times 2 = $ **32 points**

4.
$$\begin{array}{r} 1.5 \\ 3\overline{)4.5} \\ \underline{3} \\ 1\,5 \\ \underline{1\,5} \\ 0 \end{array}$$

5.
$$\begin{array}{r} 0.03 \\ 8\overline{)0.24} \\ \underline{0\,0} \\ 24 \\ \underline{24} \\ 0 \end{array}$$

6.
$$\begin{array}{r} 0.16 \\ 5\overline{)0.80} \\ \underline{5} \\ 30 \\ \underline{30} \\ 0 \end{array}$$

7. Multiples of 2
2, 4, 6, 8, 10, 12, 14, 16, 18, 20, 22, ⓐ24
Multiples of 4
4, 8, 12, 16, 20, ⓐ24, 28
Multiples of 6
6, 12, 18, ⓐ24, 30
Multiples of 8
8, 16, ⓐ24, 32, 40
LCM is **24.**

8. $\sqrt{36} = 6$

$$6 \xrightarrow{\;5 + \frac{10}{10}\;} 5\frac{10}{10}$$
$$\begin{array}{r} -\,2\frac{3}{10} \qquad\qquad -\,2\frac{3}{10} \\ \hline m = \mathbf{3\frac{7}{10}} \end{array}$$

check:
$$6 \xrightarrow{\;5 + \frac{10}{10}\;} 5\frac{10}{10}$$
$$\begin{array}{r} -\,3\frac{7}{10} \qquad\qquad -\,3\frac{7}{10} \\ \hline 2\frac{3}{10} \end{array}$$

9. $2\dfrac{2}{5} + 5\dfrac{4}{5} = 7\dfrac{6}{5} = 8\dfrac{1}{5}$

check: $\quad 7\dfrac{6}{5} - 2\dfrac{2}{5} = 5\dfrac{4}{5}$

$g = \mathbf{8\dfrac{1}{5}}$

10.
$$\begin{array}{r} \overset{1\ 1}{1.56} \\ +\ 1.44 \\ \hline 3.00 \end{array}$$
check:
$$\begin{array}{r} \overset{2\ \ 9}{\cancel{3}.\cancel{0}^1 0} \\ -\ 1.5\,6 \\ \hline 1.4\,4 \end{array}$$
$m = \mathbf{3}$

11.
$$\begin{array}{r} \overset{8\ \ 9}{\cancel{9}.\cancel{0}^1 0} \\ -\ 5.3\,9 \\ \hline 3.6\,1 \end{array}$$
check:
$$\begin{array}{r} \overset{8\ \ 9}{\cancel{9}.\cancel{0}^1 0} \\ -\ 3.6\,1 \\ \hline 5.3\,9 \end{array}$$
$n = \mathbf{3.61}$

12. $4\dfrac{3}{8} - 2\dfrac{1}{8} = 2\dfrac{2}{8} = \mathbf{2\dfrac{1}{4}}$

13. $\dfrac{8}{3} \cdot \dfrac{5}{2} = \dfrac{40}{6}$

$$6\dfrac{4}{6} = 6\dfrac{2}{3}$$
$$6\overline{)40}$$
$$\underline{36}$$
$$4$$

14. $\begin{array}{r} 700 \\ \times\ \ \ 400 \\ \hline \mathbf{280{,}000} \end{array}$

15. $\begin{array}{r} 0.7 \\ \times\ 0.6 \\ \hline 0.42 \end{array}$ $\begin{array}{r} 0.42 \\ \times\ 0.5 \\ \hline 0.210 \end{array} = \mathbf{0.21}$

16. $\begin{array}{r} 0.46 \\ \times\ 0.17 \\ \hline 322 \\ 460 \\ \hline \mathbf{0.0782} \end{array}$

17. $8 \times a = 177.6$

$$\begin{array}{r} \mathbf{22.2}\ \textbf{miles per gallon} \\ 8\overline{)177.6} \\ \underline{16} \\ 17 \\ \underline{16} \\ 1\,6 \\ \underline{1\,6} \\ 0 \end{array}$$

18. $\dfrac{3}{8} \times \dfrac{6}{1} = \dfrac{18}{8}$

$$\begin{array}{r} 2\dfrac{2}{8} = \mathbf{2\dfrac{1}{4}} \\ 8\overline{)18} \\ \underline{16} \\ 2 \end{array}$$

19. **$10; 25% of $40 is $\frac{1}{4}$ of $40. Divide $40 by 4 to find 25% of $40.**

20. $\dfrac{5}{6} \times \dfrac{2}{2} = \dfrac{10}{12}$

$\dfrac{10}{12} - \dfrac{7}{12} = \dfrac{3}{12} = \dfrac{1}{4}$

21. (a) $\sqrt{36\ \text{ft}^2} = \mathbf{6\ ft}$

(b) $6\ \text{ft} \times 4 = \mathbf{24\ ft}$

22. $\dfrac{27}{100}; \mathbf{0.27}$

23. $1\dfrac{1}{8}$ **inches**

24. $\dfrac{75}{100} = \dfrac{3}{4}$

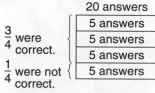

20 answers
5 answers
5 answers
5 answers
5 answers

$\frac{3}{4}$ were correct.
$\frac{1}{4}$ were not correct.

15 correct answers

25. $\dfrac{2}{3} \times \dfrac{1}{2} = \dfrac{1}{3}$

$\dfrac{1}{3} \div \dfrac{1}{2} = \dfrac{2}{3}$

$\dfrac{1}{3} \div \dfrac{2}{3} = \dfrac{1}{2}$

26. **B. 60%. Since a little more than half of the circle is shaded, a little more than 50% is shaded.**

27. (a) $\dfrac{9}{100}$

(b) **0.09**

28. $15 \div 1 = \mathbf{15}$

29. $\begin{array}{r} \overset{1}{24}\ \text{birds} \\ \times\ \ \ \ 3 \\ \hline \mathbf{72}\ \textbf{birds} \end{array}$

30. $\begin{array}{r} 2640\ \text{feet} \\ 2\overline{)5280}\ \text{feet} \\ \underline{4} \\ 12 \\ \underline{12} \\ 08 \\ \underline{8} \\ 00 \\ \underline{00} \\ 0 \end{array}$

$\begin{array}{r} \mathbf{880}\ \textbf{times} \\ 3\overline{)2640} \\ \underline{24} \\ 24 \\ \underline{24} \\ 00 \\ \underline{00} \\ 0 \end{array}$

LESSON 46, WARM-UP

a. 2100

b. 447

c. 55

d. $8.25

e. 475

f. 20

g. 7

h. 25

i. 250

j. 1.25

k. 12.5

l. False

m. True

n. $\dfrac{15}{5} = 3$

o. $\dfrac{250}{5} = 50$

Problem Solving

In order to produce a four-digit sum (with regrouping), both the hundreds digit and the tens digit in the top addend must be 9. The thousands digit in the sum must be 1, and the hundreds digit and the tens digit must both be 0. To determine the digits in the ones column, think, "What two numbers added together equal 18?" $(9 + 9)$. (We know they must equal 18 instead of 8 for regrouping to occur.)

$$
\begin{array}{r}
999 \\
+9 \\
\hline
1008
\end{array}
$$

LESSON 46, LESSON PRACTICE

a. $(2 \times 1) + \left(5 \times \dfrac{1}{100}\right)$

b. $(2 \times 10) + \left(5 \times \dfrac{1}{10}\right)$

c. $\left(2 \times \dfrac{1}{10}\right) + \left(5 \times \dfrac{1}{1000}\right)$

d. 70.8

e. 0.64

f. 3.5

g. 35

LESSON 46, MIXED PRACTICE

1. $\dfrac{30}{8} = 3\dfrac{6}{8} = 3\dfrac{3}{4}$

2. $100.2 - 98.6 = d$

$$
\begin{array}{r}
\overset{0\;\;9\;\;9}{\cancel{1}\,\cancel{0}\,\cancel{0}.{}^{1}2}°\text{F} \\
-9\,8.\,6°\text{F} \\
\hline
1.\,6°\text{F}
\end{array}
$$

3.
$$
\begin{array}{r}
20 \\
+4 \\
\hline
24
\end{array}
\qquad
\begin{array}{r}
\textbf{2 dozen} \\
12\overline{)24} \\
\underline{24} \\
0
\end{array}
$$

4. 50.607

5. $\dfrac{21}{100}$; 0.21

6. $\dfrac{20}{100} = \dfrac{1}{5}$

7.
$$
\begin{array}{r}
1.27 \\
5\overline{)6.35} \\
\underline{5} \\
1\,3 \\
\underline{1\,0} \\
35 \\
\underline{35} \\
0
\end{array}
$$

8.

$$\begin{array}{r} 0.125 \\ 4\overline{)0.500} \\ \underline{4} \\ 10 \\ \underline{8} \\ 20 \\ \underline{20} \\ 0 \end{array}$$

9.

$$\begin{array}{r} 0.125 \\ 8\overline{)1.000} \\ \underline{8} \\ 20 \\ \underline{16} \\ 40 \\ \underline{40} \\ 0 \end{array}$$

10.

$$9 \xrightarrow{\;8 + \frac{8}{8}\;} 8\frac{8}{8}$$

$$-3\frac{5}{8} \qquad -3\frac{5}{8}$$

$$\qquad\qquad 5\frac{3}{8}$$

$$x = \mathbf{5\frac{3}{8}}$$

11. $16\frac{1}{4} + 4\frac{3}{4} = 20\frac{4}{4} = 21$

$y = \mathbf{21}$

12.

$$\begin{array}{r} \overset{0}{\cancel{1}}.\,\overset{9}{\cancel{0}}\,\overset{9}{\cancel{0}}\,{}^{1}0 \\ -\;0.\,2\,3\,5 \\ \hline 0.\,7\,6\,5 \end{array}$$

$q = \mathbf{0.765}$

13.

$$\begin{array}{r} 26.9 \\ +\;12.0 \\ \hline 38.9 \end{array} \qquad \begin{array}{r} 4\,\overset{8}{\cancel{9}}.\,{}^{1}2\,5 \\ -\;3\,8.\,9 \\ \hline 1\,0.\,3\,5 \end{array}$$

$w = \mathbf{10.35}$

14.

$$\begin{array}{r} 2.5 \text{ cm} \\ \times\quad 4 \text{ cm} \\ \hline 10.0 \text{ cm}^2 \end{array} \qquad \begin{array}{r} 5\,\text{cm}^2 \\ 2\overline{)10 \text{ cm}^2} \\ \underline{10} \\ 0 \end{array}$$

15. $\dfrac{\text{dime}}{\text{quarter}} = \dfrac{\mathbf{10}}{\mathbf{25}} = \dfrac{\mathbf{2}}{\mathbf{5}}$

16.

$$\begin{array}{r} 0.25 \\ \times\quad 3.7 \\ \hline 175 \\ 750 \\ \hline \mathbf{0.925} \end{array}$$

17. $\dfrac{3}{4} \times \dfrac{3}{3} = \dfrac{9}{12} \qquad \mathbf{9}$

18. Multiples of 3

3, 6, 9, 12, 15, 18, 21, ㉔

Multiples of 4

4, 8, 12, 16, 20, ㉔, 28

Multiples of 8

8, 16, ㉔, 32, 40

LCM is **24.**

19. (a) $\dfrac{1}{10} \,\bigodot\!{=}\, 0.1$

$$\dfrac{1}{10}$$

(b) $0.1 \,\bigotimes\!{>}\, (0.1)^2$

$$\begin{array}{r} 0.1 \\ \times\;0.1 \\ \hline 0.01 \end{array}$$

20. **7**

21.

$$\begin{array}{r} 80 \\ 50\overline{)4000} \\ \underline{400} \\ 00 \\ \underline{00} \\ 0 \end{array}$$

22. $\sqrt{100 \text{ cm}^2} = 10 \text{ cm}$

$$\begin{array}{r} 10 \text{ cm} \\ \times\quad 4 \\ \hline \mathbf{40 \text{ cm}} \end{array}$$

23.

$$-2$$

24. $\dfrac{1}{2} \cdot \dfrac{4}{5} = \dfrac{4}{10} = \dfrac{\mathbf{2}}{\mathbf{5}}$

25. $\left(\dfrac{3}{4}\right)\left(\dfrac{5}{3}\right) = \dfrac{15}{12} = 1\dfrac{3}{12} = \mathbf{1\dfrac{1}{4}}$

26. **Radius, diameter, circumference**

27. $7.51; Round $6.95 to $7. Eight percent is a little less than 10%. Ten percent of $7 is 70¢. Adding 70¢ to $7 gives $7.70, so $7.51 is reasonable.

28. $\frac{6}{8}$ inch $\left(\text{or } \frac{3}{4} \text{ inch}\right)$

29. (a) $\frac{8}{3}$

(b) $\frac{8}{3} \times \frac{3}{1} = \frac{24}{3} = \frac{8}{1} = 8$

30. $\frac{1}{2} \times \frac{3}{3} = \frac{3}{6}$ $\frac{1}{3} \times \frac{2}{2} = \frac{2}{6}$

$\frac{3}{6} + \frac{2}{6} = \frac{5}{6}$

LESSON 47, WARM-UP

a. 3700

b. 261

c. 37

d. $7.75

e. $6.25

f. $2.50

g. 1

Problem Solving
$44 \times 2 = 88$
$88 - 34 = \mathbf{54}$

LESSON 47, LESSON PRACTICE

a. The product of "2 times r" in the second formula is equivalent to d in the first formula because two radii equal one diameter, so two radii times π equals one diameter times π.

b. $C = \pi d$
$C \approx (3.14)(2 \text{ in.})$
$C \approx \mathbf{6.28 \text{ in.}}$

c. $d = 3 \text{ cm} \times 2 = 6 \text{ cm}$
$C = \pi d$
$C \approx (3.14)(6 \text{ cm})$
$C \approx \mathbf{18.84 \text{ cm}}$

d. $C = \pi d$
$C \approx (3.14)(0.75 \text{ in.})$
$C \approx \mathbf{2.36 \text{ inches}}$

e. $2\frac{3}{8}$ **inches**

f. **C. 90 ft;** The radius is about 15 ft, so the diameter is about 30 ft. Using 3 as a rough approximation for π, we calculate that the circumference is about 3 × 30 ft, or about 90 ft.

g. $d = 5 \text{ in.} \times 2 = 10 \text{ in.}$
$C = \pi d$
$C \approx (3.14)(10 \text{ in.})$
$C \approx \mathbf{31.4 \text{ inches}}$

LESSON 47, MIXED PRACTICE

1. 1, 3, 5, 7, 9, 11, 13, 15, 17, ⑲

2. $\mathbf{500 \times h = 3000}$

$$\begin{array}{r} 6 \text{ hours} \\ 500\overline{)3000} \\ \underline{3000} \\ 0000 \end{array}$$

3.
$$\begin{array}{r} 6 \text{ months} \\ 2\overline{)12} \\ \underline{12} \\ 0 \end{array}$$

4. $A = bh$
$A = (8)(4)$
$A = \mathbf{32}$

5. $2 \cdot 2 \cdot 3 \cdot 3 \cdot 3 \cdot 5 \cdot 5 \cdot 7$

6. $\frac{4}{1} \times \frac{1}{2} = \frac{4}{2} = \mathbf{2}$

7. $(6 \times 1) + \left(2 \times \frac{1}{10}\right) + \left(5 \times \frac{1}{100}\right)$

8. $\frac{99}{100}$; **0.99**

9.

$$\begin{array}{r} 0.015 \\ 12\overline{)0.180} \\ \underline{0\ 0} \\ 18 \\ \underline{12} \\ 60 \\ \underline{60} \\ 0 \end{array}$$

10.

$$\begin{array}{r} 1.23 \\ 10\overline{)12.30} \\ \underline{10} \\ 2\ 3 \\ \underline{2\ 0} \\ 30 \\ \underline{30} \\ 0 \end{array}$$

11. $w \div 6 = 36$

$$\begin{array}{r} 36 \\ \times\ 6 \\ \hline 216 \end{array}$$

$w = \mathbf{216}$

12.

$$\begin{array}{r} 0.25 \\ 5\overline{)1.25} \\ \underline{1\ 0} \\ 25 \\ \underline{25} \\ 0 \end{array}$$

$y = \mathbf{0.25}$

13.

$$10 \xrightarrow{\ 9 + \frac{12}{12}\ } 9\frac{12}{12}$$

$$-\ 5\frac{11}{12} \qquad\qquad -\ 5\frac{11}{12}$$

$$\overline{\qquad\qquad\qquad\qquad 4\frac{1}{12}}$$

$n = \mathbf{4\frac{1}{12}}$

14. $6\frac{2}{5} + 3\frac{3}{5} = 9\frac{5}{5} = 10$

$\qquad m = \mathbf{10}$

15. $8\frac{3}{4} + 5\frac{3}{4} = 13\frac{6}{4} = 14\frac{2}{4} = \mathbf{14\frac{1}{2}}$

16. $\frac{5}{3} \times \frac{5}{4} = \frac{25}{12} = \mathbf{2\frac{1}{12}}$

17. $\frac{3}{4} \times \frac{5}{5} = \frac{15}{20}$

18. $\frac{3}{5} \times \frac{4}{4} = \frac{\mathbf{12}}{\mathbf{20}}$

19.

$$\begin{array}{r} \overset{1}{1}8 \\ 20 \\ 18 \\ 20 \\ +\ 20 \\ \hline 96 \end{array} \qquad \begin{array}{r} 19\frac{1}{5} \\ 5\overline{)96} \\ \underline{5} \\ 46 \\ \underline{45} \\ 1 \end{array}$$

C. 19

20. $C = \pi d$

$C \approx (3.14)(20\text{ in.})$

$C \approx \mathbf{62.8\text{ inches}}$

21. Factors of 20

①, ②, 4, ⑤, ⑩, 20

Factors of 30

①, ②, 3, ⑤, 6, ⑩, 15, 30

1, 2, 5, 10

22. **62.5; Since the problem is to multiply by 10, simply shift the decimal point in 6.25 one place to the right.**

23. $\dfrac{12.5}{5}$

$$\begin{array}{r} 2.5 \\ 5\overline{)12.5} \\ \underline{10} \\ 2\ 5 \end{array}$$

24.

$$\begin{array}{r} 0.90 \\ \times\quad 40 \\ \hline 36.00 \end{array}$$

36

25.

$$\begin{array}{r} 687 \\ -\ 365 \\ \hline \mathbf{322\text{ more days}} \end{array}$$

26.

$$\begin{array}{r} 3 \\ 200\overline{)700} \\ \underline{600} \\ 100 \end{array}$$

About 3 times

27. **Length = 1 inch**

Width = $\frac{3}{4}$ inch

28. 1 in. + 1 in. + $\frac{3}{4}$ in. + $\frac{3}{4}$ in.

2 in. + $\frac{6}{4}$ in. = 2 in. + $1\frac{2}{4}$ in.

$= 2$ in. + $1\frac{1}{2}$ in. = $3\frac{1}{2}$ **inches**

29. $\frac{2}{5} \times \frac{2}{2} = \frac{4}{10}$

$\frac{9}{10} - \frac{4}{10} = \frac{5}{10} = \frac{1}{2}$

30. The numbers 1.5 and 1.50 are equivalent. Attaching a zero to a decimal number does not shift place values. To multiply 1.5 by 10, we can move the decimal point one place to the right, which shifts the place values and makes the product 15.

LESSON 48, WARM-UP

a. 200

b. 580

c. 81

d. $5.25

e. $7.50

f. $0.25

g. 9

Problem Solving

$90 \times 3 = 270$

$270 - (85 + 85) = \mathbf{100}$

LESSON 48, LESSON PRACTICE

a.

$4\frac{1}{3} \xrightarrow{\ 3 + \frac{3}{3} + \frac{1}{3}\ } 3\frac{4}{3}$

$-\ 1\frac{2}{3} \qquad\qquad -\ 1\frac{2}{3}$

$\qquad\qquad\qquad\qquad\quad 2\frac{2}{3}$

b.

$3\frac{2}{5} \xrightarrow{\ 2 + \frac{5}{5} + \frac{2}{5}\ } 2\frac{7}{5}$

$-\ 2\frac{3}{5} \qquad\qquad -\ 2\frac{3}{5}$

$\qquad\qquad\qquad\qquad\quad \frac{4}{5}$

c.

$5\frac{2}{4} \xrightarrow{\ 4 + \frac{4}{4} + \frac{2}{4}\ } 4\frac{6}{4}$

$-\ 1\frac{3}{4} \qquad\qquad -\ 1\frac{3}{4}$

$\qquad\qquad\qquad\qquad\quad 3\frac{3}{4}$

d.

$5\frac{1}{8} \xrightarrow{\ 4 + \frac{8}{8} + \frac{1}{8}\ } 4\frac{9}{8}$

$-\ 2\frac{4}{8} \qquad\qquad -\ 2\frac{4}{8}$

$\qquad\qquad\qquad\qquad\quad 2\frac{5}{8}$

e.

$7\frac{3}{12} \xrightarrow{\ 6 + \frac{12}{12} + \frac{3}{12}\ } 6\frac{15}{12}$

$-\ 4\frac{10}{12} \qquad\qquad -\ 4\frac{10}{12}$

$\qquad\qquad\qquad\qquad\quad 2\frac{5}{12}$

f.

$6\frac{1}{4} \xrightarrow{\ 5 + \frac{4}{4} + \frac{1}{4}\ } 5\frac{5}{4}$

$-\ 2\frac{3}{4} \qquad\qquad -\ 2\frac{3}{4}$

$\qquad\qquad\qquad\qquad\quad 3\frac{2}{4} = 3\frac{1}{2}$

LESSON 48, MIXED PRACTICE

1.
$\begin{array}{r} 10 \\ \times\ 2 \\ \hline \mathbf{20} \end{array}$

2.
$\begin{array}{r} \$1.449 \\ \times\qquad 10.0 \\ \hline \$14.4900 \end{array} \longrightarrow \mathbf{\$14.49}$

3.
$\begin{array}{r} {\scriptstyle 12:\,80} \\ \cancel{1}{:}2\cancel{0}\ \text{p.m.} \\ -\ 11{:}45\ \text{a.m.} \\ \hline 1{:}35 \end{array}$

1 hour 35 minutes

4.   B. 0.2

5. $-1, 0, 0.102, 0.12, 1.02, 1.20$

6.
$$
\begin{array}{r}
\overset{1}{0.1} \\
0.2 \\
0.3 \\
+\ 0.4 \\
\hline
\mathbf{1.0}\ \text{or}\ \mathbf{1}
\end{array}
$$

7.
$$
\begin{array}{r}
0.125 \\
\times\quad 8 \\
\hline
\mathbf{1.000}\ \text{or}\ \mathbf{1}
\end{array}
$$

8. $3 - 2.1 = r$
$$
\begin{array}{r}
\overset{2}{\cancel{3}}.\overset{1}{0} \\
-\ 2.1 \\
\hline
\mathbf{0.9}\ \text{mile}
\end{array}
$$

9.
$$
\begin{array}{r}
5000 \\
8000 \\
+\ 7000 \\
\hline
\mathbf{20{,}000}
\end{array}
$$

10.
$$
\begin{array}{r}
\mathbf{0.018} \\
8\overline{)0.144} \\
\underline{0\ 0} \\
14 \\
\underline{8} \\
64 \\
\underline{64} \\
0
\end{array}
$$

11.
$$
\begin{array}{r}
\mathbf{0.15} \\
6\overline{)0.90} \\
\underline{6} \\
30 \\
\underline{30} \\
0
\end{array}
$$

12.
$$
\begin{array}{r}
\mathbf{0.225} \\
4\overline{)0.900} \\
\underline{8} \\
10 \\
\underline{8} \\
20 \\
\underline{20} \\
0
\end{array}
$$

13. $\$0.39 \times 100$

   $\$39.00$

14.   50.64

15.   Multiples of 6
   6, 12, 18, ㉔, 30, 36
   Multiples of 8
   8, 16, ㉔, 32, 40
   LCM is **24.**

16. $7\dfrac{7}{12} + 5\dfrac{5}{12} = 12\dfrac{12}{12} = 13$

   $w = \mathbf{13}$

17.
$$
12 \xrightarrow{\ 11\ +\ \frac{3}{3}\ } 11\dfrac{3}{3}
$$
$$
\begin{array}{r}
-\ 5\dfrac{2}{3} \qquad\qquad -\ 5\dfrac{2}{3} \\
\hline
6\dfrac{1}{3}
\end{array}
$$

   $m = \mathbf{6\dfrac{1}{3}}$

18.
$$
5\dfrac{1}{4} \xrightarrow{\ 4\ +\ \frac{4}{4}\ +\ \frac{1}{4}\ } 4\dfrac{5}{4}
$$
$$
\begin{array}{r}
-\ 2\dfrac{3}{4} \qquad\qquad -\ 2\dfrac{3}{4} \\
\hline
2\dfrac{2}{4} = 2\dfrac{1}{2}
\end{array}
$$

   $n = \mathbf{2\dfrac{1}{2}}$

19.
$$
\begin{array}{r}
\overset{3}{\cancel{4}}.\overset{9}{\cancel{0}}\overset{1}{0} \\
-\ 3.21 \\
\hline
0.79
\end{array}
$$

   $x = \mathbf{0.79}$

20. $\dfrac{2}{3} \times \dfrac{3}{4} = \dfrac{6}{12} = \mathbf{\dfrac{1}{2}}$

21. $3 + 5 - 12$
   $\quad 8 - 12$
   $\qquad \mathbf{-4}$

22. $C = \pi d$
   $C \approx (3.14)(2\ \text{cm})$
   $C \approx \mathbf{6.28\ cm}$
   π is a little more than 3, and 3×2 cm is 6 cm, so 6.28 cm is a reasonable answer.

23. $\dfrac{12}{8} = \mathbf{\dfrac{3}{2}}$

24.
$$\begin{array}{r} 12 \text{ inches} \\ \times \quad 4 \\ \hline 48 \text{ inches} \end{array}$$

25.
$$\begin{array}{r} 12 \text{ inches} \\ \times \ 12 \text{ inches} \\ \hline 24 \\ 120 \\ \hline 144 \text{ square inches} \end{array}$$

26. $d = rt$
$d = (60)(4)$
$d = \mathbf{240}$

27. $\dfrac{75}{100} = \dfrac{3}{4}$

$\dfrac{3}{4} \times \dfrac{32}{1} = \dfrac{96}{4} = \mathbf{24 \text{ chairs}}$

28. $\dfrac{1}{3} \times \dfrac{4}{4} = \dfrac{4}{12} \qquad \dfrac{1}{4} \times \dfrac{3}{3} = \dfrac{3}{12}$

$\dfrac{4}{12} + \dfrac{3}{12} = \dfrac{7}{12}$

29. $\dfrac{35}{7} = \mathbf{5}$

30. The server can cut one of the whole pies into sixths. Then there will be $2\frac{7}{6}$ pies on the shelf. The server can remove $1\frac{5}{6}$ pies, leaving $1\frac{2}{6}$ pies ($1\frac{1}{3}$ pies) on the shelf.

LESSON 49, WARM-UP

a. 1000

b. 340

c. 39

d. $1.75

e. $15.00

f. 40

g. 5

Problem Solving
Around the block:
$200 \text{ m} + 200 \text{ m} + 200 \text{ m} + 200 \text{ m} = 800 \text{ m}$
Nimah: $200 \text{ m} + 100 \text{ m} + 100 \text{ m} + 100 \text{ m}$
$\qquad\qquad + 100 \text{ m} + 200 \text{ m} = 800 \text{ m}$
Nimah saved **0 meters** by using her "shortcut."

LESSON 49, LESSON PRACTICE

a. 10

b. 100

c. $\dfrac{2.4}{4}$
$$\begin{array}{r} 0.6 \\ 4\overline{)2.4} \\ 2.4 \\ \hline 0 \end{array}$$

d. $\dfrac{90}{3}$
$$\begin{array}{r} 30 \\ 3\overline{)90} \\ 9 \\ \hline 00 \\ 00 \\ \hline 0 \end{array}$$

e.
$$\begin{array}{r} 50 \\ 5\overline{)250} \\ 25 \\ \hline 00 \\ 00 \\ \hline 0 \end{array}$$

f.
$$\begin{array}{r} 40 \\ 3\overline{)120} \\ 12 \\ \hline 00 \\ 00 \\ \hline 0 \end{array}$$

g. $2.4 \div 8$
$$\begin{array}{r} 0.3 \\ 8\overline{)2.4} \\ 2\ 4 \\ \hline 0 \end{array}$$

h. $30 \div 3$
$$\begin{array}{r} 10 \\ 3\overline{)30} \\ 3 \\ \hline 00 \\ 00 \\ \hline 0 \end{array}$$

i.
$$\begin{array}{r} 8 \\ 5\overline{)40} \\ \underline{40} \\ 0 \end{array}$$

j. $2 \div 4$
$$\begin{array}{r} 0.5 \\ 4\overline{)2.0} \\ \underline{2\ 0} \\ 0 \end{array}$$

8.
$$\begin{array}{r} 50 \\ 7\overline{)350} \\ \underline{35} \\ 00 \\ \underline{00} \\ 0 \end{array}$$

9.
$$\begin{array}{r} 24 \\ 5\overline{)120} \\ \underline{10} \\ 20 \\ \underline{20} \\ 0 \end{array}$$

LESSON 49, MIXED PRACTICE

1. $\dfrac{(0.2 + 0.3)}{0.5} - \dfrac{(0.2 \times 0.3)}{0.06}$

$\qquad$ **0.44**

2.

$$\frac{4}{5} \text{ of } \$1.00 \qquad \frac{1}{5} \text{ of } \$1.00$$

$1.00
20¢
20¢
20¢
20¢
20¢

80¢

3.
$$\begin{array}{r} 2.6 \text{ inches} \\ \times\ 2.2 \text{ inches} \\ \hline 52 \\ 520 \\ \hline \textbf{5.72 square inches} \end{array}$$

4.
$$\begin{array}{r} \overset{1}{2.6} \text{ in.} \\ 2.6 \text{ in.} \\ 2.2 \text{ in.} \\ +\ 2.2 \text{ in.} \\ \hline \textbf{9.6 in.} \end{array}$$

5. (a) $0.31 \; \bigcirc\!\!\!> \; 0.301$

$\qquad$ (b) $31\% \; \bigcirc\!\!\!> \; 30.1\%$

6.
$$\begin{array}{r} \overset{1\ 1}{0.67} \\ 2 \\ +\ 1.33 \\ \hline \textbf{4.00 or 4} \end{array}$$

7.
$$\begin{array}{r} 12 \\ \times\ 0.25 \\ \hline 60 \\ 240 \\ \hline \textbf{3.00 or 3} \end{array}$$

10.
$$\begin{array}{r} 0.0175 \\ 8\overline{)0.1400} \\ \underline{0\ 0} \\ 14 \\ \underline{8} \\ 60 \\ \underline{56} \\ 40 \\ \underline{40} \\ 0 \end{array}$$

11.
$$\begin{array}{r} 0.012 \\ \times\ \ \ 1.5 \\ \hline 0060 \\ 00120 \\ \hline 0.0180 \text{ or } \textbf{0.018} \end{array}$$

12. $6\dfrac{1}{8} + 4\dfrac{3}{8} = 10\dfrac{4}{8} = 10\dfrac{1}{2}$

$\qquad n = \mathbf{10\dfrac{1}{2}}$

13. $\dfrac{4}{5} \times \dfrac{20}{20} = \dfrac{80}{100}$

$\qquad x = \mathbf{80}$

14.
$$\begin{array}{r} \overset{4}{\cancel{5}}.\overset{9}{\cancel{0}}{}^{1}0 \\ -\ 1.\ 3\ 7 \\ \hline 3.\ 6\ 3 \end{array}$$

$\qquad m = \mathbf{3.63}$

15.

$$15 \xrightarrow{\ 14 + \frac{4}{4}\ } 14\frac{4}{4}$$
$$-\ 7\frac{1}{4} \qquad\qquad -\ 7\frac{1}{4}$$
$$\qquad\qquad m = \mathbf{7\frac{3}{4}}$$

16. 1.012

17. $5\frac{7}{10} + 4\frac{9}{10} = 9\frac{16}{10} = 10\frac{6}{10} = 10\frac{3}{5}$

18. $\frac{5}{2} \times \frac{5}{3} = \frac{25}{6} = 4\frac{1}{6}$

19.
$$\begin{array}{r} \$25.00 \\ \times \quad 0.4 \\ \hline \$10.000 \text{ or } \$10.00 \end{array}$$

20. (a) $\frac{8}{24} = \frac{1}{3}$

(b) $\frac{1}{3}$

(c) $\begin{array}{r} \overset{1}{2}^{1}4 \\ - \quad 8 \\ \hline 1\ 6 \end{array}$ $\quad \frac{16}{24} = \frac{2}{3}$

21. Factors of 12
①, ②, ③, 4, ⑥, 12
Factors of 18
①, ②, ③, ⑥, 9, 18
1, 2, 3, 6

22. $\begin{array}{r} \overset{1}{1}.2 \\ 1.3 \\ + \ 1.7 \\ \hline 4.2 \end{array}$ $\quad \begin{array}{r} 1.4 \\ 3\overline{)4.2} \\ \underline{3} \\ 1\ 2 \\ \underline{1\ 2} \\ 0 \end{array}$

23. **$25; Round 51% to 50% and round $49.78 to $50. Since 50% equals $\frac{1}{2}$, find $\frac{1}{2}$ of $50.**

24. (a) $\frac{4}{3}$

(b) $\frac{4}{3} \times \frac{4}{1} = \frac{16}{3} = 5\frac{1}{3}$

25. (a) y

(b) x

(c) z

26. $\frac{420}{70}$ $\quad \begin{array}{r} 6 \\ 70\overline{)420} \\ \underline{420} \\ 0 \end{array}$

27. $\frac{\$3.00}{\$0.25} \times \frac{100}{100} = \frac{\$300}{\$25} = 12$

28. Side $= \frac{3}{4}$ in.

$\frac{3}{4} \times \frac{4}{1} = \frac{12}{4} = 3$ in.

Perimeter = 3 in.

29. $C = \pi d$
$C \approx (3.14)(4\text{ cm})$
$C \approx$ **12.56 cm**

30. Yes, Sam found the correct answer. Both $\frac{10}{10}$ and $\frac{100}{100}$ are equal to 1. When we multiply by different fraction names for 1, we get numbers that are equal even though they may look different. So $\frac{2.5}{0.5}$, $\frac{25}{5}$, and $\frac{250}{50}$ are three equivalent division problems with the same quotient.

LESSON 50, WARM-UP

a. 1800

b. 315

c. 85

d. $2.44

e. $12.50

f. 500

g. 1

Problem Solving

Steps	Blocks
1	1
2	3
3	6
4	10
5	15
6	21
7	28
8	36
9	45
10	55

LESSON 50, LESSON PRACTICE

a. $\dfrac{1}{10} = \mathbf{0.1}$

b. $\dfrac{5}{10} = \mathbf{0.5}$

c. $\dfrac{9}{10} = \mathbf{0.9}$

d. $1\dfrac{2}{10} = \mathbf{1.2}$

e. $1\dfrac{6}{10} = \mathbf{1.6}$

f. $1\dfrac{8}{10} = \mathbf{1.8}$

g. $4 \div \dfrac{1}{4}$

$1 \div \dfrac{1}{4} = \dfrac{4}{1} = 4$

$4 \times 4 = \mathbf{16}$

h. $12 \div \dfrac{3}{8}$ in.

$1 \div \dfrac{3}{8} = \dfrac{8}{3}$

$12 \times \dfrac{8}{3} = \dfrac{12}{1} \times \dfrac{8}{3} = \dfrac{96}{3} = \mathbf{32\ pads}$

LESSON 50, MIXED PRACTICE

1. $1 + 3 + 5 + 7 + 9 + 11 + 13 + 15 + 17 + 19 = \mathbf{100}$

2. $6 \div \dfrac{3}{8}$ in.

$1 \div \dfrac{3}{8} = \dfrac{8}{3}$

$\dfrac{6}{1} \times \dfrac{8}{3} = \dfrac{48}{3} = \mathbf{16\ boxes}$

3.

11 rounds	33 minutes
$\times$ 3 minutes	+ 2 minutes
33 minutes	**35 minutes**

4. (a) $3.4 \ \text{>} \ 3.389$

(b) $0.60 \ \text{=} \ 0.600$

5. $9.25 + w = 10$

$$\begin{array}{r} \overset{0}{\cancel{1}}\overset{9}{0}.\overset{9}{\cancel{0}}\overset{1}{0} \\ -\ 9.25 \\ \hline 0.75 \end{array}$$

$w = \mathbf{0.75}$

6.

$$\begin{array}{r} 0.024 \\ 6\overline{)0.144} \\ \underline{0\ 0} \\ 14 \\ \underline{12} \\ 24 \\ \underline{24} \\ 0 \end{array}$$

$w = \mathbf{0.024}$

7. $\dfrac{12}{12} - \dfrac{5}{12} = \dfrac{7}{12}$

$w = \mathbf{\dfrac{7}{12}}$

8.

$6\dfrac{1}{8} \quad \xrightarrow{\ 5 + \frac{8}{8} + \frac{1}{8}\ } \quad 5\dfrac{9}{8}$

$-\ 1\dfrac{7}{8} \qquad\qquad\qquad\quad -\ 1\dfrac{7}{8}$

$\qquad\qquad\qquad\qquad\qquad 4\dfrac{2}{8} = 4\dfrac{1}{4}$

$x = \mathbf{4\dfrac{1}{4}}$

9.

$\$20.00$	$\$20.00$
$\times\quad .07$	$+\quad \$1.40$
$\$1.4000$	**$\$21.40$**

10.

$$\begin{array}{r} \overset{0}{\cancel{1}}.\overset{9}{\cancel{0}}\overset{1}{0} \\ -\ 0.97 \\ \hline \mathbf{0.03} \end{array}$$

11.

$$\begin{array}{r} 60 \\ 12\overline{)720} \\ \underline{72} \\ 00 \\ \underline{00} \\ 0 \end{array}$$

12.
$$
\begin{array}{r}
17.5 \\
4\overline{)70.0} \\
4 \\
\hline
30 \\
28 \\
\hline
2\,0 \\
2\,0 \\
\hline
0
\end{array}
$$

13.
$$
\begin{array}{r}
0.023 \\
6\overline{)0.138} \\
0\,0 \\
\hline
13 \\
12 \\
\hline
18 \\
18 \\
\hline
0
\end{array}
$$

14.
$$
\begin{array}{r}
3.75 \\
\times\ \ 2.4 \\
\hline
1500 \\
7500 \\
\hline
9.000 \text{ or } 9
\end{array}
$$

15. $\dfrac{3}{4} \times \dfrac{6}{6} = \dfrac{18}{24}$ **18**

16. **6**

17.
$$
\begin{array}{r}
100 \text{ cm} \\
\times\quad 100 \text{ cm} \\
\hline
\textbf{10,000 square centimeters}
\end{array}
$$

18. Multiples of 6
6, 12, ⑱, 24, 30, 36
Multiples of 9
9, ⑱, 27, 36, 45
LCM is **18.**

19. $6\dfrac{5}{8} + 4\dfrac{5}{8} = 10\dfrac{10}{8} = 11\dfrac{2}{8} = \mathbf{11\dfrac{1}{4}}$

20. $\dfrac{8}{3} \times \dfrac{3}{1} = \dfrac{24}{3} = \mathbf{8}$

21. $\dfrac{2}{3} \cdot \dfrac{3}{4} = \dfrac{6}{12} = \mathbf{\dfrac{1}{2}}$

22. $\dfrac{7}{1} \times \dfrac{22}{7} = \dfrac{154}{7} = \mathbf{22 \text{ cm}}$

π **is a little more than 3, and 3 × 7 cm**
is 21 cm, so 22 cm is reasonable.

23.
$$
\begin{array}{r}
\overset{1}{2}.4 \\
6.3 \\
+\ 5.7 \\
\hline
14.4
\end{array}
\qquad
\begin{array}{r}
4.8 \\
3\overline{)14.4} \\
12 \\
\hline
2\,4 \\
2\,4 \\
\hline
0
\end{array}
$$

24. Factors of 18
①, ②, ③, ⑥, 9, 18
Factors of 24
①, ②, ③, 4, ⑥, 8, 12, 24
1, 2, 3, 6

25. $5\dfrac{3}{10} = \mathbf{5.3}$

26. $\dfrac{210}{10} = \mathbf{21}$

27. **37.5**

28. $\dfrac{1}{3} \times \dfrac{2}{2} = \dfrac{2}{6}$

$\dfrac{5}{6} - \dfrac{2}{6} = \dfrac{3}{6} = \mathbf{\dfrac{1}{2}}$

29. (a) **y**

(b) **z**

(c) **x**

30. 552; No, last digit is not 5 or 0.
255; No, last digit is not even.
250; Yes, last digit is even and 5 or 0.
525; No, last digit is not even.
C. 250

INVESTIGATION 5

1.

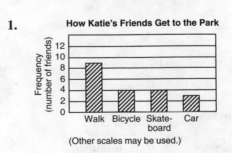

How Katie's Friends Get to the Park

(Other scales may be used.)

2.

How Katie's Friends Get to the Park

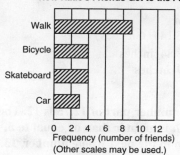

Frequency (number of friends)
(Other scales may be used.)

3.

How Katie's Friends Get to the Park

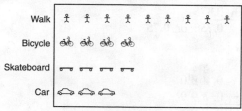

One object represents one friend.

4.

How Katie's Friends Get to the Park

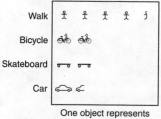

One object represents
two friends.

5.

How Katie's Friends Get to the Park

$360° ÷ 20$ friends = **18° per friend**

6. **47, 48, 49, 50, 51, 52, 53, 53, 53, 56, 56, 57, 58,**
58, 60, 60, 60, 62, 62, 63

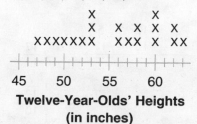

Twelve-Year-Olds' Heights
(in inches)

7. In the list of heights in increasing order (see solution to problem 6), the two 56-inch measurements are in the middle. So the median height is **56 inches.**

8. **53 inches** and **60 inches** each appear three times in the list, more than any other measurement.

9. 63 inches − 47 inches = **16 inches**

10.

Stem	Leaf
4	7 8 9
5	0 1 2 3 3 3 6 6 7 8 8
6	0 0 0 2 2 3

Extensions

a. See student work.

b. See student work.

LESSON 51, WARM-UP

a. 1000

b. 272

c. 35

d. $5.63

e. $75.00

f. $1.00

g. 20

Problem Solving

First think, "9 times 9 equals what number?" (81). Then think, "What number (with 9 in the ones place) minus 81 equals a single-digit number?" (89). Next think, "What multiple of 9 has 8 in the tens place?" (81). Finally, complete the division.

$$
\begin{array}{r}
99 \\
9)\overline{891} \\
\underline{81} \\
81 \\
\underline{81} \\
0
\end{array}
$$

LESSON 51, LESSON PRACTICE

a. $6.67

b. $0.46

c. $0.08

d. 0.1

e. 12.3

f. 2.4

g. 17

h. 5

i. 73

j.
$$\begin{array}{r} \$3.79 \\ \times \quad .06 \\ \hline \$0.2274 \\ \mathbf{\$0.23} \end{array}$$

LESSON 51, MIXED PRACTICE

1. 8, 16, ⓐ24
6, 12, 18, ⓐ24
24 − 24 = **0**

2.
$$\begin{array}{r} 3.5 \text{ miles} \\ \times \quad 2 \\ \hline \mathbf{7.0} \text{ or } \mathbf{7} \text{ miles} \end{array}$$

3. 1804 − 1769 = d

$$\begin{array}{r} 1\,\overset{7}{8}\,\overset{9}{\cancel{0}}{}^{1}4 \\ -\ 1769 \\ \hline \mathbf{3\,5} \text{ years old} \end{array}$$

4. (a)
$$\begin{array}{r} \$12.89 \\ \times \quad\quad .08 \\ \hline \$1.0312 \\ \mathbf{\$1.03} \end{array}$$

(b)
$$\begin{array}{r} \$12.\overset{1}{8}9 \\ +\ \ \$1.03 \\ \hline \mathbf{\$13.92} \end{array}$$

5. (a) 3 × 2 = **6 inches**

(b) $C = \pi d$
$C \approx (3.14)(6 \text{ inches})$
$C \approx 18.84$ inches
$C \approx \mathbf{19}$ **inches**

6. **The whole-number part of 12.75 is 12. The next digit, 7, is greater than or equal to 5, so round up to the next whole number, 13.**

7.
$$\begin{array}{r} \overset{1\,1}{0.125} \\ 0.25 \\ +\ 0.375 \\ \hline \mathbf{0.750} \text{ or } \mathbf{0.75} \end{array}$$

8.
$$\begin{array}{r} 0.\,\overset{3}{\cancel{4}}\,\overset{9}{\cancel{0}}{}^{1}0 \\ -\ 0.399 \\ \hline 0.001 \end{array}$$
$w = \mathbf{0.001}$

9.
$$\begin{array}{r} \mathbf{16} \\ 25\overline{)400} \\ \underline{25} \\ 150 \\ \underline{150} \\ 0 \end{array}$$

10.
$$\begin{array}{r} \mathbf{0.125} \\ 4\overline{)0.500} \\ \underline{4} \\ 10 \\ \underline{8} \\ 20 \\ \underline{20} \\ 0 \end{array}$$

11.
$$\begin{array}{r} \mathbf{0.325} \\ 10\overline{)3.250} \\ \underline{3\,0} \\ 25 \\ \underline{20} \\ 50 \\ \underline{50} \\ 0 \end{array}$$

12.
$$3\frac{5}{12} \xrightarrow{\ 2 + \frac{12}{12} + \frac{5}{12}\ } 2\frac{17}{12}$$
$$-1\frac{7}{12} \qquad\qquad\qquad -1\frac{7}{12}$$
$$\qquad\qquad\qquad\qquad 1\frac{10}{12} = \mathbf{1\frac{5}{6}}$$

13. $\frac{5}{8} \times \frac{3}{3} = \frac{15}{24} = \textbf{15}$

14.

$$25 \xrightarrow{\; 24 + \frac{4}{4}\;} 24\frac{4}{4}$$
$$-17\frac{3}{4} \qquad\qquad -17\frac{3}{4}$$
$$\textbf{7}\frac{\textbf{1}}{\textbf{4}}$$

15.
$$\begin{array}{r} 0.19 \\ \times\ 0.21 \\ \hline 19 \\ 380 \\ \hline \textbf{0.0399} \end{array}$$

16. $\dfrac{1}{100}$

17. **60.07**

18. $\sqrt{64\ \text{cm}^2} = 8\ \text{cm}$
8 cm × 4 = **32 cm**

19. Multiples of 2
2, 4, 6, 8, 10, ⑫, 14
Multiples of 3
3, 6, 9, ⑫, 15, 18
Multiples of 4
4, 8, ⑫, 16, 20, 24
LCM is **12.**

20. $5\frac{3}{10} + 6\frac{9}{10} = 11\frac{12}{10} = 12\frac{2}{10} = \textbf{12}\frac{\textbf{1}}{\textbf{5}}$

21. $\frac{10}{3} \times \frac{1}{2} = \frac{10}{6} = 1\frac{4}{6} = \textbf{1}\frac{\textbf{2}}{\textbf{3}}$

22. $12 \div \frac{3}{4}$ in.

$1 \div \frac{3}{4} = \frac{4}{3}$

$\frac{12}{1} \times \frac{4}{3} = \frac{48}{3} = \textbf{16 books}$

23.
$$\begin{array}{r} 50 \\ 100\overline{)5000} \\ 500 \\ \hline 00 \\ 00 \\ \hline 0 \end{array}$$

24. Factors of 16
①, ②, ④, ⑧, 16
Factors of 24
①, ②, 3, ④, 6, ⑧, 12, 24
1, 2, 4, 8

25. 12 × 4 = **48**
Rounded 11.8 to 12 and 3.89 to 4; then multiplied 12 by 4.

26. $\frac{7}{10} + 1\frac{3}{10} = 1\frac{10}{10} = 2$
$2 \div 2 = \textbf{1}$

27. $\frac{180}{60} = \textbf{3}$

28. **$7.90; Since the problem is to multiply by 10, shift the decimal point in $0.79 one place to the right.**

29. $\frac{2}{3} \times \frac{4}{4} = \frac{8}{12} \qquad \frac{3}{4} \times \frac{3}{3} = \frac{9}{12}$
$\frac{8}{12} + \frac{9}{12} = \frac{17}{12} = \textbf{1}\frac{\textbf{5}}{\textbf{12}}$

30. 75, 80, 80, 85, 85, 90, 90, 90, 100
Median: 85
Mode: 90

LESSON 52, WARM-UP

a. **1000**

b. **218**

c. **349**

d. **$22.50**

e. $\textbf{2}\frac{\textbf{1}}{\textbf{2}}$

f. **800**

g. **25**

Problem Solving
90% × 4 = 360%
360% − (88% × 3) = **96%**

LESSON 52, LESSON PRACTICE

a. 0.25

b. 0.025

c. 8.75

d. 0.875

e. 0.05

f. 0.005

g. 2.5

h. 0.25

LESSON 52, MIXED PRACTICE

1. $\frac{1}{2} \times \frac{2}{3} = \frac{2}{6} = \mathbf{\frac{1}{3}}$

2. $\begin{array}{r} 88 \\ -\ 52 \\ \hline 36 \end{array}$ black keys $\begin{array}{r} \overset{4}{5}{}^{1}2 \\ -\ 3\,6 \\ \hline \mathbf{1\,6} \end{array}$ **more white keys**

3. **28,232 feet**

4. (a) **37.5**; To multiply 3.75 by 10, shift the decimal point one place to the right.

    (b) **0.375**; To divide 3.75 by 10, shift the decimal point one place to the left.

5. $\dfrac{\text{camper}}{\text{counselor}} = \dfrac{320}{16} = \mathbf{\dfrac{20}{1}}$

6. **32**

7. $\begin{array}{r} 0.125 \\ \times\quad 4 \\ \hline 0.500 \end{array}$ or **0.5**

8. $\begin{array}{r} 12\frac{6}{12} = \mathbf{12\frac{1}{2}} \\ 12\overline{)150} \\ \underline{12} \\ 30 \\ \underline{24} \\ 6 \end{array}$

9. $\dfrac{(1\ +\ 0.2)}{(1\ -\ 0.2)} = \dfrac{1.2}{0.8}$

    $\begin{array}{r} \mathbf{0.15} \\ 8\overline{)1.20} \\ \underline{8} \\ 40 \\ \underline{40} \\ 0 \end{array}$

10. $\dfrac{5}{2} \times \dfrac{4}{1} = \dfrac{20}{2} = \mathbf{10}$

11. $\begin{array}{r} 5\frac{1}{3} \\ -\ 1\frac{2}{3} \\ \hline \end{array}$ $\xrightarrow{\ 4\ +\ \frac{3}{3}\ +\ \frac{1}{3}\ }$ $\begin{array}{r} 4\frac{4}{3} \\ -\ 1\frac{2}{3} \\ \hline m = \mathbf{3\frac{2}{3}} \end{array}$

12. $5\frac{1}{3} + 1\frac{2}{3} = 6\frac{3}{3} = 7$

     $m = \mathbf{7}$

13. $\begin{array}{r} \$\overset{0}{\cancel{1}}\overset{9}{\cancel{0}}.\overset{1}{0}\,0 \\ -\ \ \$\,0.\,1\,0 \\ \hline \$\,9.\,9\,0 \end{array}$

     $w = \mathbf{\$9.90}$

14. $\begin{array}{r} \$8.59 \\ \times\quad .06 \\ \hline \$0.5154 \\ \mathbf{52¢} \end{array}$

15. $C = \pi d$
     $C \approx (3.14)(24 \text{ inches})$
     $C \approx 75.36 \text{ inches}$
     $C \approx \mathbf{75 \text{ inches}}$

16. **0.201, 0.21, 1.02, 1.2**

Saxon Math 7/6—Homeschool

17. 16

18.
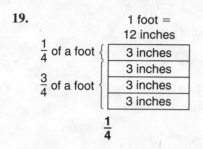

$$4\overline{)80}\text{ feet}$$
20 feet

20 tiles in a row
× 20 rows
400 square tiles

19.
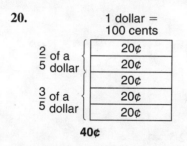

1 foot =
12 inches

$\frac{1}{4}$ of a foot { 3 inches

$\frac{3}{4}$ of a foot { 3 inches / 3 inches / 3 inches

$\frac{1}{4}$

20.

1 dollar =
100 cents

$\frac{2}{5}$ of a dollar { 20¢ / 20¢

$\frac{3}{5}$ of a dollar { 20¢ / 20¢ / 20¢

40¢

21. $12 \div \frac{3}{4}$ in.

$$1 \div \frac{3}{4} = \frac{4}{3}$$

$$\frac{12}{1} \times \frac{4}{3} = \frac{48}{3} = \textbf{16 pennies}$$

22. Multiples of 2
2, 4, 6, 8, 10, ⑫, 14
Multiples of 4
4, 8, ⑫, 16, 20, 24
Multiples of 6
6, ⑫, 18, 24, 30
LCM is **12.**

23. (a) $1 - 1 = \textbf{0}$

(b) $2 - 4 = \textbf{-2}$

24. **Answers may vary.**
Sample answer: about 2 meters

25. 1.8

26. $\frac{72}{12} = \textbf{6}$

27. $\frac{1}{2} \times \frac{3}{3} = \frac{3}{6}$ $\frac{2}{3} \times \frac{2}{2} = \frac{4}{6}$

$\frac{3}{6} + \frac{4}{6} = \frac{7}{6} = 1\frac{1}{6}$

28. $\frac{78}{100}$; **0.78**

29. **80%**; $\frac{80}{100} = \frac{4}{5}$

Approximately $\frac{4}{5}$ of the earth's atmosphere is nitrogen.

30. Sample answer:

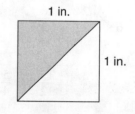

1 in.

1 in.

LESSON 53, WARM-UP

a. 1800

b. 290

c. 151

d. $12.50

e. 5

f. 40

g. 11

Problem Solving
Both the middle row and the middle column contain multiples of 7.
49

LESSON 53, LESSON PRACTICE

a. Sample answer:

$5 - 4.2$ 0.4×0.2
   line up   multiply; then count

$0.12 \div 3$ $5 \div 0.4$
   up   over, over, up

b.   **Decimal Arithmetic Reminders**

Operation	+ or −	×	÷ by whole (W)	÷ by decimal (D)
Memory cue	line up $\pm \begin{smallmatrix}\cdot\\\cdot\\\cdot\end{smallmatrix}$	×; then count $\times \dfrac{\cdot\!\!-\!\!-}{-\!\!-\!\!-}$	up $W)\overline{\cdot\ \ }$	over, over, up $D)\overline{\cdot\ \ }$

You may need to…
- Place a decimal point to the right of a whole number.
- Fill empty places with zeros.

c.

$$\dfrac{5}{6} \times \dfrac{2}{2} = \dfrac{10}{12}$$
$$+ \dfrac{5}{12} \qquad = \dfrac{5}{12}$$
$$\overline{\qquad\qquad\qquad} $$
$$\dfrac{15}{12} = \dfrac{5}{4} = 1\dfrac{1}{4}$$

d.

$$\dfrac{9}{10} \qquad = \dfrac{9}{10}$$
$$+ \dfrac{3}{5} \times \dfrac{2}{2} = \dfrac{6}{10}$$
$$\overline{\qquad\qquad\qquad}$$
$$\dfrac{15}{10} = \dfrac{3}{2} = 1\dfrac{1}{2}$$

e.

$$\dfrac{2}{3} \times \dfrac{4}{4} = \dfrac{8}{12}$$
$$+ \dfrac{7}{12} \qquad = \dfrac{7}{12}$$
$$\overline{\qquad\qquad\qquad}$$
$$\dfrac{15}{12} = \dfrac{5}{4} = 1\dfrac{1}{4}$$

LESSON 53, MIXED PRACTICE

1. We add or subtract only digits that have the same place value. When we line up the decimal points, digits with the same place value are automatically aligned.

2.
$$\overset{12:180}{\cancel{3}:\cancel{00}} \text{ p.m.}$$
$$- \ 4:45$$
$$\overline{8:135}$$
10:15 a.m.

3. 7 seconds $\div \dfrac{1}{4}$

$$1 \div \dfrac{1}{4} = \dfrac{4}{1}$$

7 seconds $\times 4 = $ **28 seconds**

4.

$$\overset{1}{47}$$
$$52$$
$$63$$
$$+ \ 66$$
$$\overline{228}$$

$$\begin{array}{r} 57 \text{ points} \\ 4)\overline{228} \\ \underline{20} \\ 28 \\ \underline{28} \\ 0 \end{array}$$

Mean = **57 points**

$$\overset{5}{\cancel{6}}\overset{1}{6}$$
$$- \ 4\ 7$$
$$\overline{1\ 9} \text{ points}$$

Range = **19 points**

5.

$$\begin{array}{r} 100 \\ 375)\overline{37500} \\ \underline{375} \\ 00 \\ \underline{00} \\ 00 \\ \underline{00} \\ 0 \end{array}$$

$x = $ **100**

6. $1.25 \times 10 = 12.5$
$$m = \textbf{12.5}$$

7. $\dfrac{1}{100}$**; 0.01**

8.
$$3.6$$
$$4$$
$$+ \ 0.39$$
$$\overline{\textbf{7.99}}$$

9.

$$\begin{array}{r} 300 \\ 12)\overline{3600} \\ \underline{36} \\ 00 \\ \underline{00} \\ 00 \\ \underline{00} \\ 0 \end{array}$$

10.

$$\begin{array}{r} 0.0375 \\ 4)\overline{0.1500} \\ \underline{12} \\ 30 \\ \underline{28} \\ 20 \\ \underline{20} \\ 0 \end{array}$$

11.

$$6\frac{1}{4} \xrightarrow{\ 5 + \frac{4}{4} + \frac{1}{4}\ } 5\frac{5}{4}$$

$$-\ 3\frac{3}{4} \qquad\qquad -\ 3\frac{3}{4}$$

$$2\frac{2}{4} = \mathbf{2\frac{1}{2}}$$

12. $\dfrac{2}{3} \times \dfrac{3}{5} = \dfrac{6}{15} = \mathbf{\dfrac{2}{5}}$

13. $5\dfrac{5}{8} + 7\dfrac{7}{8} = 12\dfrac{12}{8} = 12\dfrac{3}{2} = \mathbf{13\dfrac{1}{2}}$

14. 5

15. (a) $\begin{array}{r} \$5.20 \\ \times\ \ .08 \\ \hline \$0.4160 \end{array}$

$\qquad\qquad$ **42¢**

$\qquad$ (b) $\begin{array}{r} \$5.20 \\ +\ \$0.42 \\ \hline \mathbf{\$5.62} \end{array}$

16. B. 0.9

17. $C = \pi d$

$\quad C \approx (3.14)(40 \text{ feet})$

$\quad C \approx 125.6 \text{ feet}$

$\quad C \approx \mathbf{126\ feet}$

18. $\dfrac{3}{8}$ in. $\times\ \dfrac{4}{1} = \dfrac{12}{8} = 1\dfrac{4}{8} = \mathbf{1\dfrac{1}{2}}$ **in.**

19. $\dfrac{3}{36} = \mathbf{\dfrac{1}{12}}$

20. (a) Factors of 11: **1, 11**

$\qquad$ (b) **Prime number**

21. $16 - 2 = \mathbf{14}$

22. Multiple of 6

$\quad$ 6, 12, ⑱, 24, 30

$\quad$ Multiple of 9

$\quad$ 9, ⑱, 36, 45

$\quad$ LCM is **18.**

23. $\dfrac{3}{2} \cdot \dfrac{2}{3} = 1$

$\quad 1 \div \dfrac{2}{3} = \dfrac{3}{2}$

$\quad 1 \div \dfrac{3}{2} = \dfrac{2}{3}$

24. $\dfrac{600}{100} = \mathbf{6}$

25. $\dfrac{5}{6} \times \dfrac{4}{4} = \dfrac{20}{24} \qquad \mathbf{20}$

26. **Answers may vary. Sample answer:**

27.

$\qquad$ 350 students

30% enrolled. $\left\{ \begin{array}{|c|} \hline \text{35 students} \\ \hline \text{35 students} \\ \hline \text{35 students} \\ \hline \end{array} \right.$

70% did not enroll. $\left\{ \begin{array}{|c|} \hline \text{35 students} \\ \hline \text{35 students} \\ \hline \text{35 students} \\ \hline \text{35 students} \\ \hline \text{35 students} \\ \hline \text{35 students} \\ \hline \text{35 students} \\ \hline \end{array} \right.$

$\qquad$ **105 students**

28. $\dfrac{1}{4} \times \dfrac{3}{3} = \dfrac{3}{12} \qquad \dfrac{1}{6} \times \dfrac{2}{2} = \dfrac{2}{12}$

$\quad \dfrac{3}{12} + \dfrac{2}{12} = \dfrac{5}{12}$

29. $\begin{array}{r} 8.7 \\ -\ 7.4 \\ \hline \mathbf{1.3} \end{array}$

30. $24 \div \dfrac{3}{4}$ in.

$\quad 1 \div \dfrac{3}{4} = \dfrac{4}{3}$

$\quad \dfrac{24}{1} \times \dfrac{4}{3} = \dfrac{96}{3} = \mathbf{32\ books}$

LESSON 54, WARM-UP

a. 1500

b. 336

c. 474

d. $13.75

e. $4\frac{1}{2}$

f. 900

g. 100

Problem Solving

Walk to Natalie's house:

$1 + 1 + 1 + 1 + 1 + 1 = 6$

Walk home: $3 + 3 = 6$

The walk to Natalie's house was equal in distance to the walk home.

LESSON 54, LESSON PRACTICE

a. $\dfrac{2 \cdot 2 \cdot 5 \cdot 3}{2 \cdot 2 \cdot 5} = 1 \cdot 1 \cdot 1 \cdot 3 = \mathbf{3}$

b. $\dfrac{2 \cdot 2 \cdot 3 \cdot 3 \cdot 5}{2 \cdot 2 \cdot 3 \cdot 5 \cdot 5} = 1 \cdot 1 \cdot 1 \cdot \dfrac{3}{5} \cdot 1 = \dfrac{\mathbf{3}}{\mathbf{5}}$

c. $\dfrac{1}{2} \div \dfrac{3}{8}$

$1 \div \dfrac{3}{8} = \dfrac{8}{3}$

$\dfrac{1}{2} \times \dfrac{8}{3} = \dfrac{8}{6} = 1\dfrac{2}{6} = \mathbf{1\dfrac{1}{3}}$

d. $\dfrac{3}{8} \div \dfrac{1}{2}$

$1 \div \dfrac{1}{2} = 2$

$\dfrac{3}{8} \times \dfrac{2}{1} = \dfrac{6}{8} = \dfrac{\mathbf{3}}{\mathbf{4}}$

LESSON 54, MIXED PRACTICE

1. Decimal Arithmetic Reminders

Operation	+ or −	×	÷ by whole (W)	÷ by decimal (D)
Memory cue	line up $\dfrac{\pm\ .}{\ .}$	×; then count $\dfrac{\times\ .\underline{\ \ }}{\underline{\ \ }}$	up $W\overline{)\ .}$	over, over, up $D\overline{)\overset{.}{}}$

You may need to…
- Place a decimal point to the right of a whole number.
- Fill empty places with zeros.

2.
$$4\overline{)0.4}$$
quotient 0.1, with $\dfrac{4}{0}$

3.
$$\begin{array}{r} \overset{0\ \ {}^{1}4}{\cancel{1}\ \cancel{5}{}^{1}1,000,000} \\ -\ \ 76,000,000 \\ \hline \mathbf{75,000,000} \end{array}$$

4.
$$61\dfrac{1}{4} \quad \xrightarrow{\ 60 + \frac{4}{4} + \frac{1}{4}\ } \quad 60\dfrac{5}{4}$$
$$-\ 59\dfrac{3}{4} \qquad\qquad\qquad\qquad -\ 59\dfrac{3}{4}$$
$$\qquad\qquad\qquad\qquad\qquad 1\dfrac{2}{4} = \mathbf{1\dfrac{1}{2}\ inches}$$

5. $1000 - (100 - 1)$

$1000 - 99$

$\mathbf{901}$

6.
$$24\overline{)1000} \quad 41\dfrac{16}{24} = \mathbf{41\dfrac{2}{3}}$$
$$\begin{array}{r} 96 \\ \hline 40 \\ 24 \\ \hline 16 \end{array}$$

7.
$$\begin{array}{r} 1\,\overset{3}{\cancel{4}}{}^{1}3 \\ -\ \ 37 \\ \hline 106 \end{array} \qquad \begin{array}{r} 53 \\ 2\overline{)106} \\ \underline{10} \\ 06 \\ \underline{06} \\ 0 \end{array} \qquad \begin{array}{r} \overset{1}{3}7 \\ +\ 53 \\ \hline \mathbf{90} \end{array}$$

8.
$$\begin{array}{r} \$\overset{2}{\cancel{3}}.\overset{9}{\cancel{0}}{}^{1}0 \\ -\ \$0.24 \\ \hline \$2.76 \end{array}$$

$n = \mathbf{\$2.76}$

9.
$$6\dfrac{2}{5} \quad \xrightarrow{\ 5 + \frac{5}{5} + \frac{2}{5}\ } \quad 5\dfrac{7}{5}$$
$$-\ 3\dfrac{4}{5} \qquad\qquad\qquad\qquad -\ 3\dfrac{4}{5}$$
$$\qquad\qquad\qquad\qquad\qquad m = \mathbf{2\dfrac{3}{5}}$$

10. $4.2 \div 100 = \mathbf{0.042}$

11. $(1.2 \div 0.12)(1.2)$

$$12\overline{)120}$$
$$\underline{12}$$
$$00$$
$$\underline{00}$$
$$0$$

$10(1.2) = \mathbf{12}$

12. $\dfrac{4}{3} \times \dfrac{4}{3} = \dfrac{16}{9} = \mathbf{1\dfrac{7}{9}}$

13. $3 + 4 = \mathbf{7}$

14. **4**

15.
$$\begin{array}{r} \$289.90 \\ \times .08 \\ \hline \$23.1920 \end{array}$$
$23.19

16. Two centimeters $<$ one inch

17. $12 \times 2 = 24 - 1 = \mathbf{23}$

18.
$$\begin{array}{r} 14 \text{ ft.} \\ \times 12 \text{ ft.} \\ \hline 28 \\ 140 \\ \hline \end{array}$$
168 square feet of tile

19. (a) $\dfrac{9}{30} = \dfrac{3}{10}$

(b) $\dfrac{9}{21} = \dfrac{3}{7}$

20. (a) $\dfrac{5}{6} \times \dfrac{4}{4} = \dfrac{20}{24}$

(b) $\dfrac{5}{8} \times \dfrac{3}{3} = \dfrac{15}{24}$

21. Multiples of 3
3, 6, 9, ⑫, 15, 18
Multiples of 4
4, 8, ⑫, 16, 20, 24
Multiples of 6
6, ⑫, 18, 24, 30
LCM is **12**.

22. $\dfrac{2}{3} \div \dfrac{1}{2}$

$1 \div \dfrac{1}{2} = 2$

$\dfrac{2}{3} \times \dfrac{2}{1} = \dfrac{4}{3} = 1\dfrac{1}{3}$

23.

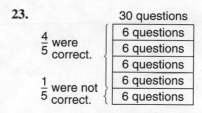

30 questions

$\dfrac{4}{5}$ were correct. { 6 questions / 6 questions / 6 questions

$\dfrac{1}{5}$ were not correct. { 6 questions / 6 questions

24 questions

24. $2.54 \text{ cm} \times 100 = \mathbf{254}$ **centimeters**

25. $\dfrac{2 \cdot 3 \cdot 5 \cdot 2 \cdot 3}{2 \cdot 3 \cdot 5 \cdot 2} = 1 \cdot 1 \cdot 1 \cdot 1 \cdot 3 = \mathbf{3}$

26. $\dfrac{2 \cdot 3 \cdot 5 \cdot 3}{2 \cdot 3 \cdot 5 \cdot 2 \cdot 2} = 1 \cdot 1 \cdot 1 \cdot \dfrac{3}{4} = \dfrac{\mathbf{3}}{\mathbf{4}}$

27. $\dfrac{2}{3} \times \dfrac{2}{2} = \dfrac{4}{6} \qquad \dfrac{1}{2} \times \dfrac{3}{3} = \dfrac{3}{6}$

$\dfrac{4}{6} + \dfrac{3}{6} = \dfrac{7}{6} = \mathbf{1\dfrac{1}{6}}$

28. $C = \pi d$
$C \approx (3.14)(15 \text{ in.})$
$C \approx 47.1 \text{ in.}$
$C \approx \mathbf{47}$ **inches**

π is a little more than 3, and 3 $\times$ 15 inches is 45 inches. So 47 inches is reasonable.

29.

$1\dfrac{1}{2}$ in.

1 in.

(a) $1\dfrac{1}{2}$ in. $+ 1\dfrac{1}{2}$ in. $+ 1$ in. $+ 1$ in.

$= 4\dfrac{2}{2}$ in. $= \mathbf{5}$ **inches**

(b) $1\dfrac{1}{2}$ in. $\times 1$ in. $= \mathbf{1\dfrac{1}{2}}$ **square inches**

30. $\left(2\dfrac{1}{2} \times 2\right) \div \left(\dfrac{1}{2} \times 2\right)$

$5 \div 1 = \mathbf{5}$

LESSON 55, WARM-UP

a. 2600

b. 379

c. 276

d. $7.50

e. 7

f. 30

g. 4

Problem Solving

1 pint = 2 cups
1 quart = 2 pints = 4 cups
1 half gallon = 2 quarts = 8 cups
1 gallon = 2 half gallons = **16 cups**

LESSON 55, LESSON PRACTICE

a.
$$\frac{1}{2} \times \frac{4}{4} = \frac{4}{8}$$
$$+ \frac{3}{8} \qquad = \frac{3}{8}$$
$$\overline{\qquad\qquad \frac{7}{8}}$$

b.
$$\frac{3}{8} \qquad = \frac{3}{8}$$
$$+ \frac{1}{4} \times \frac{2}{2} = \frac{2}{8}$$
$$\overline{\qquad\qquad \frac{5}{8}}$$

c.
$$\frac{3}{4} \times \frac{2}{2} = \frac{6}{8}$$
$$+ \frac{1}{8} \qquad = \frac{1}{8}$$
$$\overline{\qquad\qquad \frac{7}{8}}$$

d.
$$\frac{1}{2} \times \frac{2}{2} = \frac{2}{4}$$
$$- \frac{1}{4} \qquad = \frac{1}{4}$$
$$\overline{\qquad\qquad \frac{1}{4}}$$

e.
$$\frac{5}{8} \qquad = \frac{5}{8}$$
$$- \frac{1}{4} \times \frac{2}{2} = \frac{2}{8}$$
$$\overline{\qquad\qquad \frac{3}{8}}$$

f.
$$\frac{3}{4} \times \frac{2}{2} = \frac{6}{8}$$
$$- \frac{3}{8} \qquad = \frac{3}{8}$$
$$\overline{\qquad\qquad \frac{3}{8}}$$

LESSON 55, MIXED PRACTICE

1. When we divide a decimal number by a whole number using a division box, we place a decimal point in the quotient directly "up" from the decimal point in the dividend.

2. $12 \div \frac{3}{8}$ in.

$$1 \div \frac{3}{8} = \frac{8}{3}$$

$$\frac{12}{1} \times \frac{8}{3} = \frac{96}{3} = \textbf{32 CD cases}$$

3.
54 average pumpkins
$$6\overline{)324}$$
$$\underline{30}$$
$$24$$
$$\underline{24}$$
$$0$$

4.
$$\frac{1}{8} \qquad = \frac{1}{8}$$
$$+ \frac{1}{2} \times \frac{4}{4} = \frac{4}{8}$$
$$\overline{\qquad\qquad \frac{5}{8}}$$

5.
$$\frac{1}{2} \times \frac{4}{4} = \frac{4}{8}$$
$$- \frac{1}{8} \qquad = \frac{1}{8}$$
$$\overline{\qquad\qquad \frac{3}{8}}$$

6.
$$\frac{2}{3} \times \frac{2}{2} = \frac{4}{6}$$
$$- \frac{1}{6} \qquad = \frac{1}{6}$$
$$\overline{\qquad\qquad \frac{3}{6}} = \mathbf{\frac{1}{2}}$$

7. $\overset{1}{6.28}$
4
$+\;\; 0.13$
10.41

8. $9\overline{)810}$ with quotient **90**
$\underline{81}$
00
$\underline{00}$
0

9. $10\overline{)0.20}$ with quotient **0.02**
$\underline{20}$
0

10. **17**

11. $\frac{3}{4} + 3\frac{1}{4} = 3\frac{4}{4} =$ **4**

12. $\frac{5}{6} \cdot \frac{2}{3} = \frac{10}{18} = \frac{5}{9}$

13. $\frac{5}{8} \times \frac{3}{3} = \frac{15}{24}$ **15**

14. $60{,}000 + 400 + 20 =$ **60,420**

15. 0.14
$\times \;\; 0.8$
$\overline{0.112}$
0.11

16. $\frac{2}{3} \ominus \frac{2}{3} \times \frac{2}{2}$
$\frac{2}{3} \times 1$
$\frac{2}{3}$

17. $C = \pi d$
$C \approx (3.14)(5\text{ feet})$
$C \approx$ **15.7 feet**

18. $4\overline{)20\text{ feet}}$ with quotient 5 feet
$\underline{20}$
0
 5 feet
$\times$ 5 feet
25 square feet

19. $\frac{60}{100} = \frac{3}{5}$

20. Multiples of 3
3, 6, 9, ⑫, 15, 18
Multiples of 4
4, 8, ⑫, 16, 20
LCM is **12.**

21. (a) Factors of 23
1, 23

(b) **Prime number**

22. $\frac{5}{2}$

23. $9 + 16 \;\ominus\; 25$

24. $\frac{1}{2} \div \frac{2}{5}$
$1 \div \frac{2}{5} = \frac{5}{2}$
$\frac{1}{2} \times \frac{5}{2} = \frac{5}{4} = 1\frac{1}{4}$

25. $12\overline{)1212}$ with quotient **101**
$\underline{12}$
01
$\underline{00}$
12
$\underline{12}$
0

26. $\frac{36}{48} = \frac{3}{4}$

27. $\frac{1}{3}$

28. $\frac{2 \cdot 3 \cdot 2 \cdot 5 \cdot 3 \cdot 7}{2 \cdot 3 \cdot 2 \cdot 5 \cdot 5 \cdot 5}$
$= 1 \cdot 1 \cdot 1 \cdot 1 \cdot \frac{21}{25} = \frac{21}{25}$

29. $\overset{9}{\cancel{10}}:\overset{85}{\cancel{25}}$ p.m.
$-\;\; 7:45$ p.m.
$2:40$
2 hours 40 minutes

30. (a) **Triangle GIH**

(b) **Triangle DEF**

LESSON 56, WARM-UP

a. 2000

b. 112

c. 199

d. $8.25

e. $7\frac{1}{2}$

f. 1200

g. 8

Problem Solving

First think, "6 plus what number equals 10?" (4).
Then think, "5 plus 1 (from regrouping) plus
what number equals 10?" (4). Next think, "What
number plus 1 (from regrouping) equals 10?" (9).

$$\begin{array}{r} 956 \\ +44 \\ \hline 1000 \end{array}$$

LESSON 56, LESSON PRACTICE

a.
$$\begin{array}{r} \frac{2}{3} \times \frac{2}{2} = \frac{4}{6} \\ -\ \frac{1}{2} \times \frac{3}{3} = \frac{3}{6} \\ \hline \frac{1}{6} \end{array}$$

b.
$$\begin{array}{r} \frac{1}{4} \times \frac{5}{5} = \frac{5}{20} \\ +\ \frac{2}{5} \times \frac{4}{4} = \frac{8}{20} \\ \hline \frac{13}{20} \end{array}$$

c.
$$\begin{array}{r} \frac{3}{4} \times \frac{3}{3} = \frac{9}{12} \\ -\ \frac{1}{3} \times \frac{4}{4} = \frac{4}{12} \\ \hline \frac{5}{12} \end{array}$$

d.
$$\begin{array}{r} \frac{2}{3} \times \frac{4}{4} = \frac{8}{12} \\ +\ \frac{1}{4} \times \frac{3}{3} = \frac{3}{12} \\ \hline \frac{11}{12} \end{array}$$

e.
$$\begin{array}{r} \frac{1}{3} \times \frac{4}{4} = \frac{4}{12} \\ -\ \frac{1}{4} \times \frac{3}{3} = \frac{3}{12} \\ \hline \frac{1}{12} \end{array}$$

f.
$$\frac{2}{3} \oslash \frac{1}{2}$$
$$\frac{2}{3} \times \frac{2}{2} \qquad \frac{1}{2} \times \frac{3}{3}$$
$$\frac{4}{6} > \frac{3}{6}$$

g.
$$\frac{4}{6} \oslash \frac{3}{4}$$
$$\frac{4}{6} \times \frac{2}{2} \qquad \frac{3}{4} \times \frac{3}{3}$$
$$\frac{8}{12} < \frac{9}{12}$$

h.
$$\frac{2}{3} \oslash \frac{3}{5}$$
$$\frac{2}{3} \times \frac{5}{5} \qquad \frac{3}{5} \times \frac{3}{3}$$
$$\frac{10}{15} > \frac{9}{15}$$

LESSON 56, MIXED PRACTICE

1.
$$\begin{array}{r} \frac{1}{4} \times \frac{3}{3} = \frac{3}{12} \\ +\ \frac{1}{3} \times \frac{4}{4} = \frac{4}{12} \\ \hline \frac{7}{12} \end{array}$$

2.
$$\begin{array}{r} \frac{1}{2} \times \frac{3}{3} = \frac{3}{6} \\ -\ \frac{1}{3} \times \frac{2}{2} = \frac{2}{6} \\ \hline \frac{1}{6} \end{array}$$

3. (a) $\dfrac{52}{88} = \dfrac{13}{22}$

(b)
$$\begin{array}{r} 88 \\ -\ 52 \\ \hline 36 \text{ black keys} \end{array}$$

$$\frac{\text{black keys}}{\text{white keys}} = \frac{36}{52} = \frac{9}{13}$$

4.
$$\begin{array}{r} 4.5 \\ \times\ \ 2 \\ \hline \mathbf{9.0} \ \text{or}\ \mathbf{9\ apples} \end{array}$$

5.
$$\begin{array}{r} \dfrac{2}{3} \times \dfrac{4}{4} = \dfrac{8}{12} \\ -\ \dfrac{1}{4} \times \dfrac{3}{3} = \dfrac{3}{12} \\ \hline \dfrac{5}{12} \end{array}$$

6.
$$\begin{array}{r} \dfrac{1}{3} \times \dfrac{2}{2} = \dfrac{2}{6} \\ +\ \dfrac{1}{6} \quad\ \ = \dfrac{1}{6} \\ \hline \dfrac{3}{6} = \dfrac{1}{2} \end{array}$$

7.
$$\begin{array}{r} \dfrac{5}{6} \quad\ \ = \dfrac{5}{6} \\ -\ \dfrac{1}{2} \times \dfrac{3}{3} = \dfrac{3}{6} \\ \hline \dfrac{2}{6} = \dfrac{1}{3} \end{array}$$

8.
$$\begin{array}{r} \overset{\scriptscriptstyle 1}{\$3}\ \overset{\scriptscriptstyle 1}{} \\ \$1.75 \\ +\ \$0.65 \\ \hline \mathbf{\$5.40} \end{array}$$

9.
$$\begin{array}{r} 0.625 \\ \times\ \ \ 0.4 \\ \hline \mathbf{0.2500}\ \text{or}\ \mathbf{0.25} \end{array}$$

10. $36 \div 0.08$

$$\begin{array}{r} \mathbf{450} \\ 8\overline{)3600} \\ 32 \\ \hline 40 \\ 40 \\ \hline 00 \\ 00 \\ \hline 0 \end{array}$$

11.
$$3\dfrac{1}{8} \xrightarrow{\ 2 + \frac{8}{8} + \frac{1}{8}\ } 2\dfrac{9}{8}$$
$$-\ 1\dfrac{7}{8} \qquad\qquad\qquad -\ 1\dfrac{7}{8}$$
$$\overline{\qquad\qquad\qquad\qquad\qquad 1\dfrac{2}{8} = \mathbf{1\dfrac{1}{4}}}$$

12. $\dfrac{5}{8} \cdot \dfrac{2}{3} = \dfrac{10}{24} = \dfrac{5}{12}$

13. $\dfrac{40}{100} = \dfrac{2}{5};\ \mathbf{40\ questions}$

14. **80.65**

15. $3600 + 4200 = \mathbf{7800}$

16. **Yes. To estimate the distance the bike traveled, Molly found the circumference of the tire. She multiplied the diameter of the tire by 3, which is a very rough approximation for π. If Molly had used 3.14 for π, she would have calculated 6.28 ft for the circumference, which is about 6 ft 3 in.**

17.
$$\begin{array}{r} \overset{\scriptscriptstyle 1}{1.2} \\ 1.3 \\ 1.4 \\ +\ 1.5 \\ \hline 5.4 \end{array} \qquad \begin{array}{r} \mathbf{1.35} \\ 4\overline{)5.40} \\ 4 \\ \hline 1\ 4 \\ 1\ 2 \\ \hline 20 \\ 20 \\ \hline 0 \end{array}$$

18.
$$\begin{array}{r} 9\ \text{inches} \\ 4\overline{)36}\ \text{inches} \\ 36 \\ \hline 0 \end{array}$$
$9\ \text{inches} \times 9\ \text{inches} = \mathbf{81\ square\ inches}$

19. $2 \cdot 3 \cdot 5$

20. $0 \div \dfrac{2}{3} = 0$
$w = \mathbf{0}$

21. $m = \dfrac{3}{2}$

22. $0 + \dfrac{2}{3} = \dfrac{2}{3}$
$n = \mathbf{\dfrac{2}{3}}$

23.
$$\begin{array}{r} 100 \\ -\ \ 40 \\ \hline \mathbf{60\ problems} \end{array}$$

24.
$$\begin{array}{r} 45 \\ -\ 25 \\ \hline \end{array}$$
20 more problems

25. $\dfrac{25}{100} = \dfrac{1}{4}$

26. Answers will vary. Sample answer: What percent of the problems did Harpo answer correctly?

$\dfrac{40}{100}$ or **40%**

27. $\dfrac{2 \cdot 3 \cdot 5}{2 \cdot 3 \cdot 5 \cdot 7} = 1 \cdot 1 \cdot 1 \cdot \dfrac{1}{7} = \dfrac{1}{7}$

28. $\dfrac{1}{2} \div \dfrac{2}{3}$

$1 \div \dfrac{2}{3} = \dfrac{3}{2}$

$\dfrac{1}{2} \times \dfrac{3}{2} = \dfrac{3}{4}$

29. possibilities include:

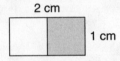

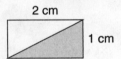

2 cm / 1 cm

30. $\dfrac{2}{3}$ $\,\bigcirc\hspace{-0.9em}<\,$ $\dfrac{5}{6}$

$\dfrac{2}{3} \times \dfrac{2}{2}$

$\dfrac{4}{6}$

LESSON 57, WARM-UP

a. 3400

b. 715

c. 101

d. $6.25

e. 9

f. 20

g. 1

Problem Solving

$15 - 2 = 13; \ 15 + 2 = 17$
13 and **17**
$15 - 4 = 11; \ 15 + 4 = 19$
11 and **19**

LESSON 57, LESSON PRACTICE

a.
$$\dfrac{1}{2} \times \dfrac{3}{3} = \dfrac{3}{6}$$
$$+\ \dfrac{1}{6} \qquad = \dfrac{1}{6}$$
$$\dfrac{4}{6} = \mathbf{\dfrac{2}{3}}$$

b.
$$\dfrac{2}{3} \times \dfrac{4}{4} = \dfrac{8}{12}$$
$$+\ \dfrac{3}{4} \times \dfrac{3}{3} = \dfrac{9}{12}$$
$$\dfrac{17}{12} = \mathbf{1\dfrac{5}{12}}$$

c.
$$\dfrac{1}{5} \times \dfrac{2}{2} = \dfrac{2}{10}$$
$$+\ \dfrac{3}{10} \qquad = \dfrac{3}{10}$$
$$\dfrac{5}{10} = \mathbf{\dfrac{1}{2}}$$

d.
$$\dfrac{5}{6} \qquad = \dfrac{5}{6}$$
$$-\ \dfrac{1}{2} \times \dfrac{3}{3} = \dfrac{3}{6}$$
$$\dfrac{2}{6} = \mathbf{\dfrac{1}{3}}$$

e.
$$\dfrac{7}{10} \qquad = \dfrac{7}{10}$$
$$-\ \dfrac{1}{2} \times \dfrac{5}{5} = \dfrac{5}{10}$$
$$\dfrac{2}{10} = \mathbf{\dfrac{1}{5}}$$

f.
$$\dfrac{5}{12} \qquad = \dfrac{5}{12}$$
$$-\ \dfrac{1}{6} \times \dfrac{2}{2} = \dfrac{2}{12}$$
$$\dfrac{3}{12} = \mathbf{\dfrac{1}{4}}$$

LESSON 57, MIXED PRACTICE

1. $\frac{1}{2} + \frac{1}{2} = \frac{2}{2} = 1$ $\frac{1}{2} \times \frac{1}{2} = \frac{1}{4}$

$\frac{4}{4} - \frac{1}{4} = \mathbf{\frac{3}{4}}$

2. $1800 - 1743 = d$

$\begin{array}{r} 1\,\overset{7}{8}\,\overset{9}{\cancel{0}}\,{}^{1}0 \\ -\ 1\,7\,4\,3 \\ \hline \mathbf{5\ 7}\ \text{years old} \end{array}$

3. $\frac{5}{6} \times \frac{2}{2} = \frac{10}{12}$

$-\ \frac{3}{4} \times \frac{3}{3} = \frac{9}{12}$
_____
$\mathbf{\frac{1}{12}}$

4. $\frac{1}{2} \times \frac{3}{3} = \frac{3}{6}$

$+\ \frac{2}{3} \times \frac{2}{2} = \frac{4}{6}$
_____
$\frac{7}{6} = \mathbf{1\frac{1}{6}}$

5. $\frac{1}{2} \times \frac{3}{3} = \frac{3}{6}$

$+\ \frac{1}{6} \quad\quad = \frac{1}{6}$
_____
$\frac{4}{6} = \mathbf{\frac{2}{3}}$

6. $\frac{5}{6} \quad\quad = \frac{5}{6}$

$+\ \frac{2}{3} \times \frac{2}{2} = \frac{4}{6}$
_____
$\frac{9}{6} = 1\frac{3}{6} = \mathbf{1\frac{1}{2}}$

7. $\frac{3}{4} \div \frac{3}{5}$

$1 \div \frac{3}{5} = \frac{5}{3}$

$\frac{3}{4} \times \frac{5}{3} = \frac{15}{12} = 1\frac{3}{12} = \mathbf{1\frac{1}{4}}$

8. **$3.25**

9. $\sqrt{4} - (1 - 0.2)$

$2 - 0.8$

1.2

10.
$\begin{array}{r} 50 \\ 12\overline{)600} \\ 60 \\ \hline 00 \\ 00 \\ \hline 0 \end{array}$

11.

$\begin{array}{r} 5\frac{3}{8} \\ -\ 2\frac{5}{8} \end{array} \xrightarrow{\ 4 + \frac{8}{8} + \frac{3}{8}\ } \begin{array}{r} 4\frac{11}{8} \\ -\ 2\frac{5}{8} \\ \hline 2\frac{6}{8} = \mathbf{2\frac{3}{4}} \end{array}$

12. $\frac{3}{4} \cdot \frac{5}{3} = \frac{15}{12} = 1\frac{3}{12} = \mathbf{1\frac{1}{4}}$

13. $\frac{50}{100} = \mathbf{\frac{1}{2}}$

$\begin{array}{r} 20\text{ mm} \\ \times\ 10\text{ mm} \\ \hline 200\text{ sq. mm} \end{array}$

$\begin{array}{r} \mathbf{100\text{ sq. mm}} \\ 2\overline{)200} \\ 2 \\ \hline 00 \\ 00 \\ \hline 00 \\ 00 \\ \hline 0 \end{array}$

14. **Thousandths**

15.
$\begin{array}{r} 0.125 \\ 4\overline{)0.500} \\ 4 \\ \hline 10 \\ 8 \\ \hline 20 \\ 20 \\ \hline 0 \end{array}$

0.1

16. **0.03, 0.3, 3.0**

17. $20 \times 2 = \mathbf{40}$

18. $\frac{4}{52} = \mathbf{\frac{1}{13}}$

19. (a)
$$\frac{1}{4} = \frac{1}{4}\text{ in.}$$
$$\frac{1}{4} = \frac{1}{4}\text{ in.}$$
$$\frac{1}{2} \times \frac{2}{2} = \frac{2}{4}\text{ in.}$$
$$+\ \frac{1}{2} \times \frac{2}{2} = \frac{2}{4}\text{ in.}$$
$$\frac{6}{4}\text{ in.} = 1\frac{2}{4}\text{ in.} = \mathbf{1\frac{1}{2}\text{ in.}}$$

(b) $\frac{1}{4}\text{ in.} \times \frac{1}{2}\text{ in.} = \mathbf{\frac{1}{8}\text{ in.}^2}$

20. $\frac{5}{8} \times \frac{80}{1} = \frac{400}{8} = \mathbf{50}$

21. Factors of 29: **1, 29**

22. Multiples of 12
12, 24, (36), 48
Multiples of 18
18, (36), 54, 72
LCM is **36.**

23.
$$\frac{5}{8} \text{ (<) } \frac{7}{10}$$
$$\frac{5}{8} \times \frac{5}{5} \qquad \frac{7}{10} \times \frac{4}{4}$$
$$\frac{25}{40} \qquad \frac{28}{40}$$

24. **−4°F**

25. **16°F**

26. $\frac{2 \cdot 2 \cdot 5 \cdot 7 \cdot 3 \cdot 3}{2 \cdot 2 \cdot 5 \cdot 7 \cdot 5 \cdot 7} = \mathbf{\frac{9}{35}}$

27. $\mathbf{\frac{1}{3}}$

28. $9 \div \frac{3}{8}$ in.
$$1 \div \frac{3}{8} = \frac{8}{3}$$
$$\frac{9}{1} \times \frac{8}{3} = \frac{72}{3} = \textbf{24 CDs}$$

29.
$$\frac{4}{5} \times \frac{2}{2} = \frac{8}{10}$$
$$-\ \frac{1}{2} \times \frac{5}{5} = \frac{5}{10}$$
$$\mathbf{\frac{3}{10}}$$

30. $C = \pi d$
$C \approx (3.14)(18\text{ inches})$
$C \approx$ **56.52 inches**

Lesson 58, Warm-Up

a. **150**

b. **165**

c. **449**

d. **$20.00**

e. $\mathbf{12\frac{1}{2}}$

f. **1000**

g. **5**

Problem Solving

Add $\frac{1}{6}$ to each term to find the next term.

$\frac{1}{6}, \frac{1}{3}, \frac{1}{2}, \mathbf{\frac{2}{3}}, \mathbf{\frac{5}{6}}, \mathbf{1}, \ldots$

Lesson 58, Lesson Practice

a. $\frac{40}{100} = \frac{4}{10} = \mathbf{\frac{2}{5}}$

$\frac{40}{100} = 0.40 = \mathbf{0.4}$

b. $1 - \frac{2}{5} = \frac{5}{5} - \frac{2}{5} = \mathbf{\frac{3}{5}}$

$1 - 0.4 = \mathbf{0.6}$

c. $\frac{3}{6} = \mathbf{\frac{1}{2}}$

d. $\mathbf{\frac{5}{6}}$

e. Red: $\mathbf{\frac{1}{2}}$

Black: $\mathbf{\frac{1}{6}}$

White: $\mathbf{\frac{1}{3}}$

f. $\dfrac{2}{3}$

g. Red: $30 \div 2 = $ **15 times**
White: $30 \div 3 = $ **10 times**
Black: $30 \div 6 = $ **5 times**

h. Answers may vary. Sample answer:

1			
2					
3				
4				
5	$\cancel{||||}\,|$
6	$\cancel{||||}\,|$

I rolled 5 and 6 more times than I expected.

LESSON 58, MIXED PRACTICE

1. $\dfrac{1}{2} \times \dfrac{3}{3} = \dfrac{3}{6}$ $\dfrac{1}{2} \times \dfrac{1}{3} = \dfrac{1}{6}$

$+\dfrac{1}{3} \times \dfrac{2}{2} = \dfrac{2}{6}$
$\phantom{+\dfrac{1}{3} \times \dfrac{2}{2} = }\dfrac{5}{6}$

$\dfrac{5}{6} - \dfrac{1}{6} = \dfrac{4}{6} = \mathbf{\dfrac{2}{3}}$

2.
$$\begin{array}{r} 12 \\ \times\ 2.5 \\ \hline 60 \\ 240 \\ \hline 30.0 \end{array}$$

30 eggs

3.
$$\begin{array}{r} \mathbf{\$14{,}800} \\ 3\overline{)\$44{,}400} \\ \underline{3} \\ 14 \\ \underline{12} \\ 2\ 4 \\ \underline{2\ 4} \\ 00 \\ \underline{00} \\ 00 \\ \underline{00} \\ 0 \end{array}$$

4. $\dfrac{5}{8}$ $\bigcirc\!\!\!>$ $\dfrac{1}{2}$

$\dfrac{1}{2} \times \dfrac{4}{4}$

$\dfrac{4}{8}$

5. $36 + 64$ $\boxed{=}$ 100
100

6.
$$\begin{array}{r} \dfrac{1}{2} \times \dfrac{4}{4} = \dfrac{4}{8} \\[6pt] -\dfrac{3}{8} = \dfrac{3}{8} \\[4pt] \hline \dfrac{1}{8} \end{array}$$

$$m = \mathbf{\dfrac{1}{8}}$$

7.
$$\begin{array}{r} \dfrac{2}{3} \times \dfrac{4}{4} = \dfrac{8}{12} \\[6pt] +\dfrac{3}{4} \times \dfrac{3}{3} = \dfrac{9}{12} \\[4pt] \hline \dfrac{17}{12} = 1\dfrac{5}{12} \end{array}$$

$$n = \mathbf{1\dfrac{5}{12}}$$

8.
$$3 \xrightarrow{\;2 + \frac{6}{6}\;} 2\dfrac{6}{6}$$
$$-\dfrac{5}{6} \qquad\qquad -\dfrac{5}{6}$$
$$\phantom{-\dfrac{5}{6}\qquad\qquad} 2\dfrac{1}{6}$$

$$f = \mathbf{2\dfrac{1}{6}}$$

9. **\$325.00**

10.
$$\begin{array}{r} 6.2 \\ \times\ 0.48 \\ \hline 496 \\ 2480 \\ \hline \mathbf{2.976} \end{array}$$

11.
$$\begin{array}{r} \mathbf{1.25} \\ 8\overline{)10.00} \\ \underline{8} \\ 2\ 0 \\ \underline{1\ 6} \\ 40 \\ \underline{40} \\ 0 \end{array}$$

12.
$$\begin{array}{r} \mathbf{240} \\ 5\overline{)1200} \\ \underline{10} \\ 20 \\ \underline{20} \\ 00 \\ \underline{00} \\ 0 \end{array}$$

13. $\dfrac{7}{8} \cdot \dfrac{8}{7} = \dfrac{56}{56} = \mathbf{1}$

14. $\dfrac{5}{6} \cdot \dfrac{3}{4} = \dfrac{15}{24} = \dfrac{\mathbf{5}}{\mathbf{8}}$

15. $15 \div 3 = \mathbf{5}$

16.
$$
\begin{array}{r} \$9.79 \\ \times \quad .07 \\ \hline \$0.6853 \end{array}
\qquad
\begin{array}{r} \overset{1\ 1}{\$9.79} \\ +\ \$0.69 \\ \hline \mathbf{\$10.48} \end{array}
$$

17. **1.0 or 1**

18. $4 \text{ cm} \div 4 = 1 \text{ cm}$
$1 \text{ cm} \times 1 \text{ cm} = \mathbf{1 \text{ cm}^2}$

19. $\dfrac{3}{4} \div \dfrac{3}{5}$

$1 \div \dfrac{3}{5} = \dfrac{5}{3}$

$\dfrac{3}{4} \times \dfrac{5}{3} = \dfrac{15}{12} = 1\dfrac{3}{12} = \mathbf{1\dfrac{1}{4}}$

20.
$$
\begin{array}{r}
100 \\
32\overline{)3200} \\
32 \\
\hline
00 \\
00 \\
\hline
00 \\
00 \\
\hline
0
\end{array}
$$

$w = \mathbf{100}$

21.
$$
\begin{array}{r}
\overset{4}{\cancel{5}}.^{1}0 \\
-\ 3.\,4 \\
\hline
1.\,6
\end{array}
$$

$x = \mathbf{1.6}$

22. $\dfrac{3}{6} = \dfrac{\mathbf{1}}{\mathbf{2}}$

23. **20 mm, 1 in., 3 cm**

24. (a) **Less than half; 50% equals $\frac{1}{2}$ and Larry answered less than 50% correctly.**

(b) $\dfrac{45}{100} = \dfrac{\mathbf{9}}{\mathbf{20}}$

25. Finding $\frac{1}{10}$ of \$12.50 is the same as dividing \$12.50 by 10. We shift the decimal point in \$12.50 one place to the left, which makes \$1.250. Then we remove the trailing zero. The answer is **\$1.25.**

26. $\dfrac{2 \cdot 5 \cdot 2 \cdot 7 \cdot 3 \cdot 3}{2 \cdot 5 \cdot 2 \cdot 7 \cdot 2 \cdot 5} = \dfrac{\mathbf{9}}{\mathbf{10}}$

27. $1\dfrac{6}{10} + 2\dfrac{4}{10} = 3\dfrac{10}{10} = \mathbf{4}$

28.

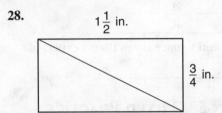

29.
$$
\begin{aligned}
1\dfrac{1}{2} \times \dfrac{2}{2} &= 1\dfrac{2}{4} \\
1\dfrac{1}{2} \times \dfrac{2}{2} &= 1\dfrac{2}{4} \\
\dfrac{3}{4} &= \dfrac{3}{4} \\
+ \quad \dfrac{3}{4} &= \dfrac{3}{4} \\
\hline
2\dfrac{10}{4} &= 4\dfrac{2}{4} = \mathbf{4\dfrac{1}{2} \text{ inches}}
\end{aligned}
$$

30. $A = lw$
$A = (1.5)(0.75)$
$A = \mathbf{1.125}$

$$
\begin{array}{r}
0.75 \\
\times\ 1.5 \\
\hline
375 \\
750 \\
\hline
1.125
\end{array}
$$

LESSON 59, WARM-UP

a. **300**

b. **629**

c. **51**

d. **\$7.25**

e. **\$3.00**

f. **20**

g. 15

Problem Solving

Count edges around the top, edges around the bottom, and edges running from top to bottom.

12 edges

LESSON 59, LESSON PRACTICE

a.
$$1\frac{1}{2} \times \frac{3}{3} = 1\frac{3}{6}$$
$$+ 1\frac{1}{3} \times \frac{2}{2} = 1\frac{2}{6}$$
$$\mathbf{2\frac{5}{6}}$$

b.
$$1\frac{1}{2} \times \frac{3}{3} = 1\frac{3}{6}$$
$$+ 1\frac{2}{3} \times \frac{2}{2} = 1\frac{4}{6}$$
$$2\frac{7}{6} = 2 + 1\frac{1}{6} = \mathbf{3\frac{1}{6}}$$

c.
$$5\frac{1}{3} \times \frac{2}{2} = 5\frac{2}{6}$$
$$+ 2\frac{1}{6} \qquad = 2\frac{1}{6}$$
$$7\frac{3}{6} = \mathbf{7\frac{1}{2}}$$

d.
$$3\frac{3}{4} \times \frac{3}{3} = 3\frac{9}{12}$$
$$+ 1\frac{1}{3} \times \frac{4}{4} = 1\frac{4}{12}$$
$$4\frac{13}{12} = 4 + 1\frac{1}{12} = \mathbf{5\frac{1}{12}}$$

e.
$$5\frac{1}{2} \times \frac{3}{3} = 5\frac{3}{6}$$
$$+ 3\frac{1}{6} \qquad = 3\frac{1}{6}$$
$$8\frac{4}{6} = \mathbf{8\frac{2}{3}}$$

f.
$$7\frac{1}{2} \times \frac{4}{4} = 7\frac{4}{8}$$
$$+ 4\frac{5}{8} \qquad = 4\frac{5}{8}$$
$$11\frac{9}{8} = 11 + 1\frac{1}{8} = \mathbf{12\frac{1}{8}}$$

LESSON 59, MIXED PRACTICE

1.
$$\begin{array}{r} 0.04 \\ \times\ \ 0.4 \\ \hline \mathbf{0.016} \end{array}$$

2.
$$\begin{array}{r} \overset{11:\ \overset{5}{\cancel{6}}\overset{10}{\cancel{0}}}{\cancel{12}:00}\ \text{a.m.} \\ -\ \ 9:45\ \text{p.m.} \\ \hline 2:15 \end{array} \qquad \begin{array}{r} 2:15 \\ +\ 8:30 \\ \hline 10:45 \end{array}$$

10 hours 45 minutes

3. 7,300,000,000 **kilometers**

4.
$$2\frac{1}{2} \times \frac{3}{3} = 2\frac{3}{6}$$
$$+ 1\frac{1}{6} \qquad - 1\frac{1}{6}$$
$$3\frac{4}{6} = \mathbf{3\frac{2}{3}}$$

5.
$$1\frac{1}{2} \times \frac{3}{3} = 1\frac{3}{6}$$
$$+ 2\frac{2}{3} \times \frac{2}{2} = 2\frac{4}{6}$$
$$3\frac{7}{6} = 3 + 1\frac{1}{6} = \mathbf{4\frac{1}{6}}$$

6.
$$\frac{1}{2} \ \text{\textcircled{<}}\ \frac{3}{5}$$
$$\frac{1}{2} \times \frac{5}{5} \qquad \frac{3}{5} \times \frac{2}{2}$$
$$\frac{5}{10} \qquad \frac{6}{10}$$

7.
$$\frac{2}{3} \ \text{\textcircled{=}}\ \frac{6}{9}$$
$$\frac{2}{3} \times \frac{3}{3}$$
$$\frac{6}{9}$$

8.
$$8\frac{1}{5} \quad \xrightarrow{7 + \frac{5}{5} + \frac{1}{5}} \quad 7\frac{6}{5}$$
$$-3\frac{4}{5} \qquad\qquad\qquad -3\frac{4}{5}$$
$$\qquad\qquad\qquad\qquad\quad \mathbf{4\frac{2}{5}}$$

9. $\dfrac{3}{4} \times \dfrac{5}{2} = \dfrac{15}{8} = \mathbf{1\dfrac{7}{8}}$

10. $\dfrac{2}{5} \div \dfrac{1}{2}$

$1 \div \dfrac{1}{2} = \dfrac{2}{1}$

$\dfrac{2}{5} \times \dfrac{2}{1} = \dfrac{4}{5}$

11.
$$\begin{array}{r} 0.875 \\ \times \quad 40 \\ \hline 35.000 \end{array} \text{ or } \mathbf{35}$$

12.
$$\begin{array}{r} \mathbf{0.0175} \\ 4\overline{)0.0700} \\ \underline{4} \\ 30 \\ \underline{28} \\ 20 \\ \underline{20} \\ 0 \end{array}$$

13.
$$\begin{array}{r} 50 \\ 6\overline{)300} \\ \underline{30} \\ 00 \\ \underline{00} \\ 0 \end{array}$$

$d = \mathbf{50}$

14.
$$\begin{array}{r} 0.24 \\ -\ 0.10 \\ \hline 0.14 \end{array} \qquad \begin{array}{r} 0.07 \\ 2\overline{)0.14} \\ \underline{14} \\ 0 \end{array}$$

$$\begin{array}{r} 0.10 \\ +\ 0.07 \\ \hline \mathbf{0.17} \end{array}$$

15. **36,000,000**

16. $8°F - 23°F = \mathbf{-15°F}$

17. $C = \pi d$
$C \approx (3.14)(10 \text{ inches})$
$C \approx 31.4 \text{ inches}$
$C \approx \mathbf{31 \text{ inches}}$

18.
$$\begin{array}{r} 12 \text{ inches} \\ \times\ 12 \text{ inches} \\ \hline 24 \\ 120 \\ \hline \mathbf{144 \text{ square inches}} \end{array}$$

19. $\dfrac{\mathbf{1}}{\mathbf{100}}$

20. (a) **62.5;** To multiply 6.25 by 10, shift the decimal point one place to the right.

(b) **0.625;** To divide 6.25 by 10, shift the decimal point one place to the left.

21. $\dfrac{2}{6} = \dfrac{\mathbf{1}}{\mathbf{3}}$

22. $(0.8)^2 \;\boxed{<}\; 0.8$
0.64

23.
$$\begin{array}{r} \overset{6}{7}{}^{1}2°F \\ -\ 6\ 4°F \\ \hline \mathbf{8°F} \end{array}$$

24. **67°F**

25. Sample answer: **How much warmer was it at noon on Wednesday than at noon on Tuesday?**
$$\begin{array}{r} 72°F \\ -\ 70°F \\ \hline \mathbf{2°F} \end{array}$$

26. $15 \div \dfrac{6}{10}$

$1 \div \dfrac{6}{10} = \dfrac{10}{6}$

$\dfrac{15}{1} \times \dfrac{10}{6} = \dfrac{150}{6} = \mathbf{25 \text{ times}}$

27.
$$\begin{array}{r} \overset{1}{12.5}\% \\ \times \quad 3 \\ \hline \mathbf{37.5\%} \text{ or } \mathbf{37\tfrac{1}{2}\%} \end{array}$$

28. **$15.99;** To multiply $1.599 by 10, shift the decimal point one place to the right.

29. $\dfrac{3}{4}$, **1**, the reciprocal of $\dfrac{3}{4}$

30. $P = 2l + 2w$
$P = 2(4) + 2(3)$
$P = 8 + 6$
$P = \mathbf{14}$

LESSON 60, WARM-UP

a. **1500**

b. **179**

c. 949

d. $11.25

e. $2.50

f. 2000

g. 25

Problem Solving

1 pint = 2 cups
1 quart = 2 pints = 4 cups
1 half gallon = 2 quarts = 8 cups
1 gallon = 2 half gallons = 16 cups
16 cups − (8 cups + 4 cups + 2 cups
    + 1 cup) = **1 cup**

LESSON 60, LESSON PRACTICE

a. **Hexagon**

b. **5 sides**

c. **Yes**

d. **Vertex**

e. **Quadrilateral**

LESSON 60, MIXED PRACTICE

1. $42 \cdot c = 1.26$

$$\begin{array}{r} \textbf{\$0.03 per ounce} \\ 42\overline{)1.26} \\ \underline{1\ 26} \\ 0\ 00 \end{array}$$

2.
$$\begin{array}{r} \$1.48 \\ \times\quad 1.1 \\ \hline 148 \\ 1480 \\ \hline \$1.628 \end{array}$$

$1.63; Amy bought only a little more than a gallon, so the cost should be only a little more than 1.47\frac{9}{10}$. The answer $1.63 is reasonable.

3. **999**

4.
$$\frac{3}{4} \times \frac{2}{2} = \frac{6}{8}$$
$$+ \frac{5}{8} \qquad = \frac{5}{8}$$
$$\overline{\qquad\qquad \frac{11}{8} = 1\frac{3}{8}}$$

5.
$$1\frac{1}{2} \times \frac{3}{3} = 1\frac{3}{6}$$
$$+\ 3\frac{1}{6} \qquad = 3\frac{1}{6}$$
$$\overline{\qquad\qquad 4\frac{4}{6} = 4\frac{2}{3}}$$

6. $1 \times (1 - 1)$
    1×0
    0

7. $\frac{3}{5} \cdot \frac{1}{3} = \frac{3}{15} = \frac{1}{5}$

8. $\frac{3}{5} \div \frac{1}{3}$

$1 \div \frac{1}{3} = 3$

$\frac{3}{5} \times \frac{3}{1} = \frac{9}{5} = 1\frac{4}{5}$

9.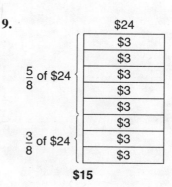

10.
$$\begin{array}{r} 0.65 \\ \times\ 0.14 \\ \hline 260 \\ 650 \\ \hline \textbf{0.0910 or 0.091} \end{array}$$

11.
$$\begin{array}{r} 1300 \\ 5\overline{)6500} \\ \underline{5} \\ 15 \\ \underline{15} \\ 00 \\ \underline{00} \\ 00 \\ \underline{00} \\ 0 \end{array}$$

12. 4 sides

13.
$$\begin{array}{r} 0.24 \\ \times\ 0.26 \\ \hline 144 \\ 480 \\ \hline .0624 \\ \mathbf{0.06} \end{array}$$

14.
$$\overset{1}{1}.3 \qquad 3\overline{)4.11}$$
$$\begin{array}{r} 2 \\ 0.81 \\ \hline 4.11 \end{array} \qquad \begin{array}{r} 1.37 \\ \hline 3 \\ 1\ 1 \\ 9 \\ \hline 21 \\ 21 \\ \hline 0 \end{array}$$

15. $1 + 3 + 5 + 7 + 9 + 11 + 13 = \mathbf{49}$

16. 3 ft $\times$ 3 ft $= \mathbf{9\ square\ feet}$

17. $\dfrac{10}{100} = \dfrac{\mathbf{1}}{\mathbf{10}}$

18. $3x = 3.6$
$$3\overline{)3.6}$$
$$\begin{array}{r} 1.2 \\ \hline 3 \\ \hline 06 \\ 06 \\ \hline 0 \end{array}$$
$$x = \mathbf{1.2}$$

19. $1 \div \dfrac{4}{3} = \dfrac{3}{4}$
$$y = \dfrac{\mathbf{3}}{\mathbf{4}}$$

20.
$$5 \xrightarrow{\quad 4 + \frac{5}{5} \quad} 4\dfrac{5}{5}$$
$$\begin{array}{r} -\ 1\dfrac{3}{5} \qquad\qquad -\ 1\dfrac{3}{5} \\ \hline 3\dfrac{2}{5} \end{array}$$
$$m = \mathbf{3\dfrac{2}{5}}$$

21.
$$6\dfrac{1}{8} \xrightarrow{\quad 5 + \frac{8}{8} + \frac{1}{8} \quad} 5\dfrac{9}{8}$$
$$\begin{array}{r} -\ 3\dfrac{5}{8} \qquad\qquad -\ 3\dfrac{5}{8} \\ \hline 2\dfrac{4}{8} = 2\dfrac{1}{2} \end{array}$$
$$w = \mathbf{2\dfrac{1}{2}}$$

22. 41, 43, 47

23. $C = \pi d$
$C \approx (3.14)(4\text{ cm})$
$C \approx 12.56\text{ cm}$
$C \approx \mathbf{13\ cm}$

24. (a) **5.4**

(b) **5**

25. $\dfrac{2}{8} = \dfrac{\mathbf{1}}{\mathbf{4}}$

26. $12 \div \dfrac{3}{8}$
$$1 \div \dfrac{3}{8} = \dfrac{8}{3}$$
$$\dfrac{12}{1} \times \dfrac{8}{3} = \dfrac{96}{3} = 32$$
about 32 years old

27. $\dfrac{12}{60} = \dfrac{\mathbf{1}}{\mathbf{5}}$

28. Possible answers:
$200 \div 8 = 25$
$100 \div 4 = 25$
$50 \div 2 = 25$

29.
$$\begin{array}{r} \$6.89 \\ \times\ \ 0.06 \\ \hline 0.4134 \end{array} \qquad \begin{array}{r} \overset{1\ \ 1}{\$6.89} \\ +\ \$0.41 \\ \hline \mathbf{\$7.30} \end{array}$$

30. (a) $3\dfrac{1}{2} \ \boxed{=}\ \dfrac{7}{2}$
$$3\dfrac{1}{2}$$

(b) $\dfrac{5}{8} \ \boxed{<}\ \dfrac{3}{4}$
$$\dfrac{3}{4} \times \dfrac{2}{2} = \dfrac{6}{8}$$

INVESTIGATION 6

1. rectangular prism

2. cylinder

3. triangular prism

4. cone

5. sphere

6. pyramid

7. 6 faces

8. 12 edges

9. 8 vertices

10. 5 faces

11. 8 edges

12. 5 vertices

13.

14.

15.

16. 5 in. × 5 in. = **25 in.²**

17. 25 in.² × 6 = **150 in.²**

18. 7 in. × 10 in. = **70 in.²**

19. 7 in. × 2 in. = **14 in.²**

20. 10 in. × 2 in. = **20 in.²**

21. (70 in.² × 2) + (14 in.² × 2)
    + (20 in.² × 2) = **208 in.²**

22. C.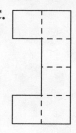

23. 6 cubes per layer × 2 layers = **12 cubes**

24. 9 cubes per layer × 3 layers = **27 cubes**

25. We count 6 cubes on the right-hand side of the figure. There are 3 such "slices" in the figure.
    6 cubes per slice × 3 slices = **18 cubes**

LESSON 61, WARM-UP

a. 3000

b. 533

c. 551

d. $5.75

e. 25

f. 30

g. 100

Problem Solving

First think, "What number times 5 equals 45?" (9). Then think, "What number minus 45 equals 1?" (46). Because the remainder is zero, we know the ones digit in the dividend must be 5, and we are able to fill in the missing digits in the last subtraction. Finally think, "What number times 5 equals 15?" (3).

$$
\begin{array}{r}
93 \\
5)\overline{465} \\
\underline{45} \\
15 \\
\underline{15} \\
0
\end{array}
$$

LESSON 61, LESSON PRACTICE

a.
$$\frac{1}{2} \times \frac{4}{4} = \frac{4}{8}$$
$$\frac{3}{4} \times \frac{2}{2} = \frac{6}{8}$$
$$+\ \frac{1}{8} \times \frac{1}{1} = \frac{1}{8}$$
$$\frac{11}{8} = \mathbf{1\frac{3}{8}}$$

b.
$$\frac{1}{2} \times \frac{3}{3} = \frac{3}{6}$$
$$\frac{1}{3} \times \frac{2}{2} = \frac{2}{6}$$
$$+\ \frac{1}{6} \times \frac{1}{1} = \frac{1}{6}$$
$$\frac{6}{6} = \mathbf{1}$$

c.
$$1\frac{1}{2} \times \frac{6}{6} = 1\frac{6}{12}$$
$$1\frac{1}{3} \times \frac{4}{4} = 1\frac{4}{12}$$
$$+\ 1\frac{1}{4} \times \frac{3}{3} = 1\frac{3}{12}$$
$$3\frac{13}{12} = \mathbf{4\frac{1}{12}}$$

d.
$$\frac{1}{2} \times \frac{3}{3} = \frac{3}{6}$$
$$\frac{2}{3} \times \frac{2}{2} = \frac{4}{6}$$
$$+\ \frac{5}{6} \times \frac{1}{1} = \frac{5}{6}$$
$$\frac{12}{6} = \mathbf{2}$$

e.
$$\frac{1}{2} \times \frac{4}{4} = \frac{4}{8}$$
$$\frac{3}{4} \times \frac{2}{2} = \frac{6}{8}$$
$$+\ \frac{7}{8} \times \frac{1}{1} = \frac{7}{8}$$
$$\frac{17}{8} = \mathbf{2\frac{1}{8}}$$

f.
$$1\frac{1}{4} \times \frac{2}{2} = 1\frac{2}{8}$$
$$1\frac{1}{8} \times \frac{1}{1} = 1\frac{1}{8}$$
$$+\ 1\frac{1}{2} \times \frac{4}{4} = 1\frac{4}{8}$$
$$\mathbf{3\frac{7}{8}}$$

LESSON 61, MIXED PRACTICE

1. $\dfrac{20}{6} = 3\dfrac{2}{6} = \mathbf{3\dfrac{1}{3}}$

2.
$$\begin{array}{r} 2.5 \\ \times\ \ \ 6 \text{ feet} \\ \hline 15.0 \ = \mathbf{15\ feet} \end{array}$$

3.
$$\begin{array}{r} \mathbf{2475\ names} \\ 3\overline{)7425} \\ \underline{6} \\ 14 \\ \underline{12} \\ 22 \\ \underline{21} \\ 15 \\ \underline{15} \\ 0 \end{array}$$

4.
$$5\frac{1}{2} \times \frac{3}{3} = 5\frac{3}{6} \xrightarrow{\ \ 4 + \frac{6}{6} + \frac{3}{6}\ \ } 4\frac{9}{6}$$
$$-\ 1\frac{2}{3} \times \frac{2}{2} = 1\frac{4}{6} \qquad\qquad\ -\ 1\frac{4}{6}$$
$$\mathbf{3\frac{5}{6}}$$

5.
$$5\frac{1}{3} \times \frac{2}{2} = 5\frac{2}{6} \xrightarrow{\ \ 4 + \frac{6}{6} + \frac{2}{6}\ \ } 4\frac{8}{6}$$
$$-\ 2\frac{1}{2} \times \frac{3}{3} = 2\frac{3}{6} \qquad\qquad\ -\ 2\frac{3}{6}$$
$$\mathbf{2\frac{5}{6}}$$

6.
$$1\frac{1}{2} \times \frac{6}{6} = 1\frac{6}{12}$$
$$2\frac{1}{3} \times \frac{4}{4} = 2\frac{4}{12}$$
$$+\ 3\frac{1}{4} \times \frac{3}{3} = 3\frac{3}{12}$$
$$6\frac{13}{12} = \mathbf{7\frac{1}{12}}$$

7.
$$3\frac{3}{4} \times \frac{3}{3} = 3\frac{9}{12}$$
$$+\ 3\frac{1}{3} \times \frac{4}{4} = 3\frac{4}{12}$$
$$6\frac{13}{12} = \mathbf{7\frac{1}{12}}$$

8. (a) $\dfrac{2}{3}$ $\ominus$ $\dfrac{3}{5}$

$\dfrac{2}{3} \times \dfrac{5}{5}$ $\dfrac{3}{5} \times \dfrac{3}{3}$

$\dfrac{10}{15}$ $\dfrac{9}{15}$

(b) 16 $\ominus$ 12

9. $\dfrac{5}{6} \times \dfrac{36}{1} = \dfrac{180}{6} = $ **30**

10. $\dfrac{3}{8} \times \dfrac{2}{3} = \dfrac{6}{24} = $ **$\dfrac{1}{4}$**

11. $\dfrac{3}{8} \div \dfrac{2}{3}$

$1 \div \dfrac{2}{3} = \dfrac{3}{2}$

$\dfrac{3}{8} \times \dfrac{3}{2} = $ **$\dfrac{9}{16}$**

12. $(4 - 0.4) \div 4$

$3.6 \div 4$

0.9

13. $4 - (0.4 \div 4)$

$4 - 0.1$

3.9

14. **6**

15. $\begin{array}{r} \$600.00 \\ + \$900.00 \\ \hline \$1500.00 \end{array}$

Round $642.23 to $600 and $861.17 to $900. Then add $600 to $900.

16. (a) $4 \text{ cm} \times 2 = $ **8 cm**

(b) $C = \pi d$

$C \approx (3.14)(8 \text{ cm})$

$C \approx $ **25.12 cm**

17. **0.1** or **$\dfrac{1}{10}$**

18. $\begin{array}{r} 3 \text{ inches} \\ 4\overline{)12 \text{ inches}} \\ \underline{12} \\ 0 \end{array}$

$3 \text{ in.} \times 3 \text{ in.} = $ **9 square inches**

19. $\dfrac{\text{dime}}{\text{quarter}} = \dfrac{10}{25} = \dfrac{2}{5}$

20. $15m = 300$

$\begin{array}{r} 20 \\ 15\overline{)300} \\ \underline{30} \\ 00 \\ \underline{00} \\ 0 \end{array}$

$m = $ **20**

21. $\dfrac{1}{10} \times \dfrac{10}{10} = \dfrac{10}{100}$

$n = $ **10**

22. $\dfrac{5}{5}$

23. $\begin{array}{r} 9\text{:}50 \text{ a.m.} \\ + \ 5\text{:}15 \\ \hline 14\text{:}65 \end{array}$

$2\text{:}65 \text{ p.m.} = $ **3:05 p.m.**

24. $\sqrt{16 \text{ in.}^2} = 4 \text{ in.}$

$4 \text{ in.} + 4 \text{ in.} + 4 \text{ in.} + 4 \text{ in.} = $ **16 inches**

25. **Pentagon**

26. $\begin{array}{r} \overset{1\ 1}{\$4.95} \\ + \$2.79 \\ \hline \$7.74 \end{array}$ $\begin{array}{r} \$7.74 \\ \times \ \ .07 \\ \hline 0.5418 \end{array}$ $\begin{array}{r} \$7.74 \\ + \$0.54 \\ \hline \textbf{\$8.28} \end{array}$

Round $4.95 to $5 and round $2.79 to $3. Add $5 and $3 to get $8. Mentally multiply 7% and $8 and get 56¢. Add 56¢ to $8 and get $8.56. Since we rounded the prices up to estimate, the answer of $8.28 is reasonable.

27. $\$37 \div 100 = $ **$0.37**

28. **One possibility:**

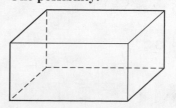

(a) **6 faces**

(b) **12 edges**

(c) **8 vertices**

29. $3 \text{ cm} \times 3 \text{ cm} = \mathbf{9 \text{ cm}^2}$

30. $9 \text{ cm}^2 \times 6 = \mathbf{54 \text{ cm}^2}$

LESSON 62, WARM-UP

a. 200

b. 1675

c. 252

d. $11.00

e. $12.00

f. $2.50

g. 2

Problem Solving

27 people per row $\times$ 2 rows $=$ 54 people
54 people $\div$ 3 rows $=$ **18 people per row**

LESSON 62, LESSON PRACTICE

a. $\dfrac{10}{5} + \dfrac{4}{5} = \dfrac{\mathbf{14}}{\mathbf{5}}$

b. $\dfrac{6}{2} + \dfrac{1}{2} = \dfrac{\mathbf{7}}{\mathbf{2}}$

c. $\dfrac{4}{4} + \dfrac{3}{4} = \dfrac{\mathbf{7}}{\mathbf{4}}$

d. $\dfrac{24}{4} + \dfrac{1}{4} = \dfrac{\mathbf{25}}{\mathbf{4}}$

e. $\dfrac{6}{6} + \dfrac{5}{6} = \dfrac{\mathbf{11}}{\mathbf{6}}$

f. $\dfrac{30}{10} + \dfrac{3}{10} = \dfrac{\mathbf{33}}{\mathbf{10}}$

g. $\dfrac{6}{3} + \dfrac{1}{3} = \dfrac{\mathbf{7}}{\mathbf{3}}$

h. $\dfrac{24}{2} + \dfrac{1}{2} = \dfrac{\mathbf{25}}{\mathbf{2}}$

i. $\dfrac{18}{6} + \dfrac{1}{6} = \dfrac{\mathbf{19}}{\mathbf{6}}$

j. $\dfrac{3}{2} \times \dfrac{10}{3} = \dfrac{30}{6} = \mathbf{5}$

LESSON 62, MIXED PRACTICE

1. (a) $1 \div \dfrac{1}{4} = \dfrac{4}{1} = \mathbf{4 \text{ quarter notes}}$

(b) $\dfrac{1}{4} \div \dfrac{1}{8}$

$1 \div \dfrac{1}{8} = 8$

$\dfrac{1}{4} \times \dfrac{8}{1} = \dfrac{8}{4} = \mathbf{2 \text{ eighth notes}}$

2.
$$\begin{array}{r} \overset{1}{1}2 \text{ inches} \\ \times \quad 5 \\ \hline 60 \text{ inches} \end{array}$$

$60 + 2\dfrac{1}{2} = \mathbf{62\dfrac{1}{2} \text{ inches}}$

3. **B. 21**

4. $\dfrac{4}{3} \cdot \dfrac{3}{2} = \dfrac{12}{6} = \mathbf{2}$

5. $100\% - 20\% = \mathbf{80\%}$

6.
$$\begin{array}{r} \$36.25 \\ \$41.50 \\ + \ \$43.75 \\ \hline \$121.50 \end{array}
\qquad
\begin{array}{r} \mathbf{\$40.50} \\ 3)\overline{\$121.50} \\ \underline{12} \\ 01 \\ \underline{00} \\ 1\ 5 \\ \underline{1\ 5} \\ 00 \\ \underline{00} \\ 0 \end{array}$$

7. $30 \div 5 = \mathbf{6}$

8.
$$\begin{aligned} 4\dfrac{3}{8} \times \dfrac{1}{1} &= 4\dfrac{3}{8} \\ + \ 3\dfrac{1}{4} \times \dfrac{2}{2} &= 3\dfrac{2}{8} \\ \hline m &= \mathbf{7\dfrac{5}{8}} \end{aligned}$$

9.

$$\frac{3}{5} \times \frac{2}{2} = \frac{6}{10}$$

$$- \frac{3}{10} = \frac{3}{10}$$

$$n = \frac{3}{10}$$

10.

$$6 \overline{)0.456} \quad 0.076$$

42

36

36

0

$$d = \mathbf{0.076}$$

11.

$$4 \overline{)150.0} \quad 37.5$$

12

30

28

2 0

2 0

0

$$w = \mathbf{37.5}$$

12.

$$\frac{1}{2} \times \frac{4}{4} = \frac{4}{8}$$

$$\frac{3}{4} \times \frac{2}{2} = \frac{6}{8}$$

$$+ \frac{5}{8} \times \frac{1}{1} = \frac{5}{8}$$

$$\frac{15}{8} = \mathbf{1\frac{7}{8}}$$

13.

$$\frac{5}{6} \times \frac{1}{1} = \frac{5}{6}$$

$$- \frac{1}{2} \times \frac{3}{3} = \frac{3}{6}$$

$$\frac{2}{6} = \mathbf{\frac{1}{3}}$$

14. $\frac{1}{2} \cdot \frac{4}{5} = \frac{4}{10} = \mathbf{\frac{2}{5}}$

15. $\frac{2}{3} \div \frac{1}{2}$

$$1 \div \frac{1}{2} = \frac{2}{1}$$

$$\frac{2}{3} \times \frac{2}{1} = \frac{4}{3} = \mathbf{1\frac{1}{3}}$$

16. $1 - (0.2 - 0.03)$

$$1 - 0.17$$

$$\mathbf{0.83}$$

17.

$$
\begin{array}{r}
0.14 \\
\times\ 0.16 \\
\hline
84 \\
140 \\
\hline
\mathbf{0.0224}
\end{array}
$$

18. $2.5 \text{ cm} \times 10 = \mathbf{25 \text{ millimeters}}$

19. Factors of 18
①, ②, ③, ⑥, 9, 18
Factors of 24
①, ②, ③, 4, ⑥, 8, 12, 24
1, 2, 3, ⑥

20. $\frac{4}{10} = \mathbf{\frac{2}{5}}$

21.

$$4 \overline{)40 \text{ mm}} \quad 10 \text{ mm} \qquad \begin{array}{r} 10 \text{ mm} \\ \times\ 10 \text{ mm} \\ \hline \mathbf{100 \text{ mm}^2} \end{array}$$

4

00

00

0

22. $14°\text{F} - (-6°\text{F}) = \mathbf{20°\text{F}}$

23. (a) $1\frac{1}{2}$ inches $\times 2 = \mathbf{3 \text{ inches}}$

(b) $C = \pi d$
$C \approx (3.14)(3 \text{ inches})$
$C \approx \mathbf{9.42 \text{ inches}}$

24.

$$
\begin{array}{r}
\overset{3}{\cancel{4}}{}^{1}0 \\
-\ 2\ 2 \\
\hline
\mathbf{1\ 8} \text{ more people}
\end{array}
$$

25. $\frac{40}{100} = \mathbf{\frac{2}{5}}$

26. **No. A majority of 100 people is at least 51 people. No sport was the favorite sport of 51 or more people.**

27. 22 out of 100 or **22%**

28.

$$
\begin{array}{r}
200 \\
\times\ 0.40 \\
\hline
80.00
\end{array}
$$

80

29. $\mathbf{2 \cdot 2 \cdot 5}$

30. 8.9, 9.0, ⑨.⑴ 9.2, 9.2

9.1

LESSON 63, WARM-UP

a. 700

b. 370

c. 252

d. $6.00

e. 15

f. $0.25

g. 0

Problem Solving

Add $\frac{1}{12}$ to each term to find the next term.

$$\frac{1}{12}, \frac{1}{6}, \frac{1}{4}, \underline{\frac{1}{3}}, \underline{\frac{5}{12}}, \underline{\frac{1}{2}}, \cdots$$

LESSON 63, LESSON PRACTICE

a.
$$5\frac{1}{2} \times \frac{3}{3} = 5\frac{3}{6}$$
$$-\ 3\frac{1}{3} \times \frac{2}{2} = 3\frac{2}{6}$$
$$2\frac{1}{6}$$

b.
$$4\frac{1}{4} \times \frac{3}{3} = \overset{3}{\cancel{4}}\overset{15}{\cancel{3}}{12}$$
$$-\ 2\frac{1}{3} \times \frac{4}{4} = 2\frac{4}{12}$$
$$1\frac{11}{12}$$

c.
$$6\frac{1}{2} \times \frac{2}{2} = \overset{5}{\cancel{6}}\overset{6}{\cancel{2}}{4}$$
$$-\ 1\frac{3}{4} \times \frac{1}{1} = 1\frac{3}{4}$$
$$4\frac{3}{4}$$

d.
$$7\frac{2}{3} \times \frac{2}{2} = \overset{6}{\cancel{7}}\overset{10}{\cancel{4}}{6}$$
$$-\ 3\frac{5}{6} \times \frac{1}{1} = 3\frac{5}{6}$$
$$3\frac{5}{6}$$

e.
$$6\frac{1}{6} \times \frac{1}{1} = \overset{5}{\cancel{6}}\overset{7}{\cancel{1}}{6}$$
$$-\ 1\frac{1}{2} \times \frac{3}{3} = 1\frac{3}{6}$$
$$4\frac{4}{6} = 4\frac{2}{3}$$

f.
$$4\frac{1}{3} \times \frac{2}{2} = \overset{3}{\cancel{4}}\overset{8}{\cancel{2}}{6}$$
$$-\ 1\frac{1}{2} \times \frac{3}{3} = 1\frac{3}{6}$$
$$2\frac{5}{6}$$

g.
$$4\frac{5}{6} \times \frac{1}{1} = 4\frac{5}{6}$$
$$-\ 1\frac{1}{3} \times \frac{2}{2} = 1\frac{2}{6}$$
$$3\frac{3}{6} = 3\frac{1}{2}$$

h.
$$6\frac{1}{2} \times \frac{3}{3} = \overset{5}{\cancel{6}}\overset{9}{\cancel{3}}{6}$$
$$-\ 3\frac{5}{6} \times \frac{1}{1} = 3\frac{5}{6}$$
$$2\frac{4}{6} = 2\frac{2}{3}$$

i.
$$8\frac{2}{3} \times \frac{4}{4} = \overset{7}{\cancel{8}}\overset{20}{\cancel{8}}{12}$$
$$-\ 5\frac{3}{4} \times \frac{3}{3} = 5\frac{9}{12}$$
$$2\frac{11}{12}$$

LESSON 63, MIXED PRACTICE

1.
$$(0.6 + 0.4) - (0.6 \times 0.4)$$
$$1 \quad - \quad 0.24$$
$$\textbf{0.76}$$

2.
$$\overset{1}{1}4,494 \text{ feet}$$
$$+\ \quad 282 \text{ feet}$$
$$\textbf{14,776 feet}$$

3.
$$\begin{array}{r} \textbf{24 feet} \\ 12\overline{)288} \\ \underline{24} \\ 48 \\ \underline{48} \\ 0 \end{array}$$

4. $\dfrac{14}{3}$

5. $\dfrac{5}{2} \cdot \dfrac{6}{5} = \dfrac{30}{10} = 3$

6.
$$
\begin{array}{r}
10{:}15 \text{ a.m.} \\
+\ 2{:}30 \\
\hline
\textbf{12{:}45 p.m.}
\end{array}
$$

7. $(30 \times 15) \div (30 - 15)$
$$450 \div 15$$
$$\textbf{30}$$

8. $\dfrac{5}{8}\ \boxed{<}\ \dfrac{2}{3}$

$$
\begin{array}{cc}
\dfrac{5}{8} \times \dfrac{3}{3} & \dfrac{2}{3} \times \dfrac{8}{8} \\[2ex]
\dfrac{15}{24} & \dfrac{16}{24}
\end{array}
$$

9.
$$
\begin{aligned}
3\dfrac{2}{3} \times \dfrac{2}{2} &= 3\dfrac{4}{6} \\
+\ 1\dfrac{1}{2} \times \dfrac{3}{3} &= 1\dfrac{3}{6} \\
\hline
4\dfrac{7}{6} &= 5\dfrac{1}{6}
\end{aligned}
$$
$$w = \mathbf{5\dfrac{1}{6}}$$

10.
$$
\begin{aligned}
\dfrac{6}{8} \times \dfrac{1}{1} &= \dfrac{6}{8} \\
-\ \dfrac{3}{4} \times \dfrac{2}{2} &= \dfrac{6}{8} \\
\hline
\dfrac{0}{8} &= \mathbf{0}
\end{aligned}
$$

11.
$$
\begin{aligned}
6\dfrac{1}{4} \times \dfrac{2}{2} &= \cancel{6}\,\overset{5}{\cancel{6}}\dfrac{\overset{10}{\cancel{2}}}{8} \\
-\ 5\dfrac{5}{8} \times \dfrac{1}{1} &= 5\dfrac{5}{8} \\
\hline
&\quad\ \mathbf{\dfrac{5}{8}}
\end{aligned}
$$

12. $\dfrac{3}{4} \times \dfrac{2}{5} = \dfrac{6}{20} = \mathbf{\dfrac{3}{10}}$

13. $\dfrac{3}{4} \div \dfrac{2}{5}$
$$1 \div \dfrac{2}{5} = \dfrac{5}{2}$$
$$\dfrac{3}{4} \times \dfrac{5}{2} = \dfrac{15}{8} = \mathbf{1\dfrac{7}{8}}$$

14. $(1 - 0.4)(1 + 0.4)$
$$(0.6)(1.4)$$
$$\mathbf{0.84}$$

15.
$$
\begin{array}{r}
\$45 \\
\times\quad 0.60 \\
\hline
\mathbf{\$27.00}
\end{array}
$$

16.
$$
\begin{array}{r}
\mathbf{0.05} \\
8\overline{)0.40} \\
\underline{40} \\
0
\end{array}
$$

17.
$$
\begin{array}{r}
\mathbf{20} \\
4\overline{)80} \\
\underline{8} \\
00 \\
\underline{00} \\
0
\end{array}
$$

18. **1.0** or **1**

19. 2, 3, 5, 7, 11, 13, 17, 19, 23, 29
29

20.
$$
\begin{aligned}
1\dfrac{1}{8}\text{ in.} \times \dfrac{1}{1} &= 1\dfrac{1}{8}\text{ in.} \\
1\dfrac{1}{8}\text{ in.} \times \dfrac{1}{1} &= 1\dfrac{1}{8}\text{ in.} \\
\dfrac{3}{4}\text{ in.} \times \dfrac{2}{2} &= \dfrac{6}{8}\text{ in.} \\
+\ \dfrac{3}{4}\text{ in.} \times \dfrac{2}{2} &= \dfrac{6}{8}\text{ in.} \\
\hline
2\dfrac{14}{8}\text{ in.} = 3\dfrac{6}{8}\text{ in.} &= \mathbf{3\dfrac{3}{4}\text{ in.}}
\end{aligned}
$$

21. (a) **5 faces**

(b) **9 edges**

(c) **6 vertices**

22. **$680; $678.25 is more than $670 but less than $680. It is closer to $680 because it is more than $675, which is halfway between $670 and $680.**

23.
$$
\begin{array}{r}
30 \text{ cm} \\
\times\ 10 \text{ cm} \\
\hline
300 \text{ cm}^2
\end{array}
\qquad
\begin{array}{r}
\mathbf{150 \text{ cm}^2} \\
2\overline{)300 \text{ cm}^2} \\
2 \\
\hline
10 \\
\underline{10} \\
00 \\
\underline{00} \\
0
\end{array}
$$

24.
$$\overset{1}{2.5}$$
$$\underline{\times\quad 2000 \text{ pounds}}$$
$$\mathbf{5000.0 \text{ pounds}}$$

25. *C*

26. (a) $C = \pi d$
$C \approx (3.14)(7 \text{ cm})$
$C \approx 21.98$
$C \approx \mathbf{22 \text{ cm}}$

(b) $C = \pi d$
$C \approx (3.14)(5 \text{ cm})$
$C \approx 15.70$
$C \approx \mathbf{16 \text{ cm}}$

27. $\dfrac{30}{100} = \dfrac{\mathbf{3}}{\mathbf{10}}$

$\dfrac{3}{10} \times \dfrac{240}{1} = \dfrac{720}{10} = \mathbf{72 \text{ hits}}$

28.
$$2\,\overset{5}{\cancel{6}}^{1}2 \text{ mi}$$
$$\underline{-\ 1\ 1.\ 5 \text{ mi}}$$
$$\mathbf{1\ 4.\ 7 \text{ mi}}$$

29.

$\overset{1\ \ 1}{\$15.49}$	$\$30.98$	$\overset{1\ 1}{\$30.98}$
$\underline{\times\qquad 2}$	$\underline{\times\quad 0.07}$	$\underline{+\quad \$2.17}$
$\$30.98$	$\$2.1686$	$\mathbf{\$33.15}$

30. $\dfrac{1}{2} \text{ tsp} \times \dfrac{2}{2} = \dfrac{2}{4} \text{ tsp}$

$+ \dfrac{3}{4} \text{ tsp} \times \dfrac{1}{1} = \dfrac{3}{4} \text{ tsp}$
$$\overline{\qquad\qquad\qquad \dfrac{5}{4} = 1\dfrac{1}{4} \text{ teaspoons}}$$

LESSON 64, WARM-UP

a. 1200

b. 4950

c. 238

d. $15.00

e. $5.00

f. $7.50

g. 8

Problem Solving
8 vertices

LESSON 64, LESSON PRACTICE

a. A quadrilateral is a four-sided polygon.

b. A parallelogram has two pairs of parallel sides; a trapezoid has only one pair of parallel sides.

c.

d.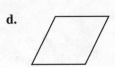

e. True

f. True

LESSON 64, MIXED PRACTICE

1. $(1.3 + 1.2) \div (1.3 - 1.2)$
$\qquad 2.5 \qquad \div \qquad 0.1$
$\qquad\qquad\qquad \mathbf{25}$

2.
$$1\,\overset{5}{\cancel{6}}^{1}1\ 6 \text{ years}$$
$$\underline{-\ 1\ 5\ 6\ 4 \text{ years}}$$
$$\mathbf{5\ 2 \text{ years}}$$

3.
$$\overset{1}{45} \text{ feet}$$
$$\underline{\times\quad 3}$$
$$\mathbf{135 \text{ feet}}$$

4. A square is a four-sided polygon, so it is a quadrilateral. The four sides of a square have the same length, and the four angles have the same measure, so a square is "regular."

5.
$$\begin{array}{r} \textbf{6 inches} \\ 6\overline{)36\text{ inches}} \\ \underline{36} \\ 00 \end{array}$$

6. $\dfrac{1}{4} \times \dfrac{25}{25} = \dfrac{25}{100}$ **25**

7. $\dfrac{8 \times 8}{8 + 8} = \dfrac{64}{16} = \dfrac{4}{1} = \textbf{4}$

8.
$$5\dfrac{2}{3} \times \dfrac{4}{4} = 5\dfrac{8}{12}$$
$$+ 3\dfrac{3}{4} \times \dfrac{3}{3} = 3\dfrac{9}{12}$$
$$8\dfrac{17}{12} = \mathbf{9\dfrac{5}{12}}$$

9.
$$\dfrac{1}{2} \times \dfrac{6}{6} = \dfrac{6}{12}$$
$$\dfrac{2}{3} \times \dfrac{4}{4} = \dfrac{8}{12}$$
$$+ \dfrac{1}{4} \times \dfrac{3}{3} = \dfrac{3}{12}$$
$$\dfrac{17}{12} = \mathbf{1\dfrac{5}{12}}$$

10.
$$\dfrac{9}{10} \times \dfrac{1}{1} = \dfrac{9}{10}$$
$$- \dfrac{1}{2} \times \dfrac{5}{5} = \dfrac{5}{10}$$
$$\dfrac{4}{10} = \mathbf{\dfrac{2}{5}}$$

11.
$$6\dfrac{1}{2} \times \dfrac{4}{4} = \cancel{6}^{5}\cancel{\dfrac{4}{8}}^{12}$$
$$- 2\dfrac{7}{8} \times \dfrac{1}{1} = 2\dfrac{7}{8}$$
$$\mathbf{3\dfrac{5}{8}}$$

12. $2 \times 0.4 \;\textcircled{<}\; 2 + 0.4$
   $\quad 0.8 \qquad\qquad 2.4$

13.
$$\begin{array}{r} 4.8 \\ \times\ 0.35 \\ \hline 240 \\ 1440 \\ \hline 1.680 = \mathbf{1.68} \end{array}$$

14.
$$\begin{array}{r} \textbf{2.5} \\ 4\overline{)10.0} \\ \underline{8} \\ 2\ 0 \\ \underline{2\ 0} \\ 0 \end{array}$$

15.
$$\begin{array}{r} \textbf{40 pencils} \\ 12\overline{)480} \\ \underline{48} \\ 00 \\ \underline{00} \\ 0 \end{array}$$

16.
$$\begin{array}{r} 0.33 \\ \times\ 0.38 \\ \hline 264 \\ 990 \\ \hline 0.1254 \end{array}$$
0.13

17. $\dfrac{1}{2}$ in. $\times \dfrac{3}{4}$ in. $= \mathbf{\dfrac{3}{8}}$ **sq. in.**

18. Yes

19. 2, 3, 5, 7, 11, 13, 17, 19, 23, 29, 31, 37
37

20. (a) $\sqrt{9\text{ cm}^2} = \mathbf{3\text{ cm}}$
   (b) 3 cm $\times$ 4 = **12 cm**

21. 6 faces

22. (a) 10 in. $\times$ 10 in. = **100 in.²**
   (b) 100 in.² $\times$ 6 = **600 in.²**

23. 2.5 $\times$ 100 cm = **250 centimeters**

24. $\dfrac{3}{2} \cdot \dfrac{5}{2} = \dfrac{15}{4} = \mathbf{3\dfrac{3}{4}}$

25. $9 = 3 \cdot 3$
   $10 = 2 \cdot 5$
   $12 = 2 \cdot 2 \cdot 3$

26. $\dfrac{75}{100}$; **0.75**

27. $\dfrac{2 \cdot 2 \cdot 3 \cdot 2 \cdot 3}{2 \cdot 2 \cdot 3 \cdot 5 \cdot 5} = 1 \cdot 1 \cdot 1 \cdot \dfrac{6}{25} = \mathbf{\dfrac{6}{25}}$

28.
$$\begin{array}{r} {}^{1}\cancel{2}\,{}^{15}\cancel{6}.{}^{1}2 \text{ miles} \\ -\ 1\,6.\,6 \text{ miles} \\ \hline 9.6 \text{ miles} \end{array}$$
$d = \textbf{9.6 miles}$

29. $4° - (-6°C)$
$$\begin{array}{r} 4°C \\ +\ 6°C \\ \hline \textbf{10°C} \end{array}$$

30. $16° - (-6°C)$
$$\begin{array}{r} {}^{1}16°C \\ +\ 6°C \\ \hline \textbf{22°C} \end{array}$$

LESSON 65, WARM-UP

a. 300

b. 536

c. 195

d. $17.50

e. $1.50

f. $0.75

g. 1

Problem Solving

Nelson ate $\frac{1}{2}$.

His sister ate $\frac{1}{2}$ of $\frac{1}{2}$, or $\frac{1}{4}$.

His little brother ate $\frac{1}{2}$ of $\frac{1}{4}$, or $\frac{1}{8}$.

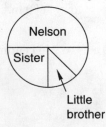

LESSON 65, LESSON PRACTICE

a. 20, 21, 22

b. $20 = 2 \cdot 2 \cdot 5$
$21 = 3 \cdot 7$
$22 = 2 \cdot 11$

c. one possibility:

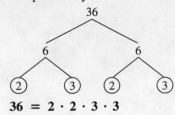

$\mathbf{36 = 2 \cdot 2 \cdot 3 \cdot 3}$

d.
$$\begin{array}{r} 1 \\ 3)\overline{3} \\ 2)\overline{6} \\ 2)\overline{12} \\ 2)\overline{24} \\ 2)\overline{48} \end{array}$$
$\mathbf{48 = 2 \cdot 2 \cdot 2 \cdot 2 \cdot 3}$

e.

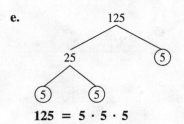

$\mathbf{125 = 5 \cdot 5 \cdot 5}$

f.

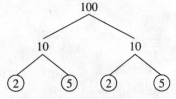

$\mathbf{10 = 2 \cdot 5}$
$\mathbf{100 = 2 \cdot 2 \cdot 5 \cdot 5}$

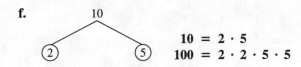

Each prime factorization contains the same prime numbers (2 and 5). Also, there are equal numbers of 2's and 5's in each prime factorization.
$\mathbf{1000 = 2 \cdot 2 \cdot 2 \cdot 5 \cdot 5 \cdot 5}$

LESSON 65, MIXED PRACTICE

1. **57,280,000 square miles**

2.
$$\begin{array}{r} 6.5 \\ \times\ 12 \text{ inches} \\ \hline 130 \\ 650 \\ \hline 78.0 \end{array}$$
or **78 inches**

3. $\dfrac{6}{10} = \dfrac{3}{5}$

$\dfrac{6}{10} \times \dfrac{10}{10} = \dfrac{60}{100} = \mathbf{60\%}$

4. One possibility:

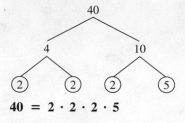

$$40 = 2 \cdot 2 \cdot 2 \cdot 5$$

5. A. **21**

6. $\dfrac{8}{3} \cdot \dfrac{3}{8} = \dfrac{24}{24} = \mathbf{1}$

7. $\dfrac{6}{10} = \dfrac{\mathbf{3}}{\mathbf{5}}$

8.
$$8\tfrac{1}{2} \times \tfrac{3}{3} = 8\tfrac{3}{6}$$
$$1\tfrac{1}{3} \times \tfrac{2}{2} = 1\tfrac{2}{6}$$
$$+\ 2\tfrac{1}{6} \times \tfrac{1}{1} = 2\tfrac{1}{6}$$
$$11\tfrac{6}{6} = \mathbf{12}$$

9.
$$\tfrac{1}{12} \times \tfrac{1}{1} = \tfrac{1}{12}$$
$$\tfrac{1}{6} \times \tfrac{2}{2} = \tfrac{2}{12}$$
$$+\ \tfrac{1}{2} \times \tfrac{6}{6} = \tfrac{6}{12}$$
$$\tfrac{9}{12} = \tfrac{\mathbf{3}}{\mathbf{4}}$$

10.
$$15\tfrac{3}{4} \times \tfrac{2}{2} = 15\tfrac{6}{8}$$
$$-\ 2\tfrac{1}{8} \times \tfrac{1}{1} = 2\tfrac{1}{8}$$
$$m = \mathbf{13\tfrac{5}{8}}$$

11. $\dfrac{4}{25} \times \dfrac{4}{4} = \dfrac{16}{100}$

$n = \mathbf{16}$

12.
$$12\overline{)0.0144} \qquad w = \mathbf{0.0012}$$
$$\underline{12}$$
$$24$$
$$\underline{24}$$
$$0$$
(quotient **0.0012**)

13. $\dfrac{3}{8} \times \dfrac{1}{3} = \dfrac{3}{24} = \dfrac{1}{8}$

$y = \dfrac{\mathbf{1}}{\mathbf{8}}$

14.
$$\tfrac{1}{2} - \tfrac{1}{3} \ \textcircled{=}\ \tfrac{2}{3} - \tfrac{1}{2}$$

$$\tfrac{1}{2} \times \tfrac{3}{3} = \tfrac{3}{6} \qquad\qquad \tfrac{2}{3} \times \tfrac{2}{2} = \tfrac{4}{6}$$
$$-\ \tfrac{1}{3} \times \tfrac{2}{2} = \tfrac{2}{6} \qquad\quad -\ \tfrac{1}{2} \times \tfrac{3}{3} = \tfrac{3}{6}$$
$$\tfrac{1}{6} \qquad\qquad\qquad\qquad \tfrac{1}{6}$$

15.
$$1 - (0.2 + 0.48)$$
$$1 - 0.68$$
$$\mathbf{0.32}$$

16.
$$50¢ \times 24 = \$12$$
$$8¢ \times 12 = 96¢$$
$$\$12 + \$0.96 = \mathbf{\$12.96}$$
Two dozen is 24. The price is 50¢ each, which is half a dollar each. So 24 erasers cost \$12. 8% of \$12 is 8¢ for each dollar. Twelve times 8¢ is 96¢. Add \$12 and 96¢ and get \$12.96.

17.
$$\begin{array}{r} \mathbf{80}\text{ pieces} \\ 25\overline{)2000} \\ \underline{200} \\ 00 \\ \underline{00} \\ 0 \end{array}$$

18. (a) **5 faces**

(b) **8 edges**

(c) **5 vertices**

19.
$$\begin{array}{r} 1 \\ 5\overline{)5} \\ 5\overline{)25} \\ 2\overline{)50} \end{array}$$
$$\mathbf{50 = 2 \cdot 5 \cdot 5}$$

20. **Hexagon; 6 vertices**

21. $\dfrac{\mathbf{25}}{\mathbf{7}}$

22. (a) $\sqrt{36 \text{ in.}^2} = \mathbf{6 \text{ in.}}$

(b) $6 \text{ in.} \times 4 = \mathbf{24 \text{ in.}}$

23. $\dfrac{16}{100} = \dfrac{4}{25}$

24. **50 mm**

25. **About 4 meters**

26.

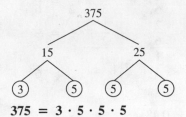

$375 = 3 \cdot 5 \cdot 5 \cdot 5$

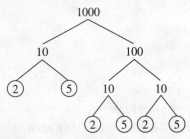

$1000 = 2 \cdot 2 \cdot 2 \cdot 5 \cdot 5 \cdot 5$

27. $\dfrac{5 \cdot 5 \cdot 5 \cdot 3}{5 \cdot 5 \cdot 5 \cdot 2 \cdot 2 \cdot 2} = 1 \cdot 1 \cdot 1 \cdot \dfrac{3}{8} = \dfrac{3}{8}$

28. $C = \pi d$
    $C \approx (3.14)(30 \text{ feet})$
    $C \approx 94.2$
    $C \approx$ **94 feet**
    The radius is 15 ft, so the diameter is 30 ft.
    π is a little more than 3, and 3 times 30 ft is
    90 ft, so 94 ft is reasonable.

29. $\dfrac{80}{100} \times \dfrac{20}{1} = \dfrac{1600}{100} =$ **16 answers**

30. **A rectangle is a four-sided polygon with four**
    right angles. Since every square is four-sided
    with four right angles, every square is a
    rectangle. (A rectangle need not be longer
    than it is wide.)

LESSON 66, WARM-UP

a. **800**

b. **475**

c. **184**

d. **$3.50**

e. **$20.00**

f. **$3.00**

g. **1**

Problem Solving

First bring down the 6 in the dividend, and think,
"What number times 6 equals 36?" (6). Then
multiply 4 (the tens digit in the quotient) by 6
and think, "What number minus 24 equals 3?"
(27). We know the hundreds digit in the quotient
must be 1 (because 2 would produce a four-digit
dividend). The hundreds digit in the dividend
must be 8.

$$
\begin{array}{r}
146 \\
6\overline{)876} \\
6 \\
\hline
27 \\
24 \\
\hline
36 \\
36 \\
\hline
0
\end{array}
$$

LESSON 66, LESSON PRACTICE

a. $\dfrac{3}{2} \times \dfrac{2}{3} = \dfrac{6}{6} =$ **1**

b. $\dfrac{5}{3} \times \dfrac{3}{4} = \dfrac{15}{12} = 1\dfrac{3}{12} = \mathbf{1\dfrac{1}{4}}$

c. $\dfrac{3}{2} \times \dfrac{5}{3} = \dfrac{15}{6} = 2\dfrac{3}{6} = \mathbf{2\dfrac{1}{2}}$

d. $\dfrac{5}{3} \times \dfrac{3}{1} = \dfrac{15}{3} =$ **5**

e. $\dfrac{5}{2} \times \dfrac{8}{3} = \dfrac{40}{6} = 6\dfrac{4}{6} = \mathbf{6\dfrac{2}{3}}$

f. $\dfrac{3}{1} \times \dfrac{7}{4} = \dfrac{21}{4} = \mathbf{5\dfrac{1}{4}}$

g. $\dfrac{10}{3} \times \dfrac{5}{3} = \dfrac{50}{9} = 5\dfrac{5}{9}$

h. $\dfrac{11}{4} \times \dfrac{2}{1} = \dfrac{22}{4} = 5\dfrac{2}{4} = 5\dfrac{1}{2}$

i. $\dfrac{2}{1} \times \dfrac{7}{2} = \dfrac{14}{2} = 7$

LESSON 66, MIXED PRACTICE

1.
$$\begin{array}{r} 6\,0 \\ \times\ 0.50 \\ \hline 30.00 \end{array}$$ or **30 questions**

2. (a)
$$\begin{array}{r} \overset{2}{\cancel{3}}{}^{1}0 \\ -\ 1\,2 \\ \hline 1\,8 \end{array}\text{ girls}$$
$\dfrac{\text{boys}}{\text{girls}} = \dfrac{12}{18} = \dfrac{2}{3}$

(b) $\dfrac{18}{30} = \dfrac{3}{5}$

3.
$$\begin{array}{r} 11 \text{ yards} \\ 3\overline{)33} \\ \underline{3} \\ 03 \\ \underline{03} \\ 0 \end{array}$$

$$\begin{array}{r} 155 \\ \times\ \ 11 \\ \hline 155 \\ 1550 \\ \hline \mathbf{1705} \text{ pounds} \end{array}$$

4. $\dfrac{3}{2} \times \dfrac{8}{3} = \dfrac{24}{6} = 4$

5. $\dfrac{8}{3} \times \dfrac{2}{1} = \dfrac{16}{3} = 5\dfrac{1}{3}$

6.
$$\begin{array}{r} \mathbf{40} \\ 5\overline{)200} \\ \underline{20} \\ 00 \\ \underline{00} \\ 0 \end{array}$$

7. $\dfrac{175}{25} = 7$

8.
$$\begin{array}{r} 1\dfrac{1}{5} \times \dfrac{2}{2} = 1\dfrac{2}{10} \\ +\ 3\dfrac{1}{2} \times \dfrac{5}{5} = 3\dfrac{5}{10} \\ \hline 4\dfrac{7}{10} \end{array}$$

9.
$$\begin{array}{r} \dfrac{1}{3} \times \dfrac{4}{4} = \dfrac{4}{12} \\ \dfrac{1}{6} \times \dfrac{2}{2} = \dfrac{2}{12} \\ +\ \dfrac{1}{12} \times \dfrac{1}{1} = \dfrac{1}{12} \\ \hline \dfrac{7}{12} \end{array}$$

10.
$$\begin{array}{r} 35\dfrac{1}{4} \times \dfrac{1}{1} = \overset{34}{\cancel{35}}\dfrac{\overset{5}{\cancel{1}}}{4} \\ -\ 12\dfrac{1}{2} \times \dfrac{2}{2} = 12\dfrac{2}{4} \\ \hline 22\dfrac{3}{4} \end{array}$$

11. $\dfrac{4}{5} \times \dfrac{1}{2} = \dfrac{4}{10} = \dfrac{2}{5}$

12. $\dfrac{4}{5} \div \dfrac{1}{2}$

$1 \div \dfrac{1}{2} = \dfrac{2}{1}$

$\dfrac{4}{5} \times \dfrac{2}{1} = \dfrac{8}{5} = 1\dfrac{3}{5}$

13.
$$\begin{array}{r} 0.05 \\ 5\overline{)0.25} \\ \underline{25} \\ 00 \end{array}$$

14.
$$\begin{array}{r} 20 \\ 25\overline{)500} \\ \underline{50} \\ 00 \\ \underline{00} \\ 0 \end{array}$$

15.
$$\begin{array}{r} 0.05 \\ \times\ \ 20 \\ \hline 1.00 \end{array}$$ or **1**

SOLUTIONS

16. $\frac{1}{2} + \frac{1}{2} = \frac{2}{2} = 1$

A. $\frac{1}{2} - \frac{1}{2} = \frac{0}{2} = 0$

B. $\frac{1}{2} \times \frac{1}{2} = \frac{1}{4}$

C. $\frac{1}{2} \div \frac{1}{2}$

$1 \div \frac{1}{2} = \frac{2}{1}$

$\frac{1}{2} \times \frac{2}{1} = \frac{2}{2} = 1$

C. $\frac{1}{2} \div \frac{1}{2}$

17. One possibility:

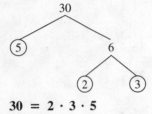

$30 = 2 \cdot 3 \cdot 5$

18.
$$\overset{1}{75}¢$$
$$\times\ 2$$
$$\overline{150¢}\ \text{or}\ \mathbf{\$1.50}$$

19. $0.075 \times \$10 = \mathbf{\$0.75}$

20.
$$0.8\ \text{meters}$$
$$\times\ 5$$
$$\overline{4.0}\ \text{or}\ \mathbf{4\ meters}$$

21. $\frac{20}{60} = \mathbf{\frac{1}{3}}$

22. $12°C - (-8°C)$
$$12°C$$
$$+\ 8°C$$
$$\overline{\mathbf{20°C}}$$

23.
$$46\ kg$$
$$-\ 41\ kg$$
$$\overline{\mathbf{5\ kg}}$$

24.
$$46\ kg \qquad \overset{\mathbf{43\ kg}}{3)\overline{129\ kg}}$$
$$42\ kg \qquad \underline{12}$$
$$\underline{+\ 41\ kg} \qquad 09$$
$$129\ kg \qquad \underline{09}$$
$$\qquad\qquad 0$$

25. Answers will vary. Sample answer: How much more than Marty does John weigh?
46 kg − 42 kg = 4 kg

26.
$$5)\overset{1}{\overline{5}}$$
$$5)\overline{25}$$
$$2)\overline{50}$$
$$2)\overline{100}$$
$$2)\overline{200}$$
$$2)\overline{400}$$

400 = 2 · 2 · 2 · 2 · 5 · 5

27. $\sqrt{144} = \mathbf{12\ tiles}$

28. 2.2 pounds $\times$ 100 = 220
about 220 pounds

29. $\frac{5 \cdot 5 \cdot 5 \cdot 7}{5 \cdot 5 \cdot 5 \cdot 2 \cdot 2 \cdot 2} = 1 \cdot 1 \cdot 1 \cdot \frac{7}{8} = \mathbf{\frac{7}{8}}$

30. B.

LESSON 67, WARM-UP

a. **1300**

b. **291**

c. **144**

d. **$8.50**

e. **$2.50**

f. **$0.30**

g. $2\frac{1}{2}$

Problem Solving

520 × 9 inches = 4680 inches
4680 inches ÷ 12 inches per foot = **390 feet**

146 *Saxon Math 7/6—Homeschool*

LESSON 67, LESSON PRACTICE

a. $\dfrac{\cancel{8}^1 \cdot \cancel{8}^1 \cdot \cancel{8}^1 \cdot 7}{2 \cdot 2 \cdot 2 \cdot \underset{1}{\cancel{8}} \cdot \underset{1}{\cancel{8}} \cdot \underset{1}{\cancel{8}}} = \dfrac{7}{8}$

b. $\dfrac{\cancel{2}^1 \cdot \cancel{2}^1 \cdot \cancel{2}^1 \cdot \cancel{2}^1 \cdot 3}{\underset{1}{\cancel{2}} \cdot \underset{1}{\cancel{2}} \cdot \underset{1}{\cancel{2}} \cdot \underset{1}{\cancel{2}} \cdot 5 \cdot 5} = \dfrac{3}{25}$

c. $\dfrac{\cancel{8}^1 \cdot \cancel{8}^1 \cdot \cancel{8}^1}{2 \cdot 2 \cdot \underset{1}{\cancel{8}} \cdot \underset{1}{\cancel{8}} \cdot \underset{1}{\cancel{8}}} = \dfrac{1}{4}$

d. $\dfrac{2 \cdot 2 \cdot \cancel{8}^1 \cdot \cancel{8}^1}{\underset{1}{\cancel{8}} \cdot \underset{1}{\cancel{8}} \cdot 3 \cdot 3} = \dfrac{4}{9}$

LESSON 67, MIXED PRACTICE

1.
$$\begin{aligned} 2 &= 2 \\ \tfrac{1}{2} \times \tfrac{2}{2} &= \tfrac{2}{4} \\ + \tfrac{3}{4} \times \tfrac{1}{1} &= \tfrac{3}{4} \\ \hline 2\tfrac{5}{4} &= 3\tfrac{1}{4} \text{ yards} \end{aligned}$$

$\dfrac{13}{4} \times \dfrac{2}{1} = \dfrac{26}{4} = 6\dfrac{2}{4} = 6\dfrac{1}{2}$

$6.50

2.
$$\begin{array}{r} \cancel{6}^{\,5}0\,8\,0\,\text{ft} \\ -\ 5\,2\,8\,0\,\text{ft} \\ \hline \textbf{About 8 0 0 feet} \end{array}$$

3. **$150 ÷ $5**

$$\begin{array}{r} \textbf{30} \text{ (not \$30 or 30¢)} \\ \$5\overline{)\$150} \\ \underline{15} \\ 00 \\ \underline{00} \\ 0 \end{array}$$

4.
$$\begin{array}{r} 11\ \text{cm} \quad k = \textbf{5 cm} \\ -\ \ 6\ \text{cm} \\ \hline 5\ \text{cm} \end{array}$$

5.
$$\begin{array}{r} 1.2 \\ 8\overline{)9.6} \qquad g = \textbf{1.2} \\ \underline{8} \\ 16 \\ \underline{16} \\ 0 \end{array}$$

6.
$$\begin{aligned} \tfrac{7}{10} \times \tfrac{1}{1} &= \tfrac{7}{10} \\ -\ \tfrac{1}{2} \times \tfrac{5}{5} &= \tfrac{5}{10} \\ \hline \tfrac{2}{10} &= \tfrac{1}{5} \end{aligned}$$

$w = \dfrac{\textbf{1}}{\textbf{5}}$

7. $\dfrac{3}{5} \times \dfrac{20}{20} = \dfrac{60}{100}$

$n = \textbf{60}$

8.
$$\begin{array}{r} \textbf{43 inches} \\ 4\overline{)172\ \text{inches}} \\ \underline{16} \\ 12 \\ \underline{12} \\ 0 \end{array}$$

9. $100.00 − ($46.75 + $9.68)
$100.00 − $56.43
$43.57

10. $(2 \times 0.3) - (0.2 \times 0.3)$
$\quad (0.6) \quad - \quad (0.06)$
0.54

11.
$$\begin{aligned} 4\tfrac{1}{4} \times \tfrac{2}{2} &= \overset{3}{\cancel{4}}\overset{10}{\cancel{2}}\tfrac{}{8} \\ -\ 2\tfrac{7}{8} \times \tfrac{1}{1} &= 2\tfrac{7}{8} \\ \hline & 1\tfrac{3}{8} \end{aligned}$$

12. $\dfrac{8}{3} \times \dfrac{3}{1} = \dfrac{24}{3} = \textbf{8}$

13.
$$\begin{aligned} 3\tfrac{1}{3} \times \tfrac{4}{4} &= 3\tfrac{4}{12} \\ +\ 2\tfrac{3}{4} \times \tfrac{3}{3} &= 2\tfrac{9}{12} \\ \hline & 5\tfrac{13}{12} = 6\tfrac{1}{12} \end{aligned}$$

14. $\dfrac{4}{3} \times \dfrac{9}{4} = \dfrac{36}{12} = \textbf{3}$

15.
$$
\begin{array}{r}
0.024 \\
60{\overline{\smash{\big)}\,1.440}} \\
\underline{1\ 20} \\
240 \\
\underline{240} \\
0
\end{array}
$$

16.
$$
\begin{array}{r}
40 \\
15{\overline{\smash{\big)}\,600}} \\
\underline{60} \\
00 \\
\underline{00} \\
0
\end{array}
$$

17.
$$
\begin{array}{r}
\$1.25 \\
4{\overline{\smash{\big)}\,\$5.00}} \\
\underline{4} \\
1\ 0 \\
\underline{8} \\
20 \\
\underline{20} \\
0
\end{array}
$$

18. $\sqrt{100 \text{ in.}^2} = 10 \text{ in.}$
$10 \text{ in.} \times 4 = \mathbf{40 \text{ in.}}$

19. $\dfrac{\cancel{3} \cdot \cancel{3} \cdot \cancel{3} \cdot 5}{2 \cdot 2 \cdot 2 \cdot \cancel{3} \cdot \cancel{3} \cdot \cancel{3}} = \dfrac{\mathbf{5}}{\mathbf{8}}$

20. $\dfrac{3}{2} \text{ in.} \times \dfrac{3}{4} \text{ in.} = \dfrac{9}{8} \text{ in.}^2 = \mathbf{1\dfrac{1}{8} \text{ in.}^2}$

21. $\dfrac{36}{88} = \dfrac{\mathbf{9}}{\mathbf{22}}$

22. One possibility:

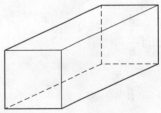

23. $\dfrac{3}{2} \times \dfrac{2}{3} = 1$

$\dfrac{\mathbf{2}}{\mathbf{3}}$

24. $2.5 \text{ km} \times 1000 = \mathbf{2500 \text{ meters}}$

25. *C*

26. $C = \pi d$
$C \approx (3.14)(12 \text{ in.})$
$C \approx 37.68$
$C \approx \mathbf{38 \text{ inches}}$

27. Sphere

28. $\dfrac{\mathbf{51}}{\mathbf{100}}$; **0.51**

29. The probability of rolling a 2, 3, or 5 is $\dfrac{3}{6}$, or $\dfrac{1}{2}$.

30. Trapezoid

LESSON 68, WARM-UP

a. **400**

b. **1775**

c. **294**

d. **$6.25**

e. **$12.00**

f. **$12.00**

g. **−1**

Problem Solving
Eliminate even numbers (92, 94, 96, 98).
Eliminate numbers divisible by 3 (93, 99).
Eliminate numbers divisible by 5 (95).
Eliminate numbers divisible by 7 (91).
97

LESSON 68, LESSON PRACTICE

a. $\dfrac{8}{5} \div \dfrac{4}{1}$

$1 \div \dfrac{4}{1} = \dfrac{1}{4}$

$\dfrac{8}{5} \times \dfrac{1}{4} = \dfrac{8}{20} = \dfrac{\mathbf{2}}{\mathbf{5}}$

b. $\dfrac{1}{4} \times \dfrac{8}{5} = \dfrac{8}{20} = \dfrac{\mathbf{2}}{\mathbf{5}}$

c. $\dfrac{12}{5} \div \dfrac{3}{1}$

$1 \div \dfrac{3}{1} = \dfrac{1}{3}$

$\dfrac{12}{5} \times \dfrac{1}{3} = \dfrac{12}{15} = \dfrac{4}{5}$

d. $\dfrac{1}{3} \times \dfrac{12}{5} = \dfrac{12}{15} = \dfrac{4}{5}$

e. $\dfrac{5}{3} \div \dfrac{5}{2}$

$1 \div \dfrac{5}{2} = \dfrac{2}{5}$

$\dfrac{5}{3} \times \dfrac{2}{5} = \dfrac{10}{15} = \dfrac{2}{3}$

f. $\dfrac{5}{2} \div \dfrac{5}{3}$

$1 \div \dfrac{5}{3} = \dfrac{3}{5}$

$\dfrac{5}{2} \times \dfrac{3}{5} = \dfrac{15}{10} = 1\dfrac{5}{10} = 1\dfrac{1}{2}$

g. $\dfrac{3}{2} \div \dfrac{3}{2}$

$1 \div \dfrac{3}{2} = \dfrac{2}{3}$

$\dfrac{3}{2} \times \dfrac{2}{3} = \dfrac{6}{6} = 1$

h. $\dfrac{7}{1} \div \dfrac{7}{4}$

$1 \div \dfrac{7}{4} = \dfrac{4}{7}$

$\dfrac{7}{1} \times \dfrac{4}{7} = \dfrac{28}{7} = 4$

LESSON 68, MIXED PRACTICE

1.

$\dfrac{1}{2} \times \dfrac{2}{2} = \dfrac{2}{4} \qquad \dfrac{1}{2} \times \dfrac{1}{4} = \dfrac{1}{8}$

$\begin{array}{r} \dfrac{2}{4} \\ + \dfrac{1}{4} = \dfrac{1}{4} \\ \hline \dfrac{3}{4} \end{array}$

$\dfrac{3}{4} \times \dfrac{2}{2} = \dfrac{6}{8}$

$\begin{array}{r} \dfrac{6}{8} \\ - \dfrac{1}{8} = \dfrac{1}{8} \\ \hline \dfrac{5}{8} \end{array}$

2. $60 \text{ seconds} \times 2 = 120 \text{ seconds}$

$\begin{array}{r} 120 \text{ seconds} \\ + \quad 55 \text{ seconds} \\ \hline \mathbf{175 \text{ seconds}} \end{array}$

3.

$\begin{array}{r} 12 \text{ in.} \\ \times \quad 4 \\ \hline 48 \text{ in.} \end{array} \qquad \begin{array}{r} \overset{1}{48} \text{ inches} \\ + \; 8.5 \text{ inches} \\ \hline \mathbf{56.5 \text{ inches}} \end{array}$

4. $\dfrac{3}{2} \div \dfrac{8}{3}$

$1 \div \dfrac{8}{3} = \dfrac{3}{8}$

$\dfrac{3}{2} \times \dfrac{3}{8} = \dfrac{\mathbf{9}}{\mathbf{16}}$

5. $\dfrac{4}{3} \div \dfrac{4}{1}$

$1 \div \dfrac{4}{1} = \dfrac{1}{4}$

$\dfrac{4}{3} \times \dfrac{1}{4} = \dfrac{4}{12} = \dfrac{\mathbf{1}}{\mathbf{3}}$

6.

$\begin{array}{r} \mathbf{18} \textbf{ points per game} \\ 6\overline{)108} \\ \underline{6} \\ 48 \\ \underline{48} \\ 0 \end{array}$

7. $\dfrac{\overset{1}{\cancel{2}} \cdot \overset{1}{\cancel{2}} \cdot \overset{1}{\cancel{2}} \cdot 3}{\underset{1}{\cancel{2}} \cdot \underset{1}{\cancel{2}} \cdot \underset{1}{\cancel{2}} \cdot 5 \cdot 5} = \dfrac{\mathbf{3}}{\mathbf{25}}$

8.

$\begin{array}{r} 5\dfrac{3}{8} \times \dfrac{2}{2} = 5\dfrac{6}{16} \\ + \; 1\dfrac{3}{16} \times \dfrac{1}{1} = 1\dfrac{3}{16} \\ \hline m = \mathbf{6\dfrac{9}{16}} \end{array}$

9.

$\begin{array}{r} 3\dfrac{3}{5} \times \dfrac{2}{2} = 3\dfrac{6}{10} \\ + \; 2\dfrac{7}{10} \times \dfrac{1}{1} = 2\dfrac{7}{10} \\ \hline 5\dfrac{13}{10} = 6\dfrac{3}{10} \end{array}$

$n = \mathbf{6\dfrac{3}{10}}$

SOLUTIONS

10.
$$\begin{array}{r} 0.015 \\ 25\overline{)0.375} \\ \underline{25} \\ 125 \\ \underline{125} \\ 0 \end{array}$$

$$d = \textbf{0.015}$$

11. $\dfrac{3}{4} \times \dfrac{25}{25} = \dfrac{75}{100}$

$w = \textbf{75}$

12.
$$5\dfrac{1}{8} \times \dfrac{1}{1} = 5\dfrac{1}{8}$$
$$-\ 1\dfrac{1}{2} \times \dfrac{4}{4} = 1\dfrac{4}{8}$$
$$\rule{3cm}{0.4pt}$$
$$3\dfrac{5}{8}$$

13. $\dfrac{10}{3} \times \dfrac{3}{2} = \dfrac{30}{6} = \textbf{5}$

14. $\dfrac{10}{3} \div \dfrac{3}{2}$

$1 \div \dfrac{3}{2} = \dfrac{2}{3}$

$\dfrac{10}{3} \times \dfrac{2}{3} = \dfrac{20}{9} = \textbf{2}\dfrac{\textbf{2}}{\textbf{9}}$

15. $\dfrac{4}{1}$ in. $\times \dfrac{7}{4}$ in. $= \dfrac{28}{4} = \textbf{7 square inches}$

16. $(3.2 + 1) - (0.6 \times 7)$
$$\quad 4.2 \quad - \quad 4.2$$
$$\textbf{0}$$

17.
$$\begin{array}{r} 31.25 \\ 4\overline{)125.00} \\ \underline{12} \\ 05 \\ \underline{4} \\ 1\,0 \\ \underline{8} \\ 20 \\ \underline{20} \\ 0 \end{array}$$

18. $3.2 \times 10 = 32$

(A) $32 \div 10 = 3.2$

(B) $320 \div 10 = 32$

(C) $0.32 \div 10 = .032$

B. 320 ÷ 10

19.
$$\begin{array}{r} 6000 \\ 6000 \\ +\ 5000 \\ \hline \textbf{17,000} \end{array}$$

20. $100 \div 3 = 33\dfrac{1}{3}$

21. (a)
$$\begin{array}{r} \textbf{0.6 meter} \\ 4\overline{)2.4 \text{ meters}} \\ \underline{24} \\ 0 \end{array}$$

(b)
$$\begin{array}{r} 0.6 \text{ m} \\ \times\ 0.6 \text{ m} \\ \hline \textbf{0.36 square meter} \end{array}$$

22.
$$\begin{array}{r} \$18,000 \\ \times\qquad .08 \\ \hline \$1440.00 \end{array}$$
$1440

23. $\dfrac{\textbf{999,999}}{\textbf{1,000,000}}$

24. **Polygons have straight sides. Since a circle is curved, it is not a polygon.**

25.
$$\dfrac{1}{3} \times 4\dfrac{1}{2} \ \ominus\ 4\dfrac{1}{2} \div 3$$
$$\dfrac{1}{3} \times \dfrac{9}{2} \qquad\qquad \dfrac{9}{2} \div 3$$
$$\dfrac{9}{6} = 1\dfrac{3}{6} = 1\dfrac{1}{2} \qquad 1 \div \dfrac{3}{1} = \dfrac{1}{3}$$
$$\dfrac{9}{2} \times \dfrac{1}{3} = \dfrac{9}{6}$$
$$= 1\dfrac{3}{6} = 1\dfrac{1}{2}$$

26. $1\dfrac{7}{8}$ **in.**

27. **∠WMX or ∠XMW**

28. $\dfrac{\textbf{3}}{\textbf{100}}$; **0.03**

29. **Rectangular prism**

30.

$$\begin{array}{r} \overset{11\,:\,60}{\cancel{12{:}00}}\ \text{p.m.} \\ -\ \ 6{:}20\ \text{a.m.} \\ \hline 5{:}40 \end{array}$$

$$\begin{array}{r} 5{:}40 \\ +\ 5{:}45 \\ \hline 10{:}85 \\ 11{:}25 \end{array}$$

11 hr 25 min

LESSON 69, WARM-UP

a. **900**

b. **480**

c. **264**

d. **$4.25**

e. **$3.50**

f. **$1.20**

g. **1**

Problem Solving

Convert 1 ft to 12 in.

Perimeter = 4 in. + 2 in. + 4 in. + 2 in.

Area = 4 in. × 2 in. = **8 sq. in.**

LESSON 69, LESSON PRACTICE

a. 26 mm + AB = 60 mm

$$\begin{array}{r} \overset{5}{\cancel{6}}{}^{1}0\ \text{mm} \\ -\ \ 2\ 6\ \text{mm} \\ \hline 3\ 4\ \textbf{mm} \end{array}$$

b.

$$\begin{array}{r} 90° \\ -\ 60° \\ \hline \mathbf{30°} \end{array}$$

c.

$$\begin{array}{r} 180° \\ -\ 60° \\ \hline \mathbf{120°} \end{array}$$

d. No. Supplementary angles total 180°, but acute angles have measures less than 90°. Two angles with measures less than 90° cannot total 180°.

e. ∠1 and ∠2

f. ∠2 and ∠3

LESSON 69, MIXED PRACTICE

1.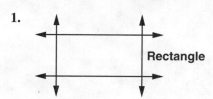

Rectangle

2. $\dfrac{1}{2} \div \dfrac{1}{8}$

$1 \div \dfrac{1}{8} = \dfrac{8}{1}$

$\dfrac{1}{2} \times \dfrac{8}{1} = \dfrac{8}{2} = \mathbf{4}$

3. 136°F − (−127°F)

$$\begin{array}{r} \overset{1}{1}36°F \\ +\ 127°F \\ \hline \mathbf{263°F} \end{array}$$

4. 6 × 1000 = 6000

$$\begin{array}{r} 500 \\ 12\overline{)6000} \\ \underline{60} \\ 00 \\ \underline{00} \\ 00 \\ \underline{00} \\ 0 \end{array}$$

About 500 feet

5. $\dfrac{\overset{1}{\cancel{3}} \cdot \overset{1}{\cancel{3}} \cdot 5}{2 \cdot 2 \cdot 2 \cdot \underset{1}{\cancel{3}} \cdot \underset{1}{\cancel{3}}} = \dfrac{5}{8}$

6. $\overline{QT}$ or $\overline{TQ}$

7. $27.50 ÷ 10 = **$2.75 per day**

8. $\dfrac{120}{15} = \mathbf{8}$

9.
$$3\frac{1}{2} \times \frac{4}{4} = 3\frac{4}{8}$$
$$2\frac{3}{4} \times \frac{2}{2} = 2\frac{6}{8}$$
$$+ \; 1\frac{5}{8} \times \frac{1}{1} = 1\frac{5}{8}$$
$$6\frac{15}{8} = \mathbf{7\frac{7}{8}}$$

10.
$$5\frac{3}{8} \times \frac{1}{1} = \cancel{5}\frac{\cancel{}^{11}}{8}$$
$$- \; 1\frac{3}{4} \times \frac{2}{2} = 1\frac{6}{8}$$
$$3\frac{5}{8}$$

$$m = \mathbf{3\frac{5}{8}}$$

11.
$$\frac{3}{4} \times \frac{3}{3} = \frac{9}{12}$$
$$- \; \frac{1}{3} \times \frac{4}{4} = \frac{4}{12}$$
$$\frac{5}{12}$$

$$f = \mathbf{\frac{5}{12}}$$

12. $w = \mathbf{\dfrac{5}{2}}$

13.
$$\frac{8}{25} \times \frac{4}{4} = \frac{32}{100}$$
$$n = \mathbf{32}$$

14.
$$\frac{5}{3} \div \frac{2}{1}$$
$$1 \div \frac{2}{1} = \frac{1}{2}$$
$$\frac{5}{3} \times \frac{1}{2} = \mathbf{\frac{5}{6}}$$

15.
$$\frac{8}{3} \times \frac{6}{5} = \frac{48}{15} = 3\frac{3}{15} = \mathbf{3\frac{1}{5}}$$

16.
$$\begin{array}{r} 30 \\ 8\overline{)240} \\ \underline{24} \\ 00 \\ \underline{00} \\ 0 \end{array}$$

17. (a)
$$\begin{array}{r} 2.5 \text{ m} \\ \times \quad 4 \\ \hline 10.0 \text{ m} = \mathbf{10 \text{ m}} \end{array}$$

(b)
$$\begin{array}{r} 2.5 \text{ m} \\ \times \; 2.5 \text{ m} \\ \hline 125 \\ 500 \\ \hline \mathbf{6.25} \text{ sq. m} \end{array}$$

18. If a counting number is divisible by a counting number other than itself or 1, then the number is composite.

19. One possibility:

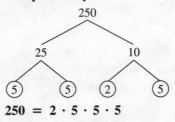

$$\mathbf{250 = 2 \cdot 5 \cdot 5 \cdot 5}$$

20. **Octagon**

21. (a) $\dfrac{12}{27} = \mathbf{\dfrac{4}{9}}$

(b) $\dfrac{15}{12} = \mathbf{\dfrac{5}{4}}$

22. $\mathbf{9 \div 3 = 3}$

23. $\mathbf{\dfrac{2}{5}}$

24. 2.25 kg $\times$ 1000 = **2250 grams**

25. **35 mm**

26.
$$\begin{array}{r} 53 \text{ mm} \\ + \; 35 \text{ mm} \\ \hline \mathbf{88 \text{ mm}} \end{array}$$

27.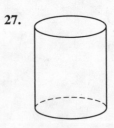

28. **−1, 0, 0.1, 1**

29.

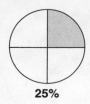

25%

30. 8 cubes

LESSON 70, WARM-UP

a. 1400

b. 575

c. 162

d. $6.25

e. $30.00

f. $25.00

g. 0

Problem Solving
Possible answers:
10 = 7 + 3;
12 = 7 + 5;
14 = 7 + 7;
16 = 5 + 11;
18 = 7 + 11;
20 = 13 + 7
Each of the sums can be further broken down.
Any combination of prime numbers is acceptable. For example, 10 = 5 + 5 or 10 = 2 + 2 + 3 + 3.

LESSON 70, LESSON PRACTICE

a. $\dfrac{3}{\cancel{4}_{1}} \cdot \dfrac{\cancel{4}^{1}}{5} = \dfrac{3}{5}$

b. $\dfrac{\cancel{2}^{1}}{\cancel{3}_{1}} \cdot \dfrac{\cancel{3}^{1}}{\cancel{4}_{2}} = \dfrac{1}{2}$

c. $\dfrac{\cancel{8}^{4}}{\cancel{9}_{1}} \cdot \dfrac{\cancel{9}^{1}}{\cancel{10}_{5}} = \dfrac{4}{5}$

d. $\dfrac{9}{\cancel{4}_{1}} \times \dfrac{\cancel{4}^{1}}{1} = \dfrac{9}{1} = 9$

e. $\dfrac{\cancel{3}^{1}}{\cancel{2}_{1}} \times \dfrac{\cancel{8}^{4}}{\cancel{3}_{1}} = \dfrac{4}{1} = 4$

f. $\dfrac{\cancel{10}^{5}}{\cancel{3}_{1}} \times \dfrac{\cancel{9}^{3}}{\cancel{4}_{2}} = \dfrac{15}{2} = 7\dfrac{1}{2}$

g. $\dfrac{2}{5} \div \dfrac{2}{3}$

$1 \div \dfrac{2}{3} = \dfrac{3}{2}$

$\dfrac{\cancel{2}^{1}}{5} \times \dfrac{3}{\cancel{2}_{1}} = \dfrac{3}{5}$

h. $\dfrac{8}{9} \div \dfrac{2}{3}$

$1 \div \dfrac{2}{3} = \dfrac{3}{2}$

$\dfrac{\cancel{8}^{4}}{\cancel{9}_{3}} \times \dfrac{\cancel{3}^{1}}{\cancel{2}_{1}} = \dfrac{4}{3} = 1\dfrac{1}{3}$

i. $\dfrac{9}{10} \div \dfrac{6}{5}$

$1 \div \dfrac{6}{5} = \dfrac{5}{6}$

$\dfrac{\cancel{9}^{3}}{\cancel{10}_{2}} \times \dfrac{\cancel{5}^{1}}{\cancel{6}_{2}} = \dfrac{3}{4}$

LESSON 70, MIXED PRACTICE

1. $7,200,000

2. $\dfrac{1}{2} \div \dfrac{1}{8}$

$1 \div \dfrac{1}{8} = \dfrac{8}{1}$

$\dfrac{1}{\cancel{2}_{1}} \times \dfrac{\cancel{8}^{4}}{1} = \dfrac{4}{1} = $ **4 eighth notes**

SOLUTIONS

3. $25 \div 5 = 5$

4. $\dfrac{\overset{1}{\cancel{8}}}{\underset{3}{\cancel{6}}} \cdot \dfrac{\overset{2}{\cancel{4}}}{\underset{1}{\cancel{8}}} = \dfrac{2}{3}$

5. $\dfrac{5}{6} \div \dfrac{5}{2}$

$1 \div \dfrac{5}{2} = \dfrac{2}{5}$

$\dfrac{\overset{1}{\cancel{5}}}{\underset{3}{\cancel{6}}} \times \dfrac{\overset{1}{\cancel{2}}}{\underset{1}{\cancel{5}}} = \dfrac{1}{3}$

6. $\dfrac{\overset{3}{\cancel{9}}}{\underset{2}{10}} \cdot \dfrac{\overset{1}{\cancel{5}}}{\underset{2}{\cancel{6}}} = \dfrac{3}{4}$

7.

$\dfrac{3}{4}$

8. $10 + 100 = \mathbf{110}$

9. $3\dfrac{2}{3} \times \dfrac{2}{2} = 3\dfrac{4}{6}$

$+\ 4\dfrac{5}{6} \times \dfrac{1}{1} = 4\dfrac{5}{6}$

$7\dfrac{9}{6} = 8\dfrac{3}{6} = \mathbf{8\dfrac{1}{2}}$

10. $7\dfrac{1}{8} \times \dfrac{1}{1} = 7\overset{9}{\cancel{\dfrac{1}{8}}}{}^{6}$

$-\ 2\dfrac{1}{2} \times \dfrac{4}{4} = 2\dfrac{4}{8}$

$\mathbf{4\dfrac{5}{8}}$

11.
$$\begin{array}{r} \overset{1}{}\overset{1}{4}.37 \\ 12.8 \\ +\ \ 6 \\ \hline \mathbf{23.17} \end{array}$$

12.
$$\begin{array}{r} \mathbf{0.092} \\ 5\overline{)0.460} \\ \underline{45} \\ 10 \\ \underline{10} \\ 0 \end{array}$$

13.
$$\begin{array}{r} 75 \\ 8\overline{)600} \\ \underline{56} \\ 40 \\ \underline{40} \\ 0 \end{array}$$

14.
$$\begin{array}{r} \overset{1}{4}.5 \\ 5 \\ +\ 5.8 \\ \hline 15.3 \end{array} \qquad \begin{array}{r} \mathbf{5.1} \\ 3\overline{)15.3} \\ \underline{15} \\ 0\ 3 \\ \underline{0\ 3} \\ 0 \end{array}$$

15. B. $150 \div 6$

16. $3.8\,L \times 1000 = \mathbf{3800\ milliliters}$

17. $\dfrac{3}{3} - \dfrac{2}{3} = \dfrac{1}{3}$

$n = \mathbf{\dfrac{1}{3}}$

18. $m = \mathbf{\dfrac{3}{2}}$

19. $\dfrac{5}{6} \times \dfrac{2}{2} = \dfrac{10}{12}$

$+\ \dfrac{3}{4} \times \dfrac{3}{3} = \dfrac{9}{12}$

$\dfrac{19}{12} = 1\dfrac{7}{12}$

$f = \mathbf{1\dfrac{7}{12}}$

20. (a) **4 faces**

(b) **6 edges**

(c) **4 vertices**

21. $\dfrac{\overset{1}{\cancel{8}}}{\underset{1}{\cancel{3}}} \times \dfrac{\overset{2}{\cancel{6}}}{\underset{1}{\cancel{8}}} = \dfrac{2}{1} = \mathbf{2}$

22. $\dfrac{8}{9} \div \dfrac{8}{3}$

$1 \div \dfrac{8}{3} = \dfrac{3}{8}$

$\dfrac{\overset{1}{\cancel{8}}}{\underset{3}{\cancel{9}}} \times \dfrac{\overset{1}{\cancel{3}}}{\underset{1}{\cancel{8}}} = \dfrac{1}{3}$

23.
$$\begin{array}{r} \overset{6}{\cancel{7}}{}^{1}0 \\ -\ 6\ 3 \\ \hline \end{array}$$
7 beats per minute more

Saxon Math 7/6—Homeschool

24.
$$\begin{array}{r} 65 \\ \times\ \ 3 \\ \hline 195 \text{ times} \end{array}$$

25. Answers will vary. Sample answer: On Wednesday John took his pulse for 2 minutes before marking the graph. How many times did his heart beat in those 2 minutes?

$$\begin{array}{r} 66 \\ \times\ \ 2 \\ \hline 132 \text{ times} \end{array}$$

26. $\dfrac{\overset{1}{\cancel{2}} \cdot \overset{1}{\cancel{2}} \cdot 2 \cdot \overset{1}{\cancel{3}} \cdot 3}{\underset{1}{\cancel{2}} \cdot \underset{1}{\cancel{2}} \cdot \underset{1}{\cancel{3}} \cdot 5 \cdot 5} = \dfrac{6}{25}$

27.
$$\begin{array}{r} \overset{2}{2.5} \text{ cm} \\ 2.5 \text{ cm} \\ 1.5 \text{ cm} \\ +\ 1.5 \text{ cm} \\ \hline 8.0 \text{ cm or } \mathbf{8\ cm} \end{array}$$

28.
$$\begin{array}{r} 2.5 \text{ cm} \\ \times\ 1.5 \text{ cm} \\ \hline 125 \\ 250 \\ \hline \mathbf{3.75\ cm^2} \end{array}$$

29. $\overline{AD}$ (or $\overline{DA}$) and $\overline{BC}$ (or $\overline{CB}$)

30.
$$\begin{array}{r} \mathbf{1.875\ cm^2} \\ 2\overline{)3.750} \\ \underline{2} \\ 1\,7 \\ \underline{1\,6} \\ 15 \\ \underline{14} \\ 10 \\ \underline{10} \\ 0 \end{array}$$

INVESTIGATION 7

1. (3, 1)

2. point C

3. (−3, −1)

4. point G

5. (−3, 1)

6. point H

7. (a) **square**

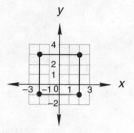

(b) 4 units + 4 units + 4 units + 4 units
= **16 units**

(c) 4 units × 4 units = **16 sq. units**

8. (a) **origin**

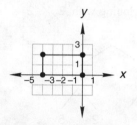

(b) 4 units + 2 units + 4 units + 2 units
= **12 units**

(c) 4 units × 2 units = **8 sq. units**

9. (a) (3, −3)

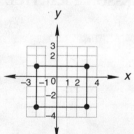

(b) 5 units + 4 units + 5 units + 4 units
= **18 units**

(c) 5 units × 4 units = **20 sq. units**

10. See student work. The figure should look like a lamp.

11. See student work. The figure should look like a space shuttle.

12. See student work.

LESSON 71, WARM-UP

a. 2400

b. 268

c. 344

d. $1.25

e. $4.50

f. $2.50

g. 1

Problem Solving

Work backward, beginning with $2 \times 2 = 4$.

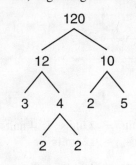

LESSON 71, LESSON PRACTICE

a. $\angle S$

b. $\angle R$

c. $\angle S$ and $\angle Q$

d. $180° - 100° = \textbf{80°}$

e. $P = 12\,\text{m} + 12\,\text{m} + 8\,\text{m} + 8\,\text{m} = \textbf{40 m}$
$A = 10\,\text{m} \times 8\,\text{m} = \textbf{80 m}^2$

f. $P = 6\,\text{in.} + 6\,\text{in.} + 8\,\text{in.} + 8\,\text{in.} = \textbf{28 in.}$
$A = 5\,\text{in.} \times 8\,\text{in.} = \textbf{40 sq. in.}$

g. $12\,\text{cm} \times 6\,\text{cm} = 72\,\text{cm}^2$
$\quad 9\,\text{cm} \times h = 72\,\text{cm}^2$
$\qquad\qquad h = 72\,\text{cm}^2 \div 9\,\text{cm}$
$\qquad\qquad h = \textbf{8 cm}$

LESSON 71, MIXED PRACTICE

1. Multiples of 6
6, 12, 18, 24, ㉚, 36
Multiples of 10
10, 20, ㉚, 40, 50
LCM is **30.**

2.
$$\begin{array}{r} \overset{1}{2}9,0\overset{1}{3}5 \text{ ft} \\ +\ \ 1,348 \text{ ft} \\ \hline \textbf{30,383 feet} \end{array}$$

3.
$$\begin{array}{r} \overset{1}{1}{:}15 \text{ p.m.} \\ +\ 1{:}45 \\ \hline 2{:}60 \text{ p.m.} \\ \textbf{3:00 p.m.} \end{array}$$

4. $\dfrac{\overset{1}{\cancel{2}}}{\cancel{3}} \cdot \dfrac{\overset{1}{\cancel{3}}}{\cancel{8}} = \dfrac{1}{4}$
$\quad {}_1 \qquad {}_4$

5. $\dfrac{5}{\cancel{4}} \cdot \dfrac{\overset{2}{\cancel{8}}}{3} = \dfrac{10}{3} = \mathbf{3\dfrac{1}{3}}$
$\quad {}_1$

6. $\dfrac{3}{4} \div \dfrac{3}{8}$

$1 \div \dfrac{3}{8} = \dfrac{8}{3}$

$\dfrac{\overset{1}{\cancel{3}}}{\cancel{4}} \cdot \dfrac{\overset{2}{\cancel{8}}}{\cancel{3}} = \dfrac{2}{1} = \mathbf{2}$
$\ {}_1 \qquad {}_1$

7. $\dfrac{9}{2} \div \dfrac{6}{1}$

$1 \div \dfrac{6}{1} = \dfrac{1}{6}$

$\dfrac{\overset{3}{\cancel{9}}}{2} \times \dfrac{1}{\cancel{6}} = \dfrac{3}{4}$
$\qquad\quad {}_2$

8.
$$6 = 6$$
$$3\dfrac{3}{4} \qquad\quad = 3\dfrac{3}{4}$$
$$+\ 2\dfrac{1}{2} \times \dfrac{2}{2} = 2\dfrac{2}{4}$$
$$\overline{\qquad\qquad 11\dfrac{5}{4} = \mathbf{12\dfrac{1}{4}}}$$

9.

$$5 \xrightarrow{\quad 4 + \frac{8}{8} \quad} 4\frac{8}{8}$$
$$-\ 3\frac{1}{8} \qquad\qquad -\ 3\frac{1}{8}$$
$$\qquad\qquad\qquad \mathbf{1\frac{7}{8}}$$

10.

$$5\frac{1}{4} \times \frac{2}{2} = \overset{4}{\cancel{5}}\overset{10}{\frac{\cancel{2}}{8}}$$
$$-\ 1\frac{7}{8} \times \frac{1}{1} = 1\frac{7}{8}$$
$$\qquad\qquad\qquad \mathbf{3\frac{3}{8}}$$

11.
$$\begin{array}{r} 3.5 \\ \times\ 3.5 \\ \hline 175 \\ 1050 \\ \hline \mathbf{12.25} \end{array}$$

12.
$$\begin{array}{r} \mathbf{\$5.00} \\ 15\overline{)\$75.00} \\ 75 \\ \hline 0\ 0 \\ 0\ 0 \\ \hline 00 \\ 00 \\ \hline 0 \end{array}$$

13. $(1 + 0.6) \div (1 - 0.6)$
$$1.6 \div 0.4$$
$$\mathbf{4}$$

14. (a)
$$\begin{array}{r} \$4.50 \\ \times\ 0.075 \\ \hline 2250 \\ 31500 \\ \hline 0.33750 \end{array} \longrightarrow \mathbf{\$0.34}$$

(b)
$$\begin{array}{r} \$4.50 \\ +\ \$0.34 \\ \hline \mathbf{\$4.84} \end{array}$$

(c)
$$\begin{array}{r} \$\overset{1}{2}\overset{9}{\cancel{0}}.\ \overset{9}{\cancel{0}}{}^1 0 \\ -\ \$4.\ 8\ 4 \\ \hline \mathbf{\$1\ 5.\ 1\ 6} \end{array}$$

15. Origin

16. (a) **Point D**

(b) **Point G**

17. (a) $\mathbf{(3, -3)}$

(b) $\mathbf{(-3, 0)}$

18.
$$\begin{array}{r} \mathbf{100} \\ 12\overline{)1200} \\ 12 \\ \hline 00 \\ 00 \\ \hline 00 \\ 00 \\ \hline 0 \end{array}$$

19.
$$\begin{array}{r} \mathbf{100} \\ 12\overline{)1200} \\ 12 \\ \hline 00 \\ 00 \\ \hline 00 \\ 00 \\ \hline 0 \end{array}$$

20. $\dfrac{\overset{1}{\cancel{2}} \cdot \overset{1}{\cancel{2}} \cdot \overset{1}{\cancel{2}} \cdot \overset{1}{\cancel{2}} \cdot \overset{1}{\cancel{2}} \cdot 2}{\underset{1}{\cancel{2}} \cdot \underset{1}{\cancel{2}} \cdot \underset{1}{\cancel{2}} \cdot \underset{1}{\cancel{2}} \cdot \underset{1}{\cancel{2}} \cdot 7} = \dfrac{2}{7}$

21.
$$\begin{array}{r} 1.6\text{ m} \\ 4\overline{)6.4} \\ 4 \\ \hline 2\ 4 \\ 2\ 4 \\ \hline 0 \end{array} \qquad \begin{array}{r} 1.6\text{ m} \\ \times\ 1.6\text{ m} \\ \hline 96 \\ 160 \\ \hline \mathbf{2.56\text{ square meters}} \end{array}$$

22. $\dfrac{6}{8} = \dfrac{3}{4}$

23. $C = \pi d$
$C \approx (3.14)(2\text{ cm})$
$C \approx \mathbf{6.28\text{ cm}}$

24. **Answers may vary but should be close to 5 cm.**

25. **4°F**

26. $\dfrac{20}{100} = \dfrac{1}{5}$

$\dfrac{1}{\underset{1}{\cancel{5}}} \cdot \dfrac{\overset{12}{\cancel{60}}}{1}\text{ min} = \dfrac{12}{1} = \mathbf{12\text{ minutes}}$

27. **Cone**

28.
$$\begin{array}{r} 6\text{ cm} \\ 4\overline{)24} \\ 24 \\ \hline 0 \end{array} \qquad 6\text{ cm} \times 3 = \mathbf{18\text{ cm}}$$

29. (a) $8\,cm + 8\,cm + 10\,cm + 10\,cm$
    $= \textbf{36 cm}$

    (b) $10\,cm \times 7\,cm = \textbf{70 cm}^2$

30. (a) $\angle AMB$ (or $\angle BMA$) and $\angle BMC$ (or $\angle CMB$); or $\angle CMD$ (or $\angle DMC$) and $\angle DMA$ (or $\angle AMD$)

    (b) $\angle BMC$ (or $\angle CMB$) and $\angle CMD$ (or $\angle DMC$)

LESSON 72, WARM-UP

a. 375

b. 325

c. $2.97

d. $10.01

e. $2.20

f. $25.00

g. 15

Problem Solving

$24 \times 2 = 48$

$48 - \left(\dfrac{1}{2} \times 24\right) = \textbf{36}$

LESSON 72, LESSON PRACTICE

a.

Fractions Chart

	+ −	× ÷
1. Shape	Write fractions with common denominators.	Write numbers in fraction form.
2. Operate	Add or subtract the numerators.	× cancel. $\dfrac{n \times n}{d \times d}$ ÷ Find reciprocal of divisor; then ...
3. Simplify	Reduce fractions. Convert improper fractions.	

b. **Step 1:** Write the fractions so the denominators are the same.
   Step 2: Add the numerators but not the denominators.
   Step 3: Simplify the answer if possible.

c. **Step 1:** Write any mixed numbers in fraction form.
   Step 2: Rewrite the division problem as a multiplication problem by changing the divisor to its reciprocal. Cancel terms; then multiply the fractions.
   Step 3: Simplify the answer if possible.

d. $\dfrac{\cancel{2}^{1}}{\cancel{3}} \cdot \dfrac{\cancel{4}^{1}}{5} \cdot \dfrac{\cancel{3}^{1}}{\cancel{8}_{2}} = \dfrac{1}{5}$ (with 1 under 3)

e. $\dfrac{\cancel{5}^{1}}{2} \cdot \dfrac{11}{\cancel{10}_{2}} \cdot \dfrac{\cancel{4}^{2}}{1} = \dfrac{11}{1} = \textbf{11}$

LESSON 72, MIXED PRACTICE

1.
   $\begin{array}{r} \overset{1}{4.2} \\ 2.61 \\ + \; 3.6 \\ \hline 10.41 \end{array}$
   $\quad$
   $\begin{array}{r} 3.47 \\ 3\overline{)10.41} \\ \underline{9} \\ 1\,4 \\ \underline{1\,2} \\ 21 \\ \underline{21} \\ 0 \end{array}$

2. $4 \times 4\,T = \textbf{16 tablespoons}$

3. $134°C - (-170°C)$
   $\begin{array}{r} 134°C \\ + \; 170°C \\ \hline \textbf{304°C} \end{array}$

4. (a) $\dfrac{4}{12} = \dfrac{1}{3}$

   (b) $\dfrac{8}{12} = \dfrac{2}{3}$

5. $C = \pi d$
   $C \approx (3.14)(100\,cm)$
   $C \approx \textbf{314 centimeters}$

6. $\dfrac{5}{100} = \dfrac{1}{20}$

7. $\dfrac{1}{2} \cdot \dfrac{\cancel{3}^{1}}{\cancel{6}_{2}} \cdot \dfrac{\cancel{3}^{1}}{\cancel{3}_{1}} = \dfrac{1}{4}$

8. $\dfrac{3}{1} \cdot \dfrac{\cancel{3}^{1}}{\cancel{2}_{1}} \cdot \dfrac{\cancel{8}^{4}}{\cancel{3}_{1}} = \dfrac{12}{1} = \textbf{12}$

9. $\dfrac{3}{4} \div \dfrac{2}{1}$

$1 \div \dfrac{2}{1} = \dfrac{1}{2}$

$\dfrac{3}{4} \cdot \dfrac{1}{2} = \mathbf{\dfrac{3}{8}}$

10. $\dfrac{3}{2} \div \dfrac{5}{3}$

$1 \div \dfrac{5}{3} = \dfrac{3}{5}$

$\dfrac{3}{2} \cdot \dfrac{3}{5} = \mathbf{\dfrac{9}{10}}$

11.
$$
\begin{array}{r}
0.12 \\
\times\ 0.24 \\
\hline
48 \\
240 \\
\hline
\mathbf{0.0288}
\end{array}
$$

12.
$$
\begin{array}{r}
\mathbf{2.4} \\
25\overline{)60.0} \\
50 \\
\hline
10\ 0 \\
10\ 0 \\
\hline
0
\end{array}
$$

13.
$$
\begin{aligned}
&\ \dfrac{3}{5} \times \dfrac{2}{2} = \dfrac{6}{10} \\
&+\ \dfrac{1}{2} \times \dfrac{5}{5} = \dfrac{5}{10} \\
&\hline
&\phantom{+\ \dfrac{1}{2}\ }\ \ \dfrac{11}{10} = 1\dfrac{1}{10}
\end{aligned}
$$

$n = \mathbf{1\dfrac{1}{10}}$

14. $\dfrac{12}{12} - \dfrac{7}{12} = \dfrac{5}{12}$

$w = \mathbf{\dfrac{5}{12}}$

15.
$$
\begin{aligned}
&\ 3\dfrac{1}{3} \times \dfrac{2}{2} = \cancel{3}\,\overset{8}{\cancel{\dfrac{2}{6}}} \\
&-\ 2\dfrac{1}{2} \times \dfrac{3}{3} = 2\dfrac{3}{6} \\
&\hline
&\phantom{-\ 2\dfrac{1}{2}\ \ }\dfrac{5}{6}
\end{aligned}
$$

$w = \mathbf{\dfrac{5}{6}}$

16.
$$
\begin{array}{r}
\overset{0}{\cancel{1}}.\,\overset{9}{\cancel{0}}{}^{1}0 \\
-\ 0.\,2\ 3 \\
\hline
0.\,7\ 7
\end{array}
$$

$w = \mathbf{0.77}$

17. 60.43

18. B. 0.1

19. 97

20. C.

21. (a) $\angle MCB$ or $\angle BCM$

(b) $\angle CMB$ (or $\angle BMC$) or $\angle DCM$ (or $\angle MCD$)

22. (a)
$$
\begin{array}{r}
\mathbf{6\ inches} \\
4\overline{)24} \\
24 \\
\hline
0
\end{array}
$$

(b) 6 in. $\times$ 6 in. = **36 square inches**

23.
$$
\begin{array}{r}
\overset{2\ \ 5}{\$3.49} \\
\$3.49 \\
\$3.29 \\
\$0.89 \\
\$1.09 \\
+\ \$1.09 \\
\hline
\mathbf{\$13.34}
\end{array}
$$

24.
$$
\begin{array}{r}
\$13.34 \\
\times\ \ 0.07 \\
\hline
\$0.9338 \longrightarrow \$0.93
\end{array}
$$

$$
\begin{array}{r}
\$13.34 \\
+\ \ \$0.93 \\
\hline
\$14.27
\end{array}
$$

$$
\begin{array}{r}
\$\overset{1}{2}\,\overset{9}{\cancel{0}}.\,\overset{9}{\cancel{0}}{}^{1}0 \\
-\ \$1\ 4.\ 2\ 7 \\
\hline
\mathbf{\$5.\ 7\ 3}
\end{array}
$$

25. **Answers will vary.**
Sample answer:

1 Pasta Salad	$\overset{2\ \ 3}{\$2.89}$
1 Taco Salad	$3.29
1 Small drink	$0.89
1 Large drink	+ $1.29
	$8.36

26. $A = (2.5)(0.4)$

$A = \mathbf{1}$

27. $\dfrac{\cancel{2} \cdot \cancel{2} \cdot \cancel{2} \cdot \cancel{3} \cdot 3}{\cancel{2} \cdot \cancel{2} \cdot \cancel{2} \cdot \cancel{3} \cdot 5} = \mathbf{\dfrac{3}{5}}$

28. (a) $(-3, 4)$

(b) $(0, -3)$

29. (a) Point E

(b) Point I

30.

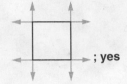

; yes

LESSON 73, WARM-UP

a. 448

b. 325

c. $3.96

d. $4.98

e. $7.00

f. $0.35

g. 21

Problem Solving

1. Shape: $\dfrac{9}{12} + \dfrac{8}{12}$

2. Operate: $\dfrac{17}{12}$

3. Simplify: $1\dfrac{5}{12}$

LESSON 73, LESSON PRACTICE

a. $10 \times 10 \times 10 \times 10 = \mathbf{10{,}000}$

b. $2 \cdot 2 \cdot 2 + 2 \cdot 2 \cdot 2 \cdot 2$
$8 + 16 = \mathbf{24}$

c. $2 \cdot 2 \cdot 5 \cdot 5 = \mathbf{100}$

d.

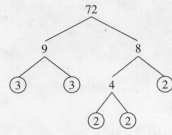

$72 = 2^3 \cdot 3^2$

e. $12\dfrac{5}{10} = 12\dfrac{1}{2}$

f. $1\dfrac{25}{100} = 1\dfrac{1}{4}$

g. $\dfrac{125}{1000} = \dfrac{1}{8}$

h. $\dfrac{5}{100} = \dfrac{1}{20}$

i. $\dfrac{24}{100} = \dfrac{6}{25}$

j. $10\dfrac{2}{10} = 10\dfrac{1}{5}$

LESSON 73, MIXED PRACTICE

1. $102°F - 98.6°F = d$

$$
\begin{array}{r}
\overset{0}{\cancel{1}}\,\overset{9}{0}\,\overset{1}{2}.0°F \\
-\quad 9\,8.\,6°F \\
\hline
3.\,4°F
\end{array}
$$

2. $180 - 42 = d$

$$
\begin{array}{r}
1\,\overset{7}{8}\,0 \\
-\quad 4\,2 \\
\hline
1\,3\,8 \text{ pages}
\end{array}
$$

3. $3p = 138$

$$
\begin{array}{r}
46 \text{ pages per day} \\
3\overline{)138} \\
\underline{12} \\
18 \\
\underline{18} \\
0
\end{array}
$$

4. $2\dfrac{5}{10} = 2\dfrac{1}{2}$

5. $\dfrac{35}{100} = \dfrac{7}{20}$

6.
$$\begin{array}{r} \$12.60 \\ \times\ \ .075 \\ \hline 6300 \\ 88200 \\ \hline \$0.94500 \end{array} \longrightarrow \$0.95$$

$$\begin{array}{r} \$12.60 \\ +\ \ \$0.95 \\ \hline \mathbf{\$13.55} \end{array}$$

7. $\dfrac{\cancel{3}^{1}}{\cancel{4}_{1}} \times \dfrac{2}{1} \times \dfrac{\cancel{4}^{1}}{\cancel{3}_{1}} = \dfrac{2}{1} = \mathbf{2}$

8. $(100 - 100) \div 25$
$0 \div 25$
0

9.
$$\begin{array}{r} 3 = 3 \\ 2\frac{1}{3} \times \frac{4}{4} = 2\frac{4}{12} \\ +\ 1\frac{3}{4} \times \frac{3}{3} = 1\frac{9}{12} \\ \hline 6\frac{13}{12} = \mathbf{7\frac{1}{12}} \end{array}$$

10.
$$\begin{array}{r} 5\frac{1}{6} \times \frac{1}{1} = \cancel{5}\,{}^{4}\cancel{1}^{7}\frac{7}{6} \\ -\ 3\frac{1}{2} \times \frac{3}{3} = 3\frac{3}{6} \\ \hline 1\frac{4}{6} = \mathbf{1\frac{2}{3}} \end{array}$$

11. $\dfrac{3}{4} \div \dfrac{3}{2}$
$1 \div \dfrac{3}{2} = \dfrac{2}{3}$
$\dfrac{\cancel{3}}{\cancel{4}_{2}} \times \dfrac{\cancel{2}^{1}}{\cancel{3}_{1}} = \mathbf{\dfrac{1}{2}}$

12.
$$\begin{array}{r} 17.5 \\ 4\overline{)70.0} \\ \underline{4} \\ 30 \\ \underline{28} \\ 2\,0 \\ \underline{2\,0} \\ 0 \end{array}$$

13. (a) $\begin{array}{c} 5^2 \\ 5 \cdot 5 \\ 25 \end{array}$ $\lessdot$ $\begin{array}{c} 2^5 \\ 2 \cdot 2 \cdot 2 \cdot 2 \cdot 2 \\ 32 \end{array}$

(b) $0.3 \ \gtrdot\ 0.125$

14. (a) $C = \pi d$
$C \approx (3.14)(2.4 \text{ cm})$
$C \approx \mathbf{7.536 \text{ cm}}$

(b) $\dfrac{1.2}{2.4} = \mathbf{\dfrac{1}{2}}$

15.
$$\begin{array}{r} 0.007 \\ 25\overline{)0.175} \\ \underline{175} \\ 0 \end{array}$$
$m = \mathbf{0.007}$

16. $5.45 + y = 7$
$$\begin{array}{r} {}^{6}7.{}^{9}\cancel{0}^{1}0 \\ -\ 5.45 \\ \hline 1.55 \end{array}$$
$y = \mathbf{1.55}$

17. **2**

18. $\mathbf{0,\ \dfrac{1}{10},\ \dfrac{1}{4},\ \dfrac{1}{2},\ 1}$

19.

$\mathbf{200 = 2^3 \cdot 5^2}$

20. (a)
$$\begin{array}{r} \$18.00 \\ \times\ \ 0.20 \\ \hline \$3.6000 \end{array}$$
\$3.60

(b)
$$\begin{array}{r} \$1\,{}^{7}\cancel{8}.{}^{1}00 \\ -\ \$3.60 \\ \hline \mathbf{\$14.40} \end{array}$$

21.
$$\begin{array}{r} {}^{4}\cancel{5}^{1}0 \text{ mm} \\ -\ 16 \text{ mm} \\ \hline \mathbf{34 \text{ mm}} \end{array}$$

22. 6 in. $\times$ 6 in. = 36 in.^{2}

$$\begin{array}{r} 18 \text{ in.}^2 \\ 2)\overline{36} \\ \underline{2} \\ 16 \\ \underline{16} \\ 0 \end{array}$$

23. Yes

24. $\dfrac{4 + 8}{2} = \dfrac{12}{2} = 6$

25. **Before we multiply fractions, we write any mixed numbers and any whole numbers as improper fractions.**

26. (a) **(−4, −3)**

(b) **(0, 3)**

27. (a) **Point J**

(b) **Point C**

28. $9 \times 9 = $ **81**

29.

30. (a) $180° - 75° = $ **105°**

(b) **75°**

Lesson 74, Warm-Up

a. 690

b. 700

c. $4.95

d. $3.02

e. $0.60

f. $12.50

g. 11

Problem Solving

The length of each side of the square is the square root of 25 sq. cm, which is 5 cm. So the perimeter of the pentagon is 5 cm $\times$ 5, which is **25 cm.**

Lesson 74, Lesson Practice

a.

$$\begin{array}{r} 0.75 \\ 4)\overline{3.00} \\ \underline{2\,8} \\ 20 \\ \underline{20} \\ 0 \end{array}$$

b.

$$\begin{array}{r} 0.2 \\ 5)\overline{1.0} \\ \underline{1\,0} \\ 0 \end{array}$$ **4.2**

c.

$$\begin{array}{r} 0.125 \\ 8)\overline{1.000} \\ \underline{8} \\ 20 \\ \underline{16} \\ 40 \\ \underline{40} \\ 0 \end{array}$$

d.

$$\begin{array}{r} 0.35 \\ 20)\overline{7.00} \\ \underline{6\,0} \\ 1\,00 \\ \underline{1\,00} \\ 0 \end{array}$$

e.

$$\begin{array}{r} 0.3 \\ 10)\overline{3.0} \\ \underline{3\,0} \\ 0 \end{array}$$ **3.3**

f.

$$\begin{array}{r} 0.28 \\ 25)\overline{7.00} \\ \underline{5\,0} \\ 2\,00 \\ \underline{2\,00} \\ 0 \end{array}$$

g. **0.6875**

h. **0.96875**

i. **3.375**

LESSON 74, MIXED PRACTICE

1. $4^3 - 5^2$
   $64 - 25$
   39

2. 10 miles $\times$ 3 = **30 miles**

3. 300×3 feet = 900 feet
   900 − 400 = d
   d = **500 feet**

4.
   $$\begin{array}{r} 0.75 \\ 4\overline{)3.00} \\ \underline{2\,8} \\ 20 \\ \underline{20} \\ 0 \end{array}$$
   2.75

5.
   $$\begin{array}{r} 0.8 \\ 5\overline{)4.0} \\ \underline{4\,0} \\ 0 \end{array}$$

6. $\dfrac{24}{100} = \dfrac{6}{25}$

7. $A = (12)(8)$
   $A = $ **96**

8. 3^2 $\bigcirc\!\!\!>$ $3 + 3$
   9 $\qquad$ 6

9. $\dfrac{1}{2} \times \dfrac{3}{3} = \dfrac{3}{6}$
   $\dfrac{2}{3} \times \dfrac{2}{2} = \dfrac{4}{6}$
   $+ \dfrac{1}{6} \times \dfrac{1}{1} = \dfrac{1}{6}$
   $\overline{\qquad\qquad\qquad}$
   $\dfrac{8}{6} = 1\dfrac{2}{6} = 1\dfrac{1}{3}$

10. $3\dfrac{1}{4} \times \dfrac{2}{2} = {}^{2}\cancel{3}\dfrac{{}^{10}\cancel{2}}{8}$
    $- 1\dfrac{7}{8} \times \dfrac{1}{1} = 1\dfrac{7}{8}$
    $\overline{\qquad\qquad\qquad}$
    $1\dfrac{3}{8}$

11. $\dfrac{\overset{1}{\cancel{5}}}{\underset{2}{\cancel{8}}} \cdot \dfrac{3}{\underset{1}{\cancel{3}}} \cdot \dfrac{\overset{1}{\cancel{4}}}{5} = \dfrac{3}{10}$

12. $\dfrac{10}{\underset{1}{\cancel{3}}} \cdot \dfrac{\overset{1}{\cancel{3}}}{1} = \dfrac{10}{1} = \mathbf{10}$

13. $\dfrac{3}{4} \div \dfrac{3}{2}$
    $1 \div \dfrac{3}{2} = \dfrac{2}{3}$
    $\dfrac{\overset{1}{\cancel{3}}}{\underset{2}{\cancel{4}}} \times \dfrac{\overset{1}{\cancel{2}}}{\underset{1}{\cancel{3}}} = \dfrac{1}{2}$

14. $(7.2) - 0.01 = $ **7.19**

15.
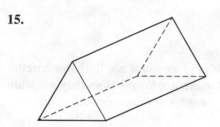

16. $12g = 10.44$
    $$\begin{array}{r} \mathbf{\$0.87} \\ 12\overline{)10.44} \\ \underline{9\,6} \\ 84 \\ \underline{84} \\ 0 \end{array}$$

17.
   $$\begin{array}{r} 80 \\ \times\ 40 \\ \hline \mathbf{3200} \end{array}$$

18.
	45 pages per day
42 pages	$4\overline{)180}$
46 pages	$\underline{16}$
35 pages	20
+ 57 pages	$\underline{20}$
180 pages	0

19. Multiples of 6
    6, 12, 18, ⓞ24, 30
    Multiples of 8
    8, 16, ⓞ24, 32
    Multiples of 12
    12, ⓞ24, 36, 48
    LCM is **24.**

20. $120 + c = 150$
    $$\begin{array}{r} 150 \\ -\ 120 \\ \hline \mathbf{30} \end{array}$$

21. $\dfrac{\overset{1}{\cancel{2}} \cdot \overset{1}{\cancel{2}} \cdot \overset{1}{\cancel{2}} \cdot 5}{\underset{1}{\cancel{2}} \cdot \underset{1}{\cancel{2}} \cdot \underset{1}{\cancel{2}} \cdot 2 \cdot 2 \cdot 3} = \dfrac{5}{12}$

22. (a) $\begin{array}{r} \textbf{10 centimeters} \\ 4\overline{)40 \text{ centimeters}} \\ \underline{4} \\ 00 \\ \underline{00} \\ 0 \end{array}$

  (b) $C = \pi d$
  $C \approx (3.14)(10 \text{ cm})$
  $C \approx \textbf{31.4 centimeters}$

23. $\dfrac{24}{36} = \dfrac{2}{3}$

24. All four sides of a square are the same length. Some rectangles are longer than they are wide, so not all the sides are the same length.

25. *B*

26. (a) $\overline{SR}$ or $\overline{RS}$

  (b) $\overline{PS}$ or $\overline{SP}$

27. (a) $3 \text{ ft} \times 3 \text{ ft} = \textbf{9 ft}^2$

  (b) $\begin{array}{r} 9 \text{ ft}^2 \\ \times\ 6 \\ \hline \textbf{54 ft}^2 \end{array}$

28. (a) $(-3, -2)$

  (b) $(0, 0)$

29. $\overline{DA}$ (or $\overline{AD}$) and $\overline{CB}$ (or $\overline{BC}$)

30. $\begin{array}{r} 300 \text{ acres} \\ \times\ 0.60 \\ \hline 180.00 \\ \textbf{180 acres} \end{array}$

LESSON 75, WARM-UP

a. 3024

b. 375

c. $5.97

d. $4.49

e. $3.20

f. $1.25

g. 1

Problem Solving

1. Shape: $\dfrac{10}{3} \div \dfrac{5}{2}$

2. Operate: $\dfrac{\overset{2}{\cancel{10}}}{3} \times \dfrac{2}{\underset{1}{\cancel{5}}} = \dfrac{4}{3}$

3. Simplify: $1\dfrac{1}{3}$

LESSON 75, LESSON PRACTICE

a. 31%

b. 1%

c. $\dfrac{1}{10} \times \dfrac{10}{10} = \dfrac{10}{100}$ **10%**

d. $\dfrac{3}{50} \times \dfrac{2}{2} = \dfrac{6}{100}$ **6%**

e. $\dfrac{7}{25} \times \dfrac{4}{4} = \dfrac{28}{100}$ **28%**

f. $\dfrac{2}{5} \times \dfrac{20}{20} = \dfrac{40}{100}$ **40%**

g. $\dfrac{12}{30} = \dfrac{2}{5}$

  $\dfrac{2}{5} \times \dfrac{20}{20} = \dfrac{40}{100}$ **40%**

h. $\dfrac{18}{20} = \dfrac{9}{10}$

  $\dfrac{9}{10} \times \dfrac{10}{10} = \dfrac{90}{100}$ **90%**

i. $\dfrac{25}{100} = \textbf{25\%}$

j. $0.30 = \dfrac{30}{100} = \textbf{30\%}$

k. $\dfrac{5}{100} = \textbf{5\%}$

l. $1.0 = \textbf{100\%}$

m. $0.70 = \dfrac{70}{100} = \textbf{70\%}$

n. $\dfrac{15}{100} = \textbf{15\%}$

LESSON 75, MIXED PRACTICE

1. $2\dfrac{3}{5} = \dfrac{13}{5}$

$\dfrac{\textbf{5}}{\textbf{13}}$

2. 2:30 p.m.
$+$ 1:35
 3:65 p.m.
 4:05 p.m.

3. $\begin{array}{r}\textbf{\$0.25 per ounce}\\ 16\overline{)\$4.00}\\ \underline{3\,2}\\ 80\\ \underline{80}\\ 0\end{array}$

4.
$\begin{array}{r}\$4.00\\ \$0.94\\ +\ \$6.35\\ \hline \$11.29\end{array}$ $\quad$ $\begin{array}{r}\$11.29\\ \times\quad\$0.08\\ \hline \$0.9032\end{array}$ $\quad$ $\begin{array}{r}\$11.29\\ +\quad\$0.90\\ \hline \textbf{\$12.19}\end{array}$

Round 94¢ to \$1 and round \$6.35 to \$6. Add \$4, \$1, and \$6 and get \$11. Tax is 8¢ per dollar, which comes to about 88¢ for the meal. So the estimated total is \$11.88. Since \$6.35 was rounded down by an amount more than 94¢ was rounded up by, the answer \$12.19 is reasonable.

5. $100\% - 50\% = \textbf{50\%}$

6.

7. (a) $\dfrac{3}{4} \times \dfrac{25}{25} = \dfrac{75}{100} = \textbf{0.75}$

(b) **75%**

8. (a) $\dfrac{3}{20} \times \dfrac{5}{5} = \dfrac{\textbf{15}}{\textbf{100}}$

(b) **15%**

9. $\dfrac{12}{100} = \dfrac{\textbf{3}}{\textbf{25}}$

$\dfrac{3}{25} \times \dfrac{4}{4} = \dfrac{12}{100} = \textbf{0.12}$

10. $\dfrac{7}{10} \times \dfrac{10}{10} = \dfrac{70}{100}$

$n = \textbf{70}$

11.
$$5 \xrightarrow{\ 4\frac{8}{8}\ } 4\frac{8}{8}$$
$$\begin{array}{r}4\frac{8}{8}\\ -\ 3\frac{1}{8}\\ \hline 1\frac{7}{8}\end{array}$$
$m = \mathbf{1\dfrac{7}{8}}$

12.
$\begin{array}{r}\overset{0}{\cancel{1}}.\overset{9}{\cancel{0}}\,{}^{1}0\\ -\ 0.9\,5\\ \hline 0.0\,5\end{array}$
$w = \textbf{0.05}$

13.
$3\dfrac{1}{6} \times \dfrac{1}{1} = \overset{2}{\cancel{3}}\,\overset{7}{\cancel{1}}\dfrac{}{6}$
$-\ 1\dfrac{2}{3} \times \dfrac{2}{2} = 1\dfrac{4}{6}$
$\qquad\qquad\qquad 1\dfrac{3}{6} = 1\dfrac{1}{2}$
$m = \mathbf{1\dfrac{1}{2}}$

14.
$\dfrac{1}{2} \times \dfrac{3}{3} = \dfrac{3}{6}$
$+\ \dfrac{1}{3} \times \dfrac{2}{2} = \dfrac{2}{6}$
$\qquad\qquad\qquad \dfrac{5}{6}$

$\dfrac{5}{6} - \dfrac{1}{6} = \dfrac{4}{6} = \dfrac{\textbf{2}}{\textbf{3}}$

15. $\dfrac{7}{\cancel{2}} \times \dfrac{\cancel{4}}{\cancel{3}} \times \dfrac{\cancel{3}}{\cancel{2}} = \dfrac{7}{1} = \textbf{7}$

16.
$$\begin{array}{r} 0.43 \\ \times\ 2.6 \\ \hline 258 \\ 860 \\ \hline 1.118 \end{array}$$

17.
$$\begin{array}{r} 0.052 \\ 5\overline{)0.260} \\ \underline{25} \\ 10 \\ \underline{10} \\ 0 \end{array}$$

18. $\dfrac{17}{20} \times \dfrac{5}{5} = \dfrac{85}{100} = \mathbf{85\%}$

19. $C = \pi d$
$C \approx (3.14)(5\text{ ft})$
$C \approx 15.7\text{ ft.}$
About 16 feet

20. **7**

21. $\dfrac{\overset{1}{\cancel{2}} \cdot \overset{1}{\cancel{3}} \cdot 3}{\cancel{2} \cdot \cancel{3} \cdot 5} = \dfrac{\mathbf{3}}{\mathbf{5}}$

22. Factors of 18
1, 2, 3, ⑥, 9, 18
Factors of 30
1, 2, 3, 5, ⑥, 10, 15, 30
GCF is **6.**

23. **B. Reciprocals**

24. **A quadrilateral is a four-sided polygon, and every rectangle has four sides.**

25. $\dfrac{8 \cdot 6}{2} = \dfrac{48}{2} = \mathbf{24}$

26. One possibility:

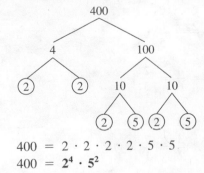

$400 = 2 \cdot 2 \cdot 2 \cdot 2 \cdot 5 \cdot 5$
$400 = \mathbf{2^4 \cdot 5^2}$

27.

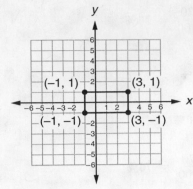

28. (a) 4 units + 2 units + 4 units
            + 2 units = **12 units**

(b) 4 units × 2 units = **8 sq. units**

29. (a) 15 cm + 20 cm + 15 cm + 20 cm
            = **70 cm**

(b)
$$\begin{array}{r} 12\text{ cm} \\ \times\ 20\text{ cm} \\ \hline 240\text{ cm}^2 \end{array}$$

30. One possibility:

; **no**

LESSON 76, WARM-UP

a. **832**

b. **535**

c. **$7.96**

d. **$5.01**

e. **$0.90**

f. **$95.00**

g. **5**

Problem Solving
Work backward, beginning with $1 \times 7 = 7$.

$$\begin{array}{r} 1 \\ 7\overline{)7} \\ 5\overline{)35} \\ 3\overline{)105} \\ 2\overline{)210} \\ 2\overline{)420} \end{array}$$

LESSON 76, LESSON PRACTICE

a. $\dfrac{3}{20}$ ⊗ $\dfrac{1}{8}$

$$
\begin{array}{r}
0.15 \\
20\overline{)3.00} \\
2\,0 \\
\hline
1\,00 \\
1\,00 \\
\hline
0
\end{array}
\qquad
\begin{array}{r}
0.125 \\
8\overline{)1.000} \\
8 \\
\hline
20 \\
16 \\
\hline
40 \\
40 \\
\hline
0
\end{array}
$$

b. $\dfrac{3}{8}$ ⊗ $\dfrac{2}{5}$

$$
\begin{array}{r}
0.375 \\
8\overline{)3.000} \\
2\,4 \\
\hline
60 \\
56 \\
\hline
40 \\
40 \\
\hline
0
\end{array}
\qquad
\begin{array}{r}
0.4 \\
5\overline{)2.0} \\
2\,0 \\
\hline
0
\end{array}
$$

c. $\dfrac{15}{25}$ ⊜ $\dfrac{3}{5}$

$$
\begin{array}{r}
0.6 \\
25\overline{)15.0} \\
15\,0 \\
\hline
0
\end{array}
\qquad
\begin{array}{r}
0.6 \\
5\overline{)3.0} \\
3\,0 \\
\hline
0
\end{array}
$$

d. 0.7 ⊗ $\dfrac{4}{5}$

$$
\begin{array}{r}
0.8 \\
5\overline{)4.0} \\
4\,0 \\
\hline
0
\end{array}
$$

e. $\dfrac{2}{5}$ ⊗ 0.5

$$
\begin{array}{r}
0.4 \\
5\overline{)2.0} \\
2\,0 \\
\hline
0
\end{array}
$$

f. $\dfrac{3}{8}$ ⊗ 0.325

$$
\begin{array}{r}
0.375 \\
8\overline{)3.000} \\
2\,4 \\
\hline
60 \\
56 \\
\hline
40 \\
40 \\
\hline
0
\end{array}
$$

LESSON 76, MIXED PRACTICE

1. $10^2 \times 2^3$
100×8
800

2.
$$
\begin{array}{r}
6.7 \\
-\ 4.5 \\
\hline
2.2
\end{array}
\qquad
\begin{array}{r}
1.1 \\
2\overline{)2.2}
\end{array}
\qquad
\begin{array}{r}
4.5 \\
+\ 1.1 \\
\hline
\mathbf{5.6}
\end{array}
$$

3. $13 \cdot 7 = y$

$$
\begin{array}{r}
13 \\
\times\ 7 \\
\hline
\mathbf{91}\ \textbf{years old}
\end{array}
$$

4. $\dfrac{2}{5}$ ⊗ $\dfrac{1}{4}$

$$
\begin{array}{r}
0.4 \\
5\overline{)2.0} \\
2\,0 \\
\hline
0
\end{array}
\qquad
\begin{array}{r}
0.25 \\
4\overline{)1.00} \\
8 \\
\hline
20 \\
20 \\
\hline
0
\end{array}
$$

5. (a) $\dfrac{3}{4}$

(b)
$$
\begin{array}{r}
\mathbf{0.75} \\
4\overline{)3.00} \\
2\,8 \\
\hline
20 \\
20 \\
\hline
0
\end{array}
$$

(c) $\dfrac{75}{100} = \mathbf{75\%}$

6.
$$
\begin{array}{r}
0.5 \\
2\overline{)1.0} \qquad \mathbf{2.5} \\
1\,0 \\
\hline
0
\end{array}
$$

7. $3\dfrac{45}{100} = \mathbf{3\dfrac{9}{20}}$

8. (a) $\dfrac{4}{100} = \mathbf{\dfrac{1}{25}}$

(b) **4%**

9.
$$
\begin{array}{r}
11\tfrac{1}{9} \\
9\overline{)100} \\
9 \\
\hline
10 \\
9 \\
\hline
1
\end{array}
$$

10.

$$6\frac{1}{3} \times \frac{4}{4} = 6\frac{4}{12}$$

$$3\frac{1}{4} \times \frac{3}{3} = 3\frac{3}{12}$$

$$+ \; 2\frac{1}{2} \times \frac{6}{6} = 2\frac{6}{12}$$

$$11\frac{13}{12} = \mathbf{12\frac{1}{12}}$$

11.

$$\frac{4}{5} \times \frac{20}{20} = \frac{80}{100}$$

$$? = \mathbf{80}$$

12.

$$\frac{\overset{1}{\cancel{8}}}{\underset{1}{\cancel{2}}} \cdot \frac{\overset{5}{\cancel{10}}}{\underset{1}{\cancel{3}}} \cdot \frac{\overset{2}{\cancel{6}}}{\underset{1}{\cancel{5}}} = \frac{10}{1} = \mathbf{10}$$

13.

$$\frac{5}{1} \div \frac{5}{2}$$

$$1 \div \frac{5}{2} = \frac{2}{5}$$

$$\frac{\overset{1}{\cancel{5}}}{1} \cdot \frac{2}{\underset{1}{\cancel{5}}} = \frac{2}{1} = \mathbf{2}$$

14. $7.18 + n = 8$

$$\begin{array}{r} \overset{7}{\cancel{8}}.\overset{9}{\cancel{0}}{}^{1}0 \\ - \; 7.18 \\ \hline 0.82 \end{array}$$

$$n = \mathbf{0.82}$$

15.

$$\begin{array}{r} \overset{0}{\cancel{1}}\,\overset{1}{\cancel{2}}.\overset{9}{\cancel{0}}{}^{1}0 \\ - \quad 4.75 \\ \hline 7.25 \end{array}$$

$$d = \mathbf{7.25}$$

16.

$$\begin{array}{r} 0.35 \\ \times \; 0.45 \\ \hline 175 \\ 1400 \\ \hline \mathbf{0.1575} \end{array}$$

17. $4.3 \div 100 = \mathbf{0.043}$

18. 0.2, 0.25, (0.27), 0.3, 0.313
0.27

19. $4000 + 5000 = \mathbf{9000}$

20. **41, 43, 47**

21. $\dfrac{12}{25} \times \dfrac{4}{4} = \dfrac{48}{100} = \mathbf{48\%}$

22. $20\,\text{mm} + 16\,\text{mm} + 12\,\text{mm} = \mathbf{48\,mm}$

23.

$$\begin{array}{r} \overset{8}{\cancel{9}}{}^{1}0° \\ - \; 5\,3° \\ \hline \mathbf{3\,7°} \end{array}$$

24. **About 45 mm**

25. (a) $20\,\text{cm} \times 10\,\text{cm} = \mathbf{200\,cm^2}$

   (b) $200\,\text{cm}^2 \div 2 = \mathbf{100\,cm^2}$

26. **12 cubes**

27. Triangle:

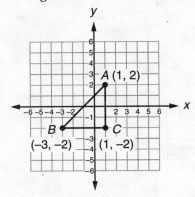

28. (a) $\overline{BC}$ or $\overline{CB}$

   (b) $\angle C$

29. $\dfrac{(12)(9)}{2} = \dfrac{108}{2} = \mathbf{54}$

30. One possibility:

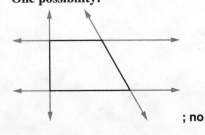

; **no**

LESSON 77, WARM-UP

a. 1555

b. 315

c. $9.95

d. **$9.49**

e. **$1.60**

f. **0.065**

g. **10**

Problem Solving

    45 pages per day $\times$ 4 days $=$ 180 pages

    180 pages $-$ 123 pages $=$ **57 pages**

LESSON 77, LESSON PRACTICE

40 engines
5 engines
5 engines
5 engines
5 engines
5 engines
5 engines
5 engines
5 engines

$\frac{3}{8}$ could climb the hill.

$\frac{5}{8}$ could not climb the hill.

a. **8 parts**

b. **5 engines**

c. **3 parts**

d. **15 engines**

e. **5 parts**

f. **25 engines**

LESSON 77, MIXED PRACTICE

1. $\dfrac{\overset{19}{\cancel{114}} \text{ pounds}}{1} \cdot \dfrac{1}{\cancel{6}_1} = \dfrac{19}{1} = $ **19 pounds**

2. **See student work.**

3. (a) $\dfrac{6}{24} = \mathbf{\dfrac{1}{4}}$

    (b) $\dfrac{1}{4} \times \dfrac{25}{25} = \dfrac{25}{100} = $ **25%**

4.

30 dogs
6 dogs
6 dogs
6 dogs
6 dogs
6 dogs

$\frac{3}{5}$ are male.

$\frac{2}{5}$ are female.

  (a) **5 parts**

  (b) **6 dogs**

  (c) **18 males**

  (d) **12 females**

5. (a) $\dfrac{4}{10} = \mathbf{\dfrac{2}{5}}$

    (b) $\dfrac{4}{10} = $ **0.4**

    (c) $\dfrac{4}{10} \times \dfrac{10}{10} = \dfrac{40}{100} = $ **40%**

6. $3\dfrac{6}{10} = \mathbf{3\dfrac{3}{5}}$

7. $\begin{array}{r} \overset{3}{\cancel{4}}.^1 5 \\ -\ 3.\ 6 \\ \hline 0.\ 5\ 5 \end{array}$

    $a = $ **0.55**

8. $x = \mathbf{\dfrac{5}{2}}$

9. **If the chance of rain is 60%, then the chance that it will not rain is 40%. It is more likely to rain because 60% is greater than 40%.**

10. $\dfrac{3}{5} \times \dfrac{20}{20} = \dfrac{60}{100} = $ **60%**

11. $32°F - (-3°F)$ $\begin{array}{r} 32°F \\ +\ \ 3°F \\ \hline \mathbf{35°F} \end{array}$

12. (a) $0.35 \;\boxed{=}\; \dfrac{7}{20}$

    $\begin{array}{r} 0.35 \\ 20\overline{)7.00} \\ \underline{6\ 0} \\ 1\ 00 \\ \underline{1\ 00} \\ 0 \end{array}$

    (b) $3^2 \;\boxed{>}\; 2^3$

       9   8

13. $\begin{aligned} \dfrac{1}{2} \times \dfrac{3}{3} &= \dfrac{3}{6} \\ +\ \dfrac{2}{3} \times \dfrac{2}{2} &= \dfrac{4}{6} \\ \hline \dfrac{7}{6} &= \mathbf{1\dfrac{1}{6}} \end{aligned}$

14.
$$\begin{array}{r} \overset{2}{\cancel{3}}\overset{\overset{6}{\cancel{1}}}{5} \\ -\ 1\frac{3}{5} \\ \hline 1\frac{3}{5} \end{array}$$

15.
$$\frac{1}{2} \times \frac{4}{4} = \frac{4}{8}$$
$$\frac{3}{4} \times \frac{2}{2} = \frac{6}{8}$$
$$+\ \frac{7}{8} \times \frac{1}{1} = \frac{7}{8}$$
$$\frac{17}{8} = 2\frac{1}{8}$$

16. $\dfrac{\overset{1}{\cancel{3}}}{1} \times \dfrac{4}{\underset{1}{\cancel{3}}} = \dfrac{4}{1} = \mathbf{4}$

17.
$$\frac{3}{1} \div \frac{4}{3}$$
$$1 \div \frac{4}{3} = \frac{3}{4}$$
$$\frac{3}{1} \times \frac{3}{4} = \frac{9}{4} = 2\frac{1}{4}$$

18.
$$\frac{4}{3} \div \frac{3}{1}$$
$$1 \div \frac{3}{1} = \frac{1}{3}$$
$$\frac{4}{3} \times \frac{1}{3} = \frac{4}{9}$$

19. 1.5 cm + 0.9 cm + 1.5 cm + 0.9 cm
$$= \textbf{4.8 cm}$$

20.
$$\begin{array}{r} 1.5\text{ cm} \\ \times\ 0.9\text{ cm} \\ \hline \textbf{1.35 cm}^2 \end{array}$$

21.

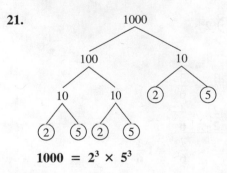

$$\mathbf{1000 = 2^3 \times 5^3}$$

22. (a)
$$\begin{array}{r} \$80 \\ \times\ 0.40 \\ \hline \$32.00 \end{array}$$

(b)
$$\begin{array}{r} \$\overset{7}{\cancel{8}}{}^{1}0 \\ -\ \$3\ 2 \\ \hline \$4\ 8 \end{array}$$

23. (a)
$$\begin{array}{r} \$38.80 \\ \times\ 0.07 \\ \hline 2.7160 \end{array} \longrightarrow \textbf{\$2.72}$$

(b)
$$\begin{array}{r} \$38.80 \\ +\ \$2.72 \\ \hline \$41.52 \end{array}$$

24. **Yes**

25.
$$\begin{array}{r} \overset{11\,:\,60}{\cancel{12:00}}\text{ p.m.} \\ -\ 1:14 \\ \hline \textbf{10:46 a.m.} \end{array}$$

26. **B. 40%**

27.

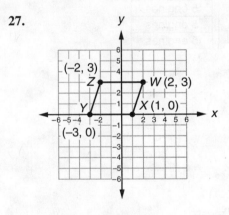

28. (a) $\overline{ZY}$ or $\overline{YZ}$

(b) $\overline{WZ}$ or $\overline{ZW}$

29. $\dfrac{\overset{1}{\cancel{2}} \cdot 3 \cdot \overset{1}{\cancel{3}} \cdot \overset{1}{\cancel{7}}}{\underset{1}{\cancel{2}} \cdot \underset{1}{\cancel{3}} \cdot 5 \cdot \underset{1}{\cancel{7}}} = \dfrac{3}{5}$

30. (a) **Sphere**

(b) $C = \pi d$
$C \approx (3.14)(2160 \text{ miles})$
$C \approx 6782.4$
$C \approx \textbf{6780 miles}$

LESSON 78, WARM-UP

a. **1300**

b. **1775**

c. $8.97

d. $17.01

e. $0.80

f. 175

g. 25

Problem Solving

Eliminate even numbers (76, 78, 80, 82, 84, 86, 88, 90, 92, 94, 96, 98).

Eliminate numbers divisible by 3 (81, 87, 93, 99).

Eliminate numbers divisible by 5 (85, 95).

Eliminate numbers divisible by 7 (77, 91).

Eliminate numbers containing 9 as a digit (79, 89, 97).

83

LESSON 78, LESSON PRACTICE

a. 1 gallon = 4 quarts

$\dfrac{1}{4}$

b. 2×1000 mL = **2000 mL**

c. $\dfrac{1}{2}$ gallon = 2 quarts

   1 quart = 4 cups

   $\dfrac{1}{2}$ gallon = **8 cups**

d. The half-gallon carton will overflow because 2 liters is a little more than half of a gallon.

LESSON 78, MIXED PRACTICE

1. $\left(\dfrac{1}{2} + \dfrac{1}{2}\right) - \left(\dfrac{1}{2} \times \dfrac{1}{2}\right)$

$\dfrac{2}{2} \times \dfrac{2}{2} = \dfrac{4}{4}$

$- \dfrac{1}{4} \times \dfrac{1}{1} = \dfrac{1}{4}$

$\qquad\qquad\quad \dfrac{3}{4}$

2. 10×10 millimeters = **100 millimeters**

3. 41, no
42, no
43, no
44, no   **48**
45, no
46, no
47, no
48, yes
49, no

4.

60 lights	
12 lights	
12 lights	(a) **5 parts**
12 lights	(b) **12 lights**
12 lights	(c) **48 lights**
12 lights	(d) **12 lights**

$\dfrac{4}{5}$ were on.

$\dfrac{1}{5}$ were off.

5. **1**

6. $m = \dfrac{5}{4}$

7. $\dfrac{5}{5} - \dfrac{4}{5} = \dfrac{1}{5}$

$w = \dfrac{1}{5}$

8. $x = \dfrac{4}{5}$

9. $1 + \dfrac{4}{5} = 1\dfrac{4}{5}$

$y = 1\dfrac{4}{5}$

10. (a) $\dfrac{4}{16} = \dfrac{1}{4}$

(b) $\dfrac{1}{4} \times \dfrac{25}{25} = \dfrac{25}{100} = $ **0.25**

(c) **25%**

11. $1\dfrac{15}{100} = 1\dfrac{3}{20}$

12. (a) $\dfrac{3}{5}$ ⊘ 0.35

$\begin{array}{r} 0.6 \\ 5\overline{)3.0} \\ 3\,0 \\ \hline 0 \end{array}$

(b) $\sqrt{100}$ ⊘ $1^4 + 2^3$

$\quad 10 \qquad\quad 1 + 8$

$\qquad\qquad\qquad 9$

13.

$$\frac{5}{6} \times \frac{1}{1} = \frac{5}{6}$$
$$- \frac{1}{2} \times \frac{3}{3} = \frac{3}{6}$$
$$\frac{2}{6} = \mathbf{\frac{1}{3}}$$

14.

$$\frac{3}{4} \times \frac{25}{25} = \frac{75}{100}$$
$$? = \mathbf{75}$$

15.

$$\frac{1}{2} \times \frac{3}{3} = \frac{3}{6}$$
$$\frac{2}{3} \times \frac{2}{2} = \frac{4}{6}$$
$$+ \frac{5}{6} \times \frac{1}{1} = \frac{5}{6}$$
$$\frac{12}{6} = \mathbf{2}$$

16. $\frac{\overset{1}{\cancel{8}}}{\underset{1}{\cancel{2}}} \times \frac{\overset{4}{\cancel{8}}}{\cancel{3}} = \frac{4}{1} = \mathbf{4}$

17.

$$\frac{3}{2} \div \frac{8}{3}$$
$$1 \div \frac{8}{3} = \frac{3}{8}$$
$$\frac{3}{2} \cdot \frac{3}{8} = \mathbf{\frac{9}{16}}$$

18.

$$\frac{8}{3} \div \frac{3}{2}$$
$$1 \div \frac{3}{2} = \frac{2}{3}$$
$$\frac{8}{3} \times \frac{2}{3} = \frac{16}{9} = \mathbf{1\frac{7}{9}}$$

19. (a) $\frac{1}{\underset{1}{\cancel{2}}} \text{ in.} \times \frac{\overset{2}{\cancel{4}}}{1} = \frac{2}{1} = \mathbf{2 \text{ in.}}$

(b) $\frac{1}{2} \text{ in.} \times \frac{1}{2} \text{ in.} = \mathbf{\frac{1}{4} \text{ in.}^2}$

20. **True**

21.

$$\begin{array}{r} \overset{1}{27} \\ + \ 25 \\ \hline 52 \end{array} \qquad \begin{array}{r} 26 \\ 2\overline{)52} \\ \underline{4} \\ 12 \\ \underline{12} \\ 0 \end{array}$$

22. **1.36**

23. $\frac{5}{\underset{1}{\cancel{2}}} \times \frac{\overset{6}{\cancel{12}} \text{ inches}}{1} = \frac{30}{1} = \mathbf{30 \text{ inches}}$

24. *C*

25. See student work. Figure should have four sides and no more than two right angles.

26. One possibility:

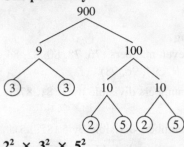

$2^2 \times 3^2 \times 5^2$

27.

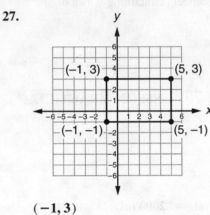

$(\mathbf{-1, 3})$

28. (a) 3 teaspoons = 1 tablespoon

$$\mathbf{\frac{1}{3}}$$

(b) 4 quarts = 1 gallon
4 quarts = 8 pints
8 pints = 16 cups
1 gallon = **16 cups**

29. B. Quart

30. 1 pint = 2 cups
2 cups = 16 ounces
1 pint = **16 ounces**

LESSON 79, WARM-UP

a. 1842

b. 580

c. **$11.96**

d. **$8.74**

e. **$48.00**

f. **12.5**

g. **8**

Problem Solving

8 in. × 2 = 16 in.
48 in. − 16 in. = 32 in.
32 in. ÷ 2 = **16 in.**

LESSON 79, LESSON PRACTICE

a. $A = \dfrac{(10\text{ ft})(6\text{ ft})}{2}$
$A = \mathbf{30\ ft^2}$

b. $A = \dfrac{(8\text{ in.})(6\text{ in.})}{2}$
$A = \mathbf{24\ in.^2}$

c. $A = \dfrac{(56\text{ mm})(15\text{ mm})}{2}$
$A = \mathbf{420\ mm^2}$

d. $A = \dfrac{(5\text{ cm})(6\text{ cm})}{2}$
$A = \mathbf{15\ cm^2}$

LESSON 79, MIXED PRACTICE

1. **Answers may vary. One method is to divide the perimeter by 2 and subtract the length from the quotient.**

2. 1 liter $\bigotimes$ 1 quart

3.
```
  1
  38
+ 33
  71 years old
```

4. (a) **False**

   (b) **True**

5. (a) $100\% - 90\% = \mathbf{10\%}$

   (b) $\dfrac{\overset{9}{\cancel{90}}}{\underset{\underset{1}{10}}{\cancel{100}}} \cdot \dfrac{\overset{3}{\cancel{30}}}{1} = \mathbf{27\ people}$

   (c) $\dfrac{3}{27} = \mathbf{\dfrac{1}{9}}$

6. (a) $\dfrac{18}{24} = \mathbf{\dfrac{3}{4}}$

   (b) $\dfrac{4}{4} - \dfrac{3}{4} = \mathbf{\dfrac{1}{4}}$

   (c) $\dfrac{1}{4} \times \dfrac{25}{25} = \dfrac{25}{100} = \mathbf{25\%}$

7. $A = 20\text{ mm} \times 12\text{ mm} = 240\text{ mm}^2$
```
     120 mm²
  2)240 mm²
     2
     04
     04
     00
     00
      0
```

8. $1000 \div 100 = \mathbf{10}$

9.
```
   1
   16.42
   12.7
 +  8
   27.12
```

10.
```
     1.2
  ×  0.12
      24
     120
   0.144
```

11.
```
      800
  8)6400
    64
    00
    00
    00
    00
     0
```

12. $\dfrac{\overset{\overset{1}{\cancel{2}}}{\cancel{10}}}{\underset{1}{\cancel{3}}} \times \dfrac{1}{\cancel{3}} \times \dfrac{\overset{1}{\cancel{3}}}{\underset{2}{\cancel{4}}} = \mathbf{\dfrac{1}{2}}$

13.
$$\frac{5}{2} \div \frac{3}{1}$$
$$1 \div \frac{3}{1} = \frac{1}{3}$$
$$\frac{5}{2} \times \frac{1}{3} = \frac{5}{6}$$

14.
$$\begin{array}{r} \overset{0}{\cancel{1}}\overset{9}{\cancel{0}}.\overset{9}{\cancel{0}}{}^{1}0 \\ - \quad 9\ .8\ 7 \\ \hline 0\ .1\ 3 \end{array}$$
$$q = \mathbf{0.13}$$

15.
$$\begin{array}{r} 0.012 \\ 24\overline{)0.288} \\ \underline{24} \\ 48 \\ \underline{48} \\ 0 \end{array}$$
$$m = \mathbf{0.012}$$

16.
$$2\frac{3}{4} \times \frac{3}{3} = 2\frac{9}{12}$$
$$+\ 3\frac{1}{3} \times \frac{4}{4} = 3\frac{4}{12}$$
$$\overline{\qquad\qquad 5\frac{13}{12} = 6\frac{1}{12}}$$
$$n = \mathbf{6\frac{1}{12}}$$

17.
$$\frac{5}{6} \times \frac{2}{2} = \frac{10}{12}$$
$$-\ \frac{1}{4} \times \frac{3}{3} = \frac{3}{12}$$
$$\overline{\qquad\qquad\quad \frac{7}{12}}$$
$$w = \mathbf{\frac{7}{12}}$$

18.
$$\begin{array}{r} 20\ \text{cm} \\ 4\overline{)80\ \text{cm}} \\ \underline{8} \\ 00 \\ \underline{00} \\ 0 \end{array} \qquad \begin{array}{r} 20\ \text{cm} \\ \times\ 20\ \text{cm} \\ \hline \mathbf{400\ sq.\ cm} \end{array}$$

19. 96.03

20.
$$C = \pi d$$
$$C \approx (3.14)(20\ \text{cm})$$
$$C \approx \mathbf{62.8\ cm}$$

21. B. 0.2

22. $7 \times 7 = \mathbf{49}$
Round 6.7 to 7 and round 7.3 to 7. Then multiply 7 by 7.

23. $3 \cdot 3 \cdot 3 \cdot 3 = \mathbf{81}$

24.
$$\begin{array}{r} 0.3 \\ +\ 0.2 \\ \hline 0.5 \end{array} \qquad \begin{array}{r} 0.25 \\ 2\overline{)0.50} \\ \underline{4} \\ 10 \\ \underline{10} \\ 0 \end{array}$$

25. 10.2

26. Trapezoid

27.

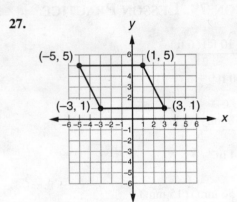

Parallelogram

28. (a) $\sqrt{100\ \text{cm}^2} = \mathbf{10\ cm}$

   (b) $10\ \text{cm} \times 4 = \mathbf{40\ cm}$

29. (a) **10 cm**

   (b) $10\ \text{cm} \times 6 = \mathbf{60\ cm}$

30.
$$\frac{\cancel{2}^{1} \cdot \cancel{2}^{1} \cdot \cancel{2}^{1} \cdot \cancel{2}^{1} \cdot 2}{\cancel{2}_{1} \cdot \cancel{2}_{1} \cdot \cancel{2}_{1} \cdot \cancel{2}_{1} \cdot 3} = \mathbf{\frac{2}{3}}$$

LESSON 80, WARM-UP

a. 1260

b. 550

c. $14.95

d. $2.01

e. $1.20

f. 0.375

g. 0

Problem Solving

For one particular prefix, there are 9999 phone numbers that end in the range 0001–9999. Also, one phone number ends with 0000, which gives us a total of **10,000 phone numbers.**

LESSON 80, LESSON PRACTICE

a.

	Ratio	Actual Count
Boys	6	b
Girls	5	60

$60 \div 5 = 12$
$6 \times 12 = 72$
72 boys

b.

	Ratio	Actual Count
Ants	8	a
Flies	3	24

$24 \div 3 = 8$
$8 \times 8 = 64$
64 ants

LESSON 80, MIXED PRACTICE

1.
$$\begin{array}{r} 96 \\ 49 \\ 68 \\ +\ 75 \\ \hline 288 \end{array} \qquad 4)\overline{288} = 72$$

Mean is **72**

$$\begin{array}{r} \overset{8}{\cancel{9}}{}^{1}6 \\ -\ 4\,9 \\ \hline 4\,7 \end{array}$$

Range is **47**

2. $\dfrac{5}{\underset{1}{\cancel{2}}} \cdot \dfrac{\overset{2640}{\cancel{5280}} \text{ ft}}{1} = \dfrac{13{,}200}{1} = \textbf{13,200 ft}$

3. $12p = 168$

$$12)\overline{168} \quad \textbf{14 players}$$
$$\begin{array}{r} 12 \\ \hline 48 \\ 48 \\ \hline 0 \end{array}$$

4. (a) 5 in. + 5 in. + 3 in. + 3 in. = **16 in.**

(b) 3 in. × 4 in. = **12 in.²**

5. (a) 3 in. + 4 in. + 5 in. = **12 in.**

(b) $A = \dfrac{3 \text{ in.} \times 4 \text{ in.}}{2} = \dfrac{12 \text{ in.}^2}{2} = \textbf{6 in.}^2$

6. **Trapezoid**

7. **True**

8. $\dfrac{4}{\underset{1}{\cancel{8}}} \cdot \dfrac{\overset{6}{\cancel{30}}}{1} = 24$ chorus members present

$30 - 24 = \textbf{6 chorus members}$ absent

9.

	Ratio	Actual Count
Dogs	2	10
Cats	5	c

$$2)\overline{10} \quad = 5$$
$$\begin{array}{r} 10 \\ \hline 0 \end{array}$$

$5 \times 5 = \textbf{25 cats}$

10. (a) $\dfrac{19}{20} \times \dfrac{5}{5} = \dfrac{95}{100} = \textbf{95\%}$

(b) $\dfrac{6}{10} \times \dfrac{10}{10} = \dfrac{60}{100} = \textbf{60\%}$

11. $\dfrac{1}{4} \times \dfrac{25}{25} = \dfrac{25}{100} = \textbf{25\%}$

12. $0.5 \ <\ \dfrac{3}{4}$

$$4)\overline{3.00} \quad = 0.75$$
$$\begin{array}{r} 2\,8 \\ \hline 20 \\ 20 \\ \hline 0 \end{array}$$

13. $4\frac{4}{10} = 4\frac{2}{5}$

14.
$$\begin{array}{r} 0.125 \\ 8\overline{)1.000} \\ \underline{8} \\ 20 \\ \underline{16} \\ 40 \\ \underline{40} \\ 0 \end{array}$$

15.
$$\begin{array}{r} \frac{5}{6} \times \frac{1}{1} = \frac{5}{6} \\ + \frac{1}{2} \times \frac{3}{3} = \frac{3}{6} \\ \hline \frac{8}{6} = 1\frac{2}{6} = 1\frac{1}{3} \end{array}$$

16.
$$\begin{array}{r} \frac{5}{8} \times \frac{1}{1} = \frac{5}{8} \\ - \frac{1}{4} \times \frac{2}{2} = \frac{2}{8} \\ \hline \frac{3}{8} \end{array}$$

17. $\dfrac{\cancel{3}^{1}}{2} \times \dfrac{\cancel{4}^{2}}{\cancel{3}_{1}} \times \dfrac{\cancel{3}^{1}}{\cancel{3}_{1}} = \dfrac{2}{1} = \mathbf{2}$

18.
$$\begin{array}{r} \overset{3}{\cancel{4}}.\overset{1}{0} \\ - 2.6 \\ \hline 1.4 \end{array}$$
$a = \mathbf{1.4}$

19.
$$\frac{3}{2} \div \frac{3}{1}$$
$$1 \div \frac{3}{1} = \frac{1}{3}$$
$$\frac{\cancel{3}^{1}}{2} \cdot \frac{1}{\cancel{3}_{1}} = \frac{1}{2}$$
$$n = \mathbf{\frac{1}{2}}$$

20.
$$\begin{array}{r} 0.072 \\ 5\overline{)0.360} \\ \underline{35} \\ 10 \\ \underline{10} \\ 0 \end{array}$$
$x = \mathbf{0.072}$

21.
$$\begin{array}{r} 70 \\ 9\overline{)630} \\ \underline{63} \\ 00 \\ \underline{00} \\ 0 \end{array}$$
$y = \mathbf{70}$

22. **0.43**

23. **About 35 grams**

24. 40 grams ÷ 2 = **20 grams**

25. Answers will vary. Sample answer: **About how many grams of sugar, per 100 grams of cereal, does Goodmeal contain?**
About 10 grams

26. 1 quart = 2 pints
2 pints = 4 cups
1 quart = 4 cups
4 cups − 1 cup = **3 cups**

27.
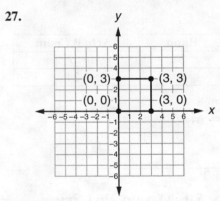

(a) **(0, 0)**

(b) 3 units × 3 units = **9 units²**

28. 90° − 30° = 60°
A. ∠

29. $A = \frac{1}{2}(6)(8)$
$A = \frac{48}{2}$
$A = \mathbf{24}$

30. Possibilities include:

 and ; **yes**

INVESTIGATION 8

Activity 1

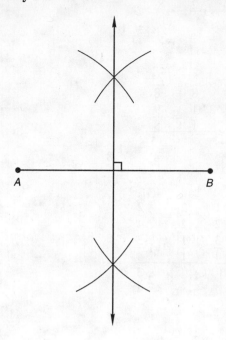

Activity 2

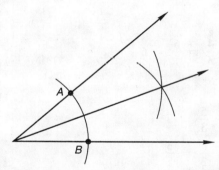

Activity 3
Student work should include a different perpendicular bisector and a different angle bisector than those in Activities 1 and 2.

LESSON 81, WARM-UP

a. **2177**

b. **750**

c. **$39.96**

d. **$12.49**

e. **$11.00**

f. **7.5**

g. **9**

Problem Solving
Since the remainder is 2, the divisor must be greater than 2. We know the divisor times 3 equals a single-digit number, so the divisor must be 3. Think, "What number times 3 equals 9?" (3). Then think, "9 plus 2 equals what number?" (11). So the ones digit in the dividend must be 1. Then think, "1 plus what single-digit number equals a two-digit number?" (9). What is the two-digit number? (10).

$$\begin{array}{r} 33 \\ 3\overline{)101} \\ \underline{9} \\ 11 \\ \underline{9} \\ 2 \end{array}$$

LESSON 81, LESSON PRACTICE

a.   24 in. $-$ 12 in. $=$ **12 in.**

b.   2 ft $\times$ 4 ft $=$ $\underbrace{2 \cdot 4}_{8}$ $\underbrace{\text{ft} \cdot \text{ft}}_{\text{ft}^2}$

 8 ft²

c. $\dfrac{\overset{4}{\cancel{12}}}{\underset{1}{\cancel{3}}}$ $\dfrac{\text{cm} \cdot \text{cm}}{\text{cm}}$

 4 cm

d. $\dfrac{\overset{60}{\cancel{300}}}{\underset{1}{\cancel{5}}}$ $\dfrac{\text{mi}}{\text{hr}}$

 60 $\dfrac{\text{mi}}{\text{hr}}$

LESSON 81, MIXED PRACTICE

1.   1 gallon $=$ 4 quarts
 2 gallons $=$ 8 quarts
 8 quarts $-$ 2 quarts $=$ **6 quarts**

2.   1 quart $\textcircled{<}$ 1 liter
 945 mL · 1000 mL

3. $\frac{5}{2}$ inches $\cdot \frac{3}{1} = \frac{15}{2} = 7\frac{1}{2}$ inches

4. $\frac{\overset{400}{\cancel{1200}} \text{ miles}}{\underset{1}{\cancel{3}} \text{ hours}} = 400 \frac{\textbf{miles}}{\textbf{hour}}$

5. $\frac{2 \cdot \overset{1}{\cancel{3}} \cdot \overset{1}{\cancel{3}} \cdot \overset{1}{\cancel{3}}}{\underset{1}{\cancel{3}} \cdot \underset{1}{\cancel{3}} \cdot \underset{1}{\cancel{3}} \cdot 5} = \frac{2}{5}$

6. $\frac{60}{100} = \frac{3}{5}$

    $\frac{3}{\underset{1}{\cancel{5}}} \cdot \frac{\overset{16}{\cancel{80}}}{1} = 48$ points

7. $\begin{array}{r} 25 \text{ m} \\ \times\ 24 \text{ m} \\ \hline 100 \\ 500 \\ \hline \textbf{600 m}^2 \end{array}$

8. $26 \text{ m} + 26 \text{ m} + 24 \text{ m} + 24 \text{ m} = \textbf{100 m}$

9. **False**

10.
	Ratio	Actual Count
Red	3	r
Blue	4	24

    $\begin{array}{r} 6 \\ 4\overline{)24} \\ \underline{24} \\ 0 \end{array}$ $6 \times 3 = 18$

    18 red marbles

11. $\frac{1}{5}$, 0.4, $\frac{1}{2}$

12. (a) $\frac{4}{25} \times \frac{4}{4} = \frac{16}{100} = \textbf{0.16}$

    (b) **16%**

13. $\begin{array}{r} 9.9 \\ \times\ 0.1 \\ \hline \textbf{0.99} \end{array}$

14. $\begin{array}{r} 1.7 \\ 2\overline{)3.4} \\ \underline{2} \\ 1\ 4 \\ \underline{1\ 4} \\ 0 \end{array}$

15. $\begin{array}{r} \frac{5}{8} \times \frac{1}{1} = \frac{5}{8} \\ + \frac{3}{4} \times \frac{2}{2} = \frac{6}{8} \\ \hline \frac{11}{8} = 1\frac{3}{8} \end{array}$

16. $\begin{array}{r} \frac{3}{1} \times \frac{8}{8} = \frac{24}{8} \\ - \frac{9}{8} \times \frac{1}{1} = \frac{9}{8} \\ \hline \frac{15}{8} = 1\frac{7}{8} \end{array}$

17. $\begin{array}{r} 4\frac{1}{2} \times \frac{2}{2} = \overset{3}{\cancel{4}}\overset{6}{\cancel{\frac{2}{4}}} \\ - 1\frac{3}{4} \times \frac{1}{1} = 1\frac{3}{4} \\ \hline 2\frac{3}{4} \end{array}$

18. $\frac{\overset{1}{\cancel{5}}}{\underset{2}{\cancel{6}}} \cdot \frac{\overset{1}{\cancel{4}}}{\underset{1}{\cancel{5}}} \cdot \frac{\overset{1}{\cancel{3}}}{\underset{2}{\cancel{8}}} = \frac{1}{4}$

19. $\frac{\overset{3}{\cancel{9}}}{\underset{1}{\cancel{2}}} \times \frac{\overset{2}{\cancel{4}}}{\underset{1}{\cancel{3}}} = \frac{6}{1} = 6$

20. $\frac{10}{3} \div \frac{5}{3}$

    $1 \div \frac{5}{3} = \frac{3}{5}$

    $\frac{\overset{2}{\cancel{10}}}{\underset{1}{\cancel{3}}} \cdot \frac{\overset{1}{\cancel{3}}}{\underset{1}{\cancel{5}}} = \frac{2}{1} = 2$

21. $\begin{array}{r} 50 \text{ cm} \\ 4\overline{)200 \text{ cm}} \\ \underline{20} \\ 00 \\ \underline{00} \\ 0 \end{array}$

22. (a) **50 cm**

    (b) $C = \pi d$
        $C \approx (3.14)(50 \text{ cm})$
        $C \approx \textbf{157 cm}$

23. $\begin{array}{r} \$12.80 \\ \times\ \ 0.06 \\ \hline \$0.7680 \end{array} \longrightarrow \textbf{77¢}$

24.   10:40 a.m.
   + 2:30
   ‾‾‾‾‾‾‾‾
   12:70 p.m.
   1:10 p.m.

25. $2\frac{10}{16}$ in. $= 2\frac{5}{8}$ in.

26.

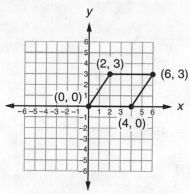

$A = bh$
$A = (4 \text{ units})(3 \text{ units})$
$A = \textbf{12 sq. units}$

27. Cone

28. $\sqrt{1 \text{ sq. foot}} = 1$ ft
   1 ft $\times$ 4 = **4 ft**

29. (a) 2 yd + 1 yd = **3 yd**

   (b) 5 m $\cdot$ 3 m = **15 m²**

   (c) $\dfrac{\overset{6}{\cancel{36}}}{\underset{1}{\cancel{6}}}$ $\dfrac{\text{ft} \times \text{ft}}{\text{ft}} = \textbf{6 ft}$

   (d) $\dfrac{\overset{20}{\cancel{400}}}{\underset{1}{\cancel{20}}} \dfrac{\text{miles}}{\text{gallons}} = \textbf{20}\dfrac{\textbf{miles}}{\textbf{gallon}}$

30. One possibility:

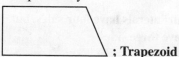

   ; Trapezoid

LESSON 82, WARM-UP

a. 400

b. 134

c. $4.98

d. $2.50

e. 25

f. 20

g. 3

Problem Solving
   $(110 \times 3) + 115 + 120 = 565$
   $565 \div 5 = \textbf{113}$

LESSON 82, LESSON PRACTICE

a.   4 $\times$ 4 = 16 sugar cubes
   16 $\times$ 4 = **64 sugar cubes**

b. $V = lwh$
   $V = (5\text{ ft})(3\text{ ft})(2\text{ ft})$
   $V = \textbf{30 ft}^3$

c. 10 $\times$ 6 = **60 cubes**
   60 $\times$ 4 = **240 cubes**

LESSON 82, MIXED PRACTICE

1. 21.05

2. 12 $\div$ 3 = 4 cans
   $2.49
   $\times$ 4
   ‾‾‾‾‾‾‾
   $9.96

3.   18 in.   6 feet
   $\times$ 4 in.   12)‾7‾2‾
   ‾‾‾‾‾‾‾‾‾   72
   72 inches   ‾‾
                      0

4. (a) **7%**

   (b) $\dfrac{7}{10} \times \dfrac{10}{10} = \dfrac{70}{100} = \textbf{70\%}$

5. $\dfrac{90}{100} = \dfrac{9}{10}$
   0.9

6. $\dfrac{23}{50} \times \dfrac{2}{2} = \dfrac{46}{100} = \textbf{46\%}$

7. $\dfrac{9}{25} \times \dfrac{4}{4} = \dfrac{36}{100} = \textbf{36\%}$

8. Rectangular prism

9.
$$3\frac{5}{6} \times \frac{1}{1} = 3\frac{5}{6}$$
$$+ \ 2\frac{1}{3} \times \frac{2}{2} = 2\frac{2}{6}$$
$$5\frac{7}{6} = 6\frac{1}{6}$$
$$w = 6\frac{1}{6}$$

10.
$$3\frac{1}{4} \times \frac{2}{2} = 3\frac{\overset{10}{2}}{8}$$
$$- \ 1\frac{5}{8} \times \frac{1}{1} = 1\frac{5}{8}$$
$$1\frac{5}{8}$$
$$y = 1\frac{5}{8}$$

11.
$$6\overline{\smash{\big)}0.12} \quad \overset{0.02}{}$$
$$\underline{12}$$
$$0$$
$$n = 0.02$$

12.
$$12\overline{\smash{\big)}600} \quad \overset{50}{}$$
$$\underline{60}$$
$$00$$
$$\underline{00}$$
$$0$$
$$m = 50$$

13.
$$5\overline{\smash{\big)}100} \quad \overset{20}{}$$
$$\underline{10}$$
$$00$$
$$\underline{00}$$
$$0$$
$$n = 20$$

14.
$$\frac{6}{1} \div \frac{3}{2}$$
$$1 \div \frac{3}{2} = \frac{2}{3}$$
$$\frac{\overset{2}{\cancel{6}}}{1} \cdot \frac{2}{\underset{1}{\cancel{3}}} = \frac{4}{1} = 4$$
$$w = 4$$

15. (a) $\dfrac{6}{10} = \dfrac{3}{5}$

(b) $\dfrac{3}{5} \times \dfrac{20}{20} = \dfrac{60}{100} = \mathbf{60\%}$

16. $0.5 + 1 + 0.25 = \mathbf{1.75}$

17.
$$\frac{1}{2} \times \frac{5}{5} = \frac{5}{10}$$
$$\frac{1}{5} \times \frac{2}{2} = \frac{2}{10}$$
$$+ \ \frac{1}{10} \times \frac{1}{1} = \frac{1}{10}$$
$$\frac{8}{10} = \frac{4}{5}$$

18. $\dfrac{\overset{3}{\cancel{9}}}{\underset{1}{\cancel{3}}} \times \dfrac{\overset{1}{\cancel{3}}}{\underset{1}{\cancel{3}}} = \dfrac{3}{1} = \mathbf{3}$

19. **6**

20. $40 \times 40 = \mathbf{1600}$

21. (a) $\dfrac{12}{36} = \dfrac{1}{3}$

(b) $\dfrac{12}{48} = \dfrac{1}{4}$

22.
$$\begin{array}{r} 25 \text{ mm} \\ \times \ 20 \text{ mm} \\ \hline \mathbf{500 \text{ mm}^2} \end{array}$$

23. $25 \text{ mm} + 25 \text{ mm} + 22 \text{ mm} + 22 \text{ mm}$
$$= \mathbf{94 \text{ mm}}$$

24.

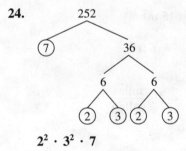

$$2^2 \cdot 3^2 \cdot 7$$

25. **False. Quadrilaterals have four sides, but triangles have three sides.**

26.

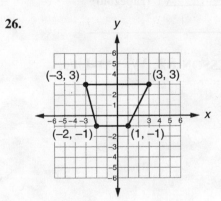

Trapezoid

27. $V = lwh$
$V = (2 \text{ in.})(2 \text{ in.})(2 \text{ in.})$
$V = \textbf{8 in.}^3$

28. (a) 2 pints = 1 quart
3 quarts + 1 quart = **4 quarts**

(b) $\dfrac{\overset{7}{\cancel{49}} \, \cancel{m} \cdot m}{\underset{1}{\cancel{7}} \, \cancel{m}} = \textbf{7 m}$

(c) $\dfrac{\overset{50}{\cancel{400}} \text{ miles}}{\underset{1}{\cancel{8}} \text{ hours}} = \textbf{50} \, \dfrac{\textbf{miles}}{\textbf{hour}}$

29. $\dfrac{3}{12} = \dfrac{1}{4} \times \dfrac{25}{25} = \dfrac{25}{100} = \textbf{25\%}$

30. 1 gallon = 4 quarts
4 quarts = 8 pints
1 gallon = 8 pints
About 8 pounds

LESSON 83, WARM-UP

a. **900**

b. **875**

c. **$11.97**

d. **7**

e. **0.025**

f. **680**

g. **3**

Problem Solving
<; Each of the three mixed numbers on the left is less than the respective whole number on the right. So the sum on the left must be less than the sum on the right.

LESSON 83, LESSON PRACTICE

a. C. $\dfrac{15}{6}$

b. $\dfrac{6}{8} = \dfrac{9}{12}$

c. $\dfrac{4}{3} = \dfrac{12}{n}$
$\dfrac{4}{3} \times \dfrac{3}{3} = \dfrac{12}{9}$
$n = \textbf{9}$

d. $\dfrac{6}{9} = \dfrac{n}{36}$
$\dfrac{6}{9} \times \dfrac{4}{4} = \dfrac{24}{36}$
$n = \textbf{24}$

LESSON 83, MIXED PRACTICE

1. $(0.2 + 0.2) \times (0.2 - 0.2)$
$\quad (0.4) \quad \times \quad (0)$
$\qquad\qquad \textbf{0}$

2. $\dfrac{3}{2} \times \dfrac{3}{1} = \dfrac{9}{2} = \textbf{4}\,\dfrac{\textbf{1}}{\textbf{2}}$ **miles per hour**

3.
$\begin{array}{r} 4{:}45 \text{ p.m.} \\ - \ 1{:}45 \text{ p.m.} \\ \hline 3{:}00 \end{array}$ $\begin{array}{r} \$4 \text{ per hour} \\ \times \ \ 3 \text{ hours} \\ \hline \$12 \end{array}$
3 hours

4. $\dfrac{55}{100} = \dfrac{\textbf{11}}{\textbf{20}}$

5. (a) **9%**

(b) $\dfrac{9}{10} \times \dfrac{10}{10} = \dfrac{90}{100} = \textbf{90\%}$

6. **100%**

7. (a) $\dfrac{10}{100} = \dfrac{\textbf{1}}{\textbf{10}}$

(b) $\dfrac{10}{100} = \textbf{10\%}$

8. (a) $\dfrac{48}{100} = \dfrac{\textbf{12}}{\textbf{25}}$

(b) $\dfrac{48}{100} = \textbf{48\%}$

9. $\begin{array}{r} \textbf{0.875} \\ 8\overline{)7.000} \\ \underline{6\ 4} \\ 60 \\ \underline{56} \\ 40 \\ \underline{40} \\ 0 \end{array}$

10.
$$1\tfrac{1}{3} \times \tfrac{2}{2} = 1\tfrac{2}{6}$$
$$+\ 1\tfrac{1}{6} \times \tfrac{1}{1} = 1\tfrac{1}{6}$$
$$\overline{\qquad\qquad 2\tfrac{3}{6} = 2\tfrac{1}{2}}$$

$$2\tfrac{1}{2} \times \tfrac{3}{3} = 2\tfrac{9}{6}$$
$$-\ 1\tfrac{2}{3} \times \tfrac{2}{2} = 1\tfrac{4}{6}$$
$$\overline{\qquad\qquad\qquad \tfrac{5}{6}}$$

11. $\dfrac{\cancel{3}}{\cancel{2}} \cdot \dfrac{\cancel{3}}{1} \cdot \dfrac{\cancel{10}}{\cancel{9}} = \dfrac{5}{1} = \textbf{5}$

12. $\dfrac{14}{3} \div \dfrac{7}{6}$

$1 \div \dfrac{7}{6} = \dfrac{6}{7}$

$\dfrac{\cancel{14}}{\cancel{3}} \cdot \dfrac{\cancel{6}}{\cancel{7}} = \dfrac{4}{1} = \textbf{4}$

13. $0.1 + 0.99$
1.09

14. $\dfrac{3}{4} = \dfrac{9}{n}$

$\dfrac{3}{4} \times \dfrac{3}{3} = \dfrac{9}{12}$

$n = \textbf{12}$

15. C. $\dfrac{12}{20}$

16. **80,420**

17. $2^4 \boxed{=} 4^2$
$\quad 16 \qquad 16$

18. $\dfrac{\cancel{2} \cdot \cancel{2} \cdot \cancel{2} \cdot 3}{\cancel{2} \cdot \cancel{2} \cdot \cancel{2} \cdot 2 \cdot 2} = \dfrac{3}{4}$

19. Factors of 24
1, 2, 3, 4, 6, ⑧, 12, 24
Factors of 32
1, 2, 4, ⑧, 16, 32
GCF is **8**.

20. $C = \pi d$
$C \approx (3.14)(4\text{ ft})$
$C \approx 12.56\text{ ft}.$
C. $12\tfrac{1}{2}$ ft.

21. 10 mm + 13 mm + 20 mm + 13 mm
$\quad = \textbf{56 mm}$

22. $V = lwh$
$V = (3\text{ cm})(3\text{ cm})(3\text{ cm})$
$V = \textbf{27 cm}^3$

23. (a) $A = bh$
$\quad A = (6\text{ in.})(4\text{ in.})$
$\quad A = \textbf{24 in.}^2$

(b) $24\text{ in.}^2 \div 2 = \textbf{12 in.}^2$

24. $\dfrac{1}{\cancel{4}} \cdot \dfrac{\cancel{120}^{30}}{1} = $ 30 campers chopped wood.

$$\begin{array}{r} 1\,2\,0 \\ -\ \ \ 3\,0 \\ \hline \textbf{9\,0 campers} \end{array}$$ did not chop wood.

25. 2.5×10 millimeters = **25 millimeters**

26. **Pyramid**

27. (a) 1 quart = 2 pints
3 quarts = 6 pints
6 pints + 2 pints = **8 pints**

(b) $\dfrac{\cancel{64}^{8}}{\cancel{8}} \dfrac{\cancel{\text{cm}} \cdot \text{cm}}{\cancel{\text{cm}}} = \textbf{8 cm}$

(c) $\dfrac{\cancel{60}^{20}}{\cancel{3}} \dfrac{\text{newspapers}}{\text{stacks}}$

$= \textbf{20 newspapers per stack}$

28. $\dfrac{20}{25} \times \dfrac{4}{4} = \dfrac{80}{100} = \textbf{80\%}$

29. One possibility:

30. (a) $\dfrac{15 \text{ nickels}}{5)\overline{75}}$

$\dfrac{5}{25}$

$\dfrac{25}{0}$

(b) $\dfrac{2}{3} = \dfrac{d}{15}$

$\dfrac{2}{3} \times \dfrac{5}{5} = \dfrac{10}{15}$

10 dimes

(c) $\quad \$0.75$

$\quad + \$1.00$

$\quad \overline{\mathbf{\$1.75}}$

LESSON 84, WARM-UP

a. 1600

b. 844

c. $8.98

d. $2.50

e. 750

f. 24

g. 2

Problem Solving

$\dfrac{5 \text{ cm} \times 4 \text{ cm}}{2} = \textbf{10 sq. cm}$

$\dfrac{5 \text{ cm} \times 6 \text{ cm}}{2} = \textbf{15 sq. cm}$

$\dfrac{5 \text{ cm} \times 10 \text{ cm}}{2} = \textbf{25 sq. cm}$

LESSON 84, LESSON PRACTICE

a. $5 + 5 \times 5 - 5 \div 5$

$5 + 25 - 1 = \textbf{29}$

b. $2(10) + 2(6)$

$20 + 12 = \textbf{32}$

c. $5 + 4 \times 3 \div 2 - 1$

$5 + 12 \div 2 - 1$

$5 + 6 - 1 = \textbf{10}$

d. $32 + 1.8(20)$

$32 + 36 = \textbf{68}$

e. $3 + 3 \times 3 - 3 \div 3$

$3 + 9 - 1 = \textbf{11}$

LESSON 84, MIXED PRACTICE

1. $\dfrac{\text{Prime numbers}}{\text{Composite numbers}} = \dfrac{\mathbf{4}}{\mathbf{5}}$

2. 1 gallon $=$ 4 quarts

4 quarts $=$ 8 pints

8 pints $=$ 16 cups

$\frac{1}{2}$ gallon $=$ 8 cups

8 cups $-$ 4 cups $= \textbf{4 cups}$

3. $6 + 6 \times 6 - 6 \div 6$

$6 + 36 - 1 = \textbf{41}$

4. $\dfrac{30}{100} = \dfrac{\mathbf{3}}{\mathbf{10}};\ \mathbf{0.3}$

5. $A = \dfrac{9 \text{ cm} \times 4 \text{ cm}}{2}$

$A = \dfrac{36 \text{ cm}^2}{2}$

$A = \textbf{18 cm}^2$

6. $A = \dfrac{6 \text{ cm} \times 6 \text{ cm}}{2}$

$A = \dfrac{36 \text{ cm}^2}{2}$

$A = \textbf{18 cm}^2$

7. (a) $\dfrac{1}{20} \times \dfrac{5}{5} = \dfrac{5}{100} = \textbf{0.05}$

(b) **5%**

8. True

9. $A = 16 \text{ cm} \times 24 \text{ cm}$

$A = \textbf{384 cm}^2$

10. $16 \text{ cm} + 16 \text{ cm} + 25 \text{ cm} + 25 \text{ cm} = \textbf{82 cm}$

11.
$$3\frac{1}{8} \times \frac{1}{1} = 3\frac{1}{8}$$
$$+\ 2\frac{1}{4} \times \frac{2}{2} = 2\frac{2}{8}$$
$$= 5\frac{3}{8}$$

$$5\frac{3}{8} \times \frac{1}{1} = \overset{4}{\cancel{5}}\overset{11}{\cancel{3}}\frac{}{8}$$
$$-\ 1\frac{1}{2} \times \frac{4}{4} = 1\frac{4}{8}$$
$$\mathbf{3\frac{7}{8}}$$

12. $\dfrac{5}{\underset{3}{\cancel{6}}} \times \dfrac{\overset{4}{\cancel{8}}}{\underset{1}{\cancel{3}}} \times \dfrac{\overset{1}{\cancel{3}}}{1} = \dfrac{20}{3} = \mathbf{6\dfrac{2}{3}}$

13. $\dfrac{25}{3} \div \dfrac{100}{1}$

$1 \div \dfrac{100}{1} = \dfrac{1}{100}$

$\dfrac{\overset{1}{\cancel{25}}}{3} \times \dfrac{1}{\underset{4}{\cancel{100}}} = \mathbf{\dfrac{1}{12}}$

14. $0.8 \div 10 = \mathbf{0.08}$

15. $0.5 \times 0.5 + 0.5 \div 0.5$
$0.25 + 1$
$\mathbf{1.25}$

16.
$$\begin{array}{r} \mathbf{200} \\ 4\overline{)800} \\ \underline{8} \\ 00 \\ \underline{00} \\ 00 \\ \underline{00} \\ 0 \end{array}$$

17. **4**

18. The mixed number $5\frac{1}{8}$ is more than 5 but less than 6. Since $\frac{1}{8}$ is less than $\frac{1}{2}$, $5\frac{1}{8}$ is closer to 5 than to 6. Thus $5\frac{1}{8}$ rounds to 5.

19.

$\mathbf{2^2 \cdot 5^2 \cdot 7}$

20. $8m = 16$

$$\begin{array}{r} 2 \\ 8\overline{)16} \\ \underline{16} \\ 0 \end{array}$$

$m = \mathbf{2}$

21.
$$\begin{array}{r} \mathbf{25\ cm} \\ 4\overline{)100\ cm} \\ \underline{8} \\ 20 \\ \underline{20} \\ 0 \end{array}$$

22. (a) $\dfrac{9}{45} = \mathbf{\dfrac{1}{5}}$

(b) $\dfrac{1}{5} \times \dfrac{20}{20} = \dfrac{20}{100} = \mathbf{20\%}$

23.
$$\begin{array}{r} \overset{1}{9}{:}30\ \text{p.m.} \\ +\ 5{:}30 \\ \hline 15{:}00 \end{array}$$
$\mathbf{3{:}00\ a.m.}$

24. $\mathbf{\dfrac{6}{4}} = \mathbf{\dfrac{n}{8}}$

$\dfrac{6}{4} \times \dfrac{2}{2} = \dfrac{12}{8}$

$n = \mathbf{12}$

25. $m\angle A = 70°$

$m\angle B = 180° - 70° = \mathbf{110°}$

$m\angle C = m\angle A = \mathbf{70°}$

$m\angle D = m\angle B = \mathbf{110°}$

26. $40 \times 1\ cm^3 = \mathbf{40\ cm^3}$

27. $24\ in. + 24\ in. = \mathbf{48\ in.}$

28. (a) $\dfrac{\overset{10}{\cancel{100}}}{\underset{1}{\cancel{10}}} \dfrac{cm \cdot cm}{cm} = \mathbf{10\ cm}$

(b) $\dfrac{\overset{45}{\cancel{180}}}{\underset{1}{\cancel{4}}} \dfrac{pages}{days} = \mathbf{45\ pages\ per\ day}$

29.

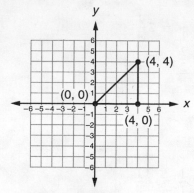

(0, 0)   (4, 4)

(4, 0)

(a) **6 full squares**

(b) **4 half squares**

30.

$$\overset{\overset{1}{2}{}^{1}4}{\underline{-\ 1\ 6}}$$
$$8 \text{ men}$$

$$\frac{\text{men}}{\text{women}} = \frac{8}{16} = \frac{1}{2}$$

LESSON 85, WARM-UP

a. 2500

b. 375

c. $15.96

d. $2.50

e. 0.75

f. 700

g. 30

Problem Solving

The third row contains multiples of 8, and the second column contains multiples of 9. The multiple of 9 that follows 63 is 72, so **81** is the wrong number.

LESSON 85, LESSON PRACTICE

a. 66 ⟵ $\frac{6}{9}$ ✕ $\frac{7}{11}$ ⟶ 63

No

b. 72 ⟵ $\frac{6}{8}$ ✕ $\frac{9}{12}$ ⟶ 72

Yes

c.
$$\frac{6}{10} = \frac{9}{x}$$
$$9 \cdot 10 = 6x$$
$$\frac{\overset{3}{\cancel{9}} \cdot \overset{5}{\cancel{10}}}{\underset{\underset{1}{2}}{\cancel{6}}} = x$$
$$15 = x$$

d.
$$\frac{12}{16} = \frac{y}{20}$$
$$12 \cdot 20 = 16y$$
$$\frac{\overset{3}{\cancel{12}} \cdot \overset{5}{\cancel{20}}}{\underset{\underset{1}{\cancel{4}}}{\cancel{16}}} = y$$
$$15 = y$$

e.
$$\frac{10}{15} = \frac{30}{x}$$
$$15 \cdot 30 = 10x$$
$$\frac{\overset{3}{\cancel{15}} \cdot \overset{15}{\cancel{30}}}{\underset{\underset{1}{2}}{\cancel{10}}} = x$$
$$45 = x$$

LESSON 85, MIXED PRACTICE

1. $\frac{21}{25} \times \frac{4}{4} = \frac{84}{100} = \mathbf{84\%}$

2. $17°F - (-6°F)$
$$\overset{1}{1}7°F$$
$$\underline{+\ 6°F}$$
$$\mathbf{23°F}$$

3.
$$\$1.50$$
$$\underline{+\ \$4.00}$$
$$\mathbf{\$5.50}$$

4.
$$\frac{5}{7} = \frac{350}{x}$$
$$7 \cdot 350 = 5x$$
$$\frac{7 \cdot \overset{70}{\cancel{350}}}{\underset{1}{\cancel{5}}} = x$$

490 walkers

5. $90° - 55° = \mathbf{35°}$

6. (a) $\begin{array}{r} \$55 \\ \times\ 0.20 \\ \hline \mathbf{\$11.00} \end{array}$

(b) $\begin{array}{r} \$55 \\ -\ \$11 \\ \hline \mathbf{\$44} \end{array}$

7. (a) $\begin{array}{r} \$39.60 \\ \times\ 0.08 \\ \hline \$3.1680 \end{array}$

$\mathbf{\$3.17}$

(b) $\begin{array}{r} \overset{1}{\$39.60} \\ +\ \$\ 3.17 \\ \hline \mathbf{\$42.77} \end{array}$

8. (a) $\dfrac{1}{25} \times \dfrac{4}{4} = \dfrac{4}{100}$ **0.04**

(b) **4%**

9.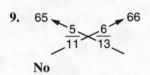

No

10. $\dfrac{4}{6} = \dfrac{10}{w}$

$6 \cdot 10 = 4w$

$\dfrac{\overset{3}{\cancel{6}} \cdot \overset{5}{\cancel{10}}}{\underset{\underset{1}{\cancel{2}}}{\cancel{4}}} = w$

$\mathbf{15} = w$

11. $\dfrac{10}{1} \div \dfrac{5}{2}$

$1 \div \dfrac{5}{2} = \dfrac{2}{5}$

$\dfrac{\overset{2}{\cancel{10}}}{1} \cdot \dfrac{2}{\underset{1}{\cancel{5}}} = \dfrac{4}{1} = \mathbf{4}$

12. $\begin{array}{r} 6.5\ -\ 3.68 \\ \mathbf{2.82} \end{array}$

13. $\begin{array}{r} 6.25 \\ \times\ 1.6 \\ \hline 3750 \\ 6250 \\ \hline 10.000 \\ \mathbf{10} \end{array}$

14. $\begin{array}{r} \mathbf{0.005} \\ 12\overline{)0.060} \\ \underline{60} \\ 0 \end{array}$

15. $3\dfrac{1}{4} \times \dfrac{1}{1} = \overset{2}{\cancel{3}}\dfrac{\overset{5}{\cancel{1}}}{4}$

$-\ 2\dfrac{1}{2} \times \dfrac{2}{2} = 2\dfrac{2}{4}$

$\rule{2cm}{0.4pt}$

$\dfrac{3}{4}$

$x = \dfrac{\mathbf{3}}{\mathbf{4}}$

16. $4\dfrac{1}{8} \times \dfrac{1}{1} = \overset{3}{\cancel{4}}\dfrac{\overset{9}{\cancel{1}}}{8}$

$-\ 1\dfrac{1}{2} \times \dfrac{4}{4} = 1\dfrac{4}{8}$

$\rule{2cm}{0.4pt}$

$2\dfrac{5}{8}$

$y = \mathbf{2\dfrac{5}{8}}$

17. $\dfrac{9}{12} = \dfrac{n}{20}$

$9 \cdot 20 = 12n$

$\dfrac{\overset{3}{\cancel{9}} \cdot \overset{5}{\cancel{20}}}{\underset{\underset{1}{\cancel{1}}}{\cancel{12}}} = n$

$\mathbf{15} = n$

18. $\mathbf{30\%,\ 0.4,\ \dfrac{1}{2}}$

19. $\dfrac{300}{15} = \dfrac{\mathbf{20}}{\mathbf{1}}$

20. **The probability of rolling a 4 or a 6 is $\dfrac{2}{6}$, or $\dfrac{1}{3}$.**

21. $\dfrac{1}{\underset{1}{\cancel{4}}} \cdot \dfrac{\overset{8}{\cancel{32}}}{1} = 8$ burned

$\begin{array}{r} \overset{2}{\cancel{3}}{}^{1}2 \\ -\ \ \ 8 \\ \hline \mathbf{2\ 4}\ \textbf{marshmallows} \end{array}$

22.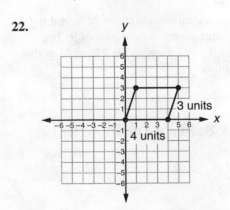

$4 \text{ units} \times 3 \text{ units} = \mathbf{12 \text{ sq. units}}$

23. $2 + 2 \times 2 - 2 \div 2$
$2 + 4 - 1 = 5$

24. $\overset{8:69}{\cancel{9:09}}$ a.m.
$- \ 8:22$ a.m.
47 min

25. $10 \times 0.621 = 6.21 \rightarrow$ **6 miles**

26. (a) $4 \times 5 =$ **20 boxes**

(b) $4 \times 5 \times 3 =$ **60 boxes**

27. (a) 2 ft $+ \ 2$ ft $=$ **4 ft**

(b) 3 yd $\cdot \ 3$ yd $=$ **9 yd^2**

28. 1 gallon $= 4$ quarts
$\dfrac{1}{4} \times \dfrac{25}{25} = \dfrac{25}{100} =$ **25%**

29. (a) 10 cm $\times \ 10$ cm $=$ **100 cm^2**

(b) 100 cm$^2 \div 2 =$ **50 cm^2**

30. **An educated guess might be "about 75 cm^2." Calculating the area of a circle will be covered in Lesson 86.**

LESSON 86, WARM-UP

a. **3600**

b. **680**

c. **$13.98**

d. **$0.50**

e. **8**

f. **3**

g. **11**

Problem Solving

Think, "What two-digit number times 8 equals a two-digit number?" The top factor must be 10, 11, or 12. Only 12 does not result in repeated digits.

$$\begin{array}{r} 12 \\ \times \ 8 \\ \hline 96 \end{array}$$

LESSON 86, LESSON PRACTICE

a. **About 50 sq. units**

b. $A = \pi r^2$
$A \approx (3.14)(4 \text{ cm})^2$
$A \approx$ **50.24 cm^2**

c. $A = \pi r^2$
$A \approx (3.14)(2 \text{ ft})^2$
$A \approx$ **12.56 ft^2**

d. $A = \pi r^2$
$A \approx (3.14)(1 \text{ ft})^2$
$A \approx$ **3.14 ft^2**

e. $A = \pi r^2$
$A \approx (3.14)(5 \text{ in.})^2$
$A \approx$ **78.5 in.^{2}**

LESSON 86, MIXED PRACTICE

1. $\begin{array}{r} 265 \\ 4\overline{)1060} \\ 8 \ \ \\ \hline 26 \\ 24 \\ \hline 20 \\ 20 \\ \hline 0 \end{array}$

2. $\overset{12:75}{\cancel{1:15}}$ p.m
$- \ 3:00$
$\overline{9:75}$ a.m.
10:15 a.m.

3. $\begin{array}{r} 12 \\ \times \ \$0.75 \\ \hline 60 \\ 840 \\ \hline \$9.00 \end{array}$ $\quad \begin{array}{r} \$10.00 \\ - \ \$9.00 \\ \hline \textbf{\$1.00} \end{array}$

4. $32 + 1.8(50)$
$32 + \ \ 90$
122

5. **16 in.^{3}**

6.
$$\frac{5}{2} = \frac{600}{n}$$

$$2 \cdot 600 = 5n$$

$$\frac{2 \cdot \overset{120}{\cancel{600}}}{\underset{1}{\cancel{5}}} = n$$

240 paperbacks

7. $\dfrac{3}{20} \times \dfrac{5}{5} = \dfrac{15}{100} = \mathbf{15\%}$

8.
$$\begin{array}{r} 0.015 \\ \times\ \$2000 \\ \hline \$30.000 \end{array}$$

$30

9. (a) $\dfrac{4}{5} \times \dfrac{20}{20} = \dfrac{80}{100} = \mathbf{0.8}$

(b) **80%**

10. $\dfrac{1}{4}$

11.
$$5\frac{1}{2} \times \frac{4}{4} = 5\frac{4}{8}$$
$$+\ 3\frac{7}{8} \times \frac{1}{1} = 3\frac{7}{8}$$
$$\overline{\qquad\qquad\qquad 8\frac{11}{8} = \mathbf{9\frac{3}{8}}}$$

12.
$$3\frac{1}{4} \times \frac{2}{2} = \overset{2}{\cancel{3}}\overset{10}{\cancel{\frac{2}{8}}}$$
$$-\ \frac{5}{8} \times \frac{1}{1} = \frac{5}{8}$$
$$\overline{\qquad\qquad\qquad 2\frac{5}{8}}$$

13. $\dfrac{\overset{3}{\cancel{6}}}{\underset{1}{\cancel{2}}} \cdot \dfrac{\overset{1}{\cancel{2}}}{\underset{1}{\cancel{3}}} = \dfrac{3}{1} = \mathbf{3}$

14. $\dfrac{25}{2} \div \dfrac{100}{1}$

$$1 \div \frac{100}{1} = \frac{1}{100}$$

$$\frac{\overset{1}{\cancel{25}}}{2} \cdot \frac{1}{\underset{4}{\cancel{100}}} = \mathbf{\frac{1}{8}}$$

15. $\dfrac{5}{1} \div \dfrac{3}{2}$

$$1 \div \frac{3}{2} = \frac{2}{3}$$

$$\frac{5}{1} \cdot \frac{2}{3} = \frac{10}{3} = \mathbf{3\frac{1}{3}}$$

16. $\dfrac{5}{\underset{1}{\cancel{6}}} \cdot \dfrac{\overset{5}{\cancel{30}}}{1} = \dfrac{25}{1} = \mathbf{\$25}$

17. $16.72 + n = 50.4$

$$\begin{array}{r} \overset{4}{\cancel{5}}\overset{9}{\cancel{0}}.\overset{{}^{13}}{\cancel{4}}{}^{1}0 \\ -\ 1\,6.\,7\,2 \\ \hline 3\,3.\,6\,8 \end{array}$$

$$n = \mathbf{33.68}$$

18.
$$\begin{array}{r} \$\overset{0}{\cancel{1}}\overset{9}{\cancel{0}}.\overset{9}{\cancel{0}}{}^{1}0 \\ -\ \$\ 9.\,8\,7 \\ \hline \$\ 0.\,1\,3 \end{array}$$

$$m = \mathbf{\$0.13}$$

19.
$$\begin{array}{r} 0.16 \\ 3\overline{)0.48} \\ \underline{3} \\ 18 \\ \underline{18} \\ 0 \end{array}$$

$$n = \mathbf{0.16}$$

20.
$$\frac{w}{8} = \frac{25}{20}$$

$$25 \cdot 8 = 20w$$

$$\frac{\overset{5}{\cancel{25}} \cdot \overset{2}{\cancel{8}}}{\underset{1}{\underset{\cancel{4}}{\cancel{20}}}} = w$$

$$\mathbf{10} = w$$

21. 121, 144, 169

22. (a) $A = 15\text{ cm} \times 10\text{ cm}$
$$A = \mathbf{150\text{ cm}^2}$$

(b)
$$\begin{array}{r} \mathbf{75\text{ cm}^2} \\ 2\overline{)150\text{ cm}^2} \\ \underline{14} \\ 10 \\ \underline{10} \\ 0 \end{array}$$

23. $A = \pi r^2$
$$A \approx (3.14)(10\text{ cm})^2$$
$$A \approx \mathbf{314\text{ cm}^2}$$

24.

$$\begin{array}{r} \overset{0\ \ 9}{\cancel{1}\ \cancel{0}^1 0\%} \\ -\quad 8\ 4\% \\ \hline 1\ 6 \text{ percentage points} \end{array}$$

25. 84%, 88%, (92%), 96%, 100%

92%

26.

$$\begin{array}{cc} & \mathbf{92\%} \\ \overset{2}{}84\% & 5\overline{)460} \\ 88\% & \underline{45} \\ 96\% & 10 \\ \overset{3}{}92\% & \underline{10} \\ +\ 100\% & 0 \\ \hline 460\% & \end{array}$$

27. Answers may vary. Sample answer: How
many percentage points did Bonnie improve
from test 1 to test 3?

$$\begin{array}{r} 96\% \\ -\ 84\% \\ \hline \mathbf{12 \text{ percentage points}} \end{array}$$

28. m∠Y = 180° − 110° = **70°**

m∠Z = **110°**

29. $C = \pi d$

$C \approx (3.14)(10 \text{ in.})$

$C \approx 31.4$

$C \approx \mathbf{31 \text{ inches}}$

30.

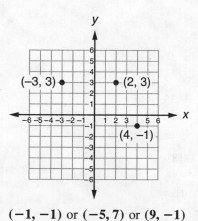

(−1, −1) or **(−5, 7)** or **(9, −1)**

LESSON 87, WARM-UP

a. 4900

b. 625

c. $24.95

d. $1.70

e. 0.625

f. 900

g. 10

Problem Solving

$(85 \times 3) + (95 \times 2) = 445$

$445 \div 5 = \mathbf{89}$

LESSON 87, LESSON PRACTICE

a.

$$\begin{array}{l} 3\frac{3}{6} = 3\frac{1}{2} \\ 6\overline{)21} \\ \underline{18} \\ 3 \end{array}$$

$w = \mathbf{3\frac{1}{2}}$

b.

$$\begin{array}{l} 16\frac{2}{3} \\ 3\overline{)50} \\ \underline{3} \\ 20 \\ \underline{18} \\ 2 \end{array}$$

$f = \mathbf{16\frac{2}{3}}$

c.

$$\begin{array}{l} 7\frac{1}{5} \\ 5\overline{)36} \\ \underline{35} \\ 1 \end{array}$$

$n = \mathbf{7\frac{1}{5}}$

d.

$$\begin{array}{l} 0.8 \\ 3\overline{)2.4} \\ \underline{2\ 4} \\ 0 \end{array}$$

$t = \mathbf{0.8}$

e.

$$\begin{array}{l} 0.4 \\ 8\overline{)3.2} \\ \underline{3\ 2} \\ 0 \end{array}$$

$m = \mathbf{0.4}$

f.

$$\begin{array}{l} 1.6 \\ 5\overline{)8.0} \\ \underline{5} \\ 3\ 0 \\ \underline{3\ 0} \\ 0 \end{array}$$

$x = \mathbf{1.6}$

LESSON 87, MIXED PRACTICE

1.
$$\begin{array}{r} 12 \\ \times\ 24 \\ \hline 48 \\ 240 \\ \hline \mathbf{288} \end{array}$$

2. $\dfrac{\overset{35}{\cancel{140}}\text{ tons}}{1} \cdot \dfrac{1}{\underset{1}{\cancel{4}}} = \dfrac{35}{1} = \mathbf{35\text{ tons}}$

3.
$$\begin{array}{r} \overset{2}{\cancel{3}}{}^{1}2 \\ -\ 1\,4 \\ \hline 1\,8 \text{ cooled cookies} \end{array}$$

$\dfrac{\text{warm cookies}}{\text{cooled cookies}} = \dfrac{14}{18} = \dfrac{\mathbf{7}}{\mathbf{9}}$

4.
$$\begin{array}{r} 0.9 \\ 3\overline{)2.7} \\ \underline{2\ 7} \\ 0 \end{array} \qquad m = \mathbf{0.9}$$

5.
$$\begin{array}{r} 6\frac{1}{5} \\ 5\overline{)31} \\ \underline{30} \\ 1 \end{array} \qquad n = \mathbf{6\frac{1}{5}}$$

6.
$$\begin{array}{r} 12 \\ 3\overline{)36} \\ \underline{3} \\ 06 \\ \underline{06} \\ 0 \end{array} \qquad n = \mathbf{12}$$

7.
$$\begin{array}{r} 0.0875 \\ 4\overline{)0.3500} \\ \underline{32} \\ 30 \\ \underline{28} \\ 20 \\ \underline{20} \\ 0 \end{array} \qquad n = \mathbf{0.0875}$$

8. $\dfrac{25}{100} = \dfrac{1}{4}$

$$\begin{array}{r} 3\frac{1}{4} \\ +\ \frac{1}{4} \\ \hline 3\frac{2}{4} = \mathbf{3\frac{1}{2}} \end{array}$$

9. $\dfrac{3}{5} \times \dfrac{20}{20} = \dfrac{60}{100} = 0.60$

$$\begin{array}{r} \overset{1}{6.5} \\ +\ 0.6 \\ \hline \mathbf{7.1} \end{array}$$

10. $\dfrac{1}{50} \times \dfrac{2}{2} = \dfrac{2}{100} = \mathbf{0.02;\ 2\%}$

11.
$$\begin{array}{r} \overset{11}{\cancel{12}}\overset{6}{\cancel{1}}\frac{1}{5} \\ -\ 3\frac{4}{5} \\ \hline \mathbf{8\frac{2}{5}} \end{array}$$

12. $\dfrac{\overset{4}{\cancel{20}}}{\underset{1}{\cancel{3}}} \times \dfrac{\overset{2}{\cancel{6}}}{\underset{1}{\cancel{5}}} = \dfrac{8}{1} = \mathbf{8}$

13. $\dfrac{100}{9} \div \dfrac{100}{1}$

$1 \div \dfrac{100}{1} = \dfrac{1}{100}$

$\dfrac{\overset{1}{\cancel{100}}}{9} \cdot \dfrac{1}{\underset{1}{\cancel{100}}} = \dfrac{\mathbf{1}}{\mathbf{9}}$

14.
$$\begin{array}{r} \overset{1}{4.75} \\ 12.6 \\ +\ 10 \\ \hline \mathbf{27.35} \end{array}$$

15. $35 - (0.35 \times 100)$

$35 - 35$

$\mathbf{0}$

16. $\dfrac{12}{m} = \dfrac{18}{9}$

$12 \cdot 9 = 18m$

$\dfrac{\overset{6}{\cancel{12}} \cdot \overset{1}{\cancel{9}}}{\underset{\underset{1}{2}}{\cancel{18}}} = m$

$\mathbf{6} = m$

17. 12.05

18. $V = lwh$
$V = 10 \text{ in.} \times 5 \text{ in.} \times 5 \text{ in.}$
$V = \mathbf{250 \text{ in.}^3}$

19. $2(15) - 5$

$\quad 30 - 5$

$\qquad$ **25**

20. $A = (25\,\text{mm})(18\,\text{mm})$

$A = \textbf{450 mm}^2$

21. $25\,\text{mm} + 25\,\text{mm} + 20\,\text{mm} + 20\,\text{mm}$

$\qquad = \textbf{90 mm}$

22. True

23. $\dfrac{1}{\cancel{10}_{1}} \cdot \dfrac{\cancel{100}^{10}}{1} = 10$ shillings

$100 - 10 = \textbf{90 shillings}$

24. $18°\text{F} - (-19°\text{F})$

$\qquad\qquad\qquad \overset{1}{1}8°\text{F}$

$\qquad\qquad\qquad \underline{-\ 19°\text{F}}$

$\qquad\qquad\qquad \textbf{37°F}$

25. 4 cm

26. $2\,\text{liter} = 2000\,\text{milliliters}$

$2000\,\text{mL} - 500\,\text{mL} = \textbf{1500 mL}$

27. Triangular prism

28. (a) $2\,\text{m} + 1\,\text{m} = \textbf{3 m}$

(b) $2\,\text{m} \times 4\,\text{m} = \textbf{8 m}^2$

29. $4 + 4 \times 4 - 4 \div 4$

$\quad 4 + 16 - 1$

$\qquad \textbf{19}$

30.

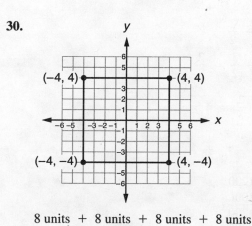

$8\,\text{units} + 8\,\text{units} + 8\,\text{units} + 8\,\text{units}$

$\qquad = \textbf{32 units}$

LESSON 88, WARM-UP

a. 6400

b. 570

c. $3.96

d. $0.50

e. 15

f. 3

g. 11

Problem Solving

Multiples of both 5 and 6: 30, 60

Since 60 is a multiple of 4, the answer is **30**.

LESSON 88, LESSON PRACTICE

a.

	Ratio	Actual Count
Damsels	5	d
Knights	4	60

$\dfrac{5}{4} = \dfrac{d}{60}$

$5 \cdot 60 = 4d$

$\dfrac{5 \cdot \cancel{60}^{15}}{\cancel{4}_{1}} = d$

$75 = d$

75 damsels

b.

	Ratio	Actual Count
Boys	5	30
Girls	3	g

$\dfrac{5}{3} = \dfrac{30}{g}$

$5g = 3 \cdot 30$

$g = \dfrac{3 \cdot \cancel{30}^{6}}{\cancel{5}_{1}}$

$g = 18$

18 girls

SOLUTIONS

LESSON 88, MIXED PRACTICE

1.
$$\frac{12}{20} = \frac{3}{5}$$
$$\frac{3}{5} \times \frac{20}{20} = \frac{60}{100} = \mathbf{60\%}$$

2.
$$\frac{4}{1} \div \frac{1}{4}$$
$$1 \div \frac{1}{4} = \frac{4}{1}$$
$$\frac{4}{1} \times \frac{4}{1} = \frac{16}{1}$$
16 lies

3.
$$\begin{array}{r} \$\overset{4}{\cancel{5}}{}^{1}0 \\ -\ \$1\,4 \\ \hline \$3\,6 \end{array}$$
$$\begin{array}{r} 6 \\ \$6\overline{)\$36} \\ \underline{36} \\ 0 \end{array}$$
6 hours

4. $A = \dfrac{\overset{8}{\cancel{16}} \cdot 8}{\underset{1}{\cancel{2}}}$

$A = \mathbf{64\ mm^2}$

5.

	Ratio	Actual Count
Skaters	3	b
Bikers	5	15

$$\frac{3}{5} = \frac{b}{15}$$
$$3 \cdot 15 = 5b$$
$$\frac{3 \cdot \overset{3}{\cancel{15}}}{\underset{1}{\cancel{5}}} = b$$
$$9 = b$$
9 skaters

6. $V = 10\ \text{in.} \times 6\ \text{in.} \times 8\ \text{in.}$
$V = \mathbf{480\ in.^3}$

7.
$$\begin{array}{r} 2\frac{5}{10} = 2\frac{1}{2} \\ 10\overline{)25} \\ \underline{20} \\ 5 \end{array}$$
$$w = \mathbf{2\frac{1}{2}}$$

8.
$$\begin{array}{r} 2\frac{2}{9} \\ 9\overline{)20} \\ \underline{18} \\ 2 \end{array}$$
$$m = \mathbf{2\frac{2}{9}}$$

9. (a) $10\ \text{in.} + 6\ \text{in.} + 8\ \text{in.} = 24\ \text{in.}$

(b) $A = \dfrac{8\ \text{in.} \times 6\ \text{in.}}{2}$

$A = \dfrac{48\ \text{in.}^2}{2}$

$A = \mathbf{24\ in.^2}$

10. (a) $\dfrac{5}{100} = \mathbf{0.05}$

(b) $\dfrac{5}{100} = \mathbf{\dfrac{1}{20}}$

11.
$$\begin{array}{r} 0.4 \\ 5\overline{)2.0} \\ \underline{2\ 0} \\ 0 \end{array}$$
$$\begin{array}{r} 2.5 \\ \times\ 0.4 \\ \hline 1.00 \end{array}$$
1

12.
$$\frac{2}{3} + \frac{3}{2} \ \textcircled{>}\ \frac{2}{3} \cdot \frac{3}{2}$$
$$\frac{4}{6} + \frac{9}{6} = \frac{13}{6} \qquad \frac{6}{6} = 1$$
$$2\frac{1}{6}$$

13. $\dfrac{1}{3} \times \dfrac{100}{1} = \dfrac{100}{3} = \mathbf{33\frac{1}{3}}$

14.
$$\frac{6}{1} \div \frac{3}{2}$$
$$1 \div \frac{3}{2} = \frac{2}{3}$$
$$\frac{\overset{2}{\cancel{6}}}{1} \cdot \frac{2}{\underset{1}{\cancel{3}}} = \mathbf{4}$$

15.
$$\begin{array}{r} 48 \\ 25\overline{)1200} \\ \underline{100} \\ 200 \\ \underline{200} \\ 0 \end{array}$$

192

16. $0.025 \times 100 = $ **2.5**

17.
$$\begin{array}{r} \$24.90 \\ \times\ \ 0.07 \\ \hline 1.7430 \end{array} \longrightarrow \textbf{\$1.74}$$

18. $4 \cdot 9 \cdot 25 = $ **900**

19. C. **81**

20. $1.23 \longrightarrow$ **1.2**

21. $\dfrac{\overset{5}{\cancel{60}}}{1} \cdot \dfrac{7}{\underset{1}{\cancel{12}}} = 35$ cookies

$$\begin{array}{r} \overset{5}{\cancel{6}}\overset{1}{0} \\ -\ 3\ 5 \\ \hline \textbf{2 5 cookies} \end{array}$$

22. $6 \times 3 - 6 \div 3$
$$18 - 2$$
$$\textbf{16}$$

23. 4 milliliters $\times 1000 = $ **4000 milliliters**

24.

$$A \qquad\qquad B \qquad\qquad C$$

$$\frac{9}{4} \div \frac{2}{1}$$

$$1 \div \frac{2}{1} = \frac{1}{2}$$

$$\frac{9}{4} \cdot \frac{1}{2} = \frac{9}{8} = 1\frac{1}{8} \text{ in.}$$

$$AB = BC = 1\frac{1}{8} \text{ in.}$$

25.

$A = 6$ units $\times\ 4$ units

$A = $ **24 square units**

26. $\dfrac{\text{length}}{\text{width}} = \dfrac{6}{4} = \dfrac{3}{2}$

27. **One method is to multiply the base by the height of the triangle and then divide the product by 2.**

28. (a) $180° - 45° = $ **135°**

(b) $\dfrac{45}{135} = \dfrac{1}{3}$

29. (a) $20\text{ cm} \times 20\text{ cm} = $ **400 cm²**

(b) $A = \pi r^2$
$A \approx (3.14)(10\text{ cm})^2$
$A \approx $ **314 cm²**

30.
$$\frac{6}{8} = \frac{w}{100}$$
$$6 \cdot 100 = 8w$$
$$\frac{\overset{3}{\cancel{6}} \cdot \overset{25}{\cancel{100}}}{\underset{\underset{1}{\cancel{A}}}{\cancel{8}}} = w$$
$$\textbf{75} = w$$

LESSON 89, WARM-UP

a. 8100

b. 595

c. $47.94

d. $54.00

e. 0.875

f. 720

g. 3

h. about 2 m

Problem Solving
$$1\text{ m} = 100\text{ cm}$$
$$100\text{ cm} - (21\text{ cm} \times 2) = 58\text{ cm}$$
$$58\text{ cm} \div 2 = \textbf{29 cm}$$

LESSON 89, LESSON PRACTICE

a. $\sqrt{169}$

$$
\begin{array}{r}
13 \\
\times\ 13 \\
\hline
39 \\
130 \\
\hline
169 \\
\end{array}
$$

13

b. $\sqrt{484}$

$$
\begin{array}{r}
22 \\
\times\ 22 \\
\hline
44 \\
440 \\
\hline
484 \\
\end{array}
$$

22

c. $\sqrt{961}$

$$
\begin{array}{r}
31 \\
\times\ 31 \\
\hline
31 \\
930 \\
\hline
961 \\
\end{array}
$$

31

d. The number 2 is between the perfect squares 1 and 4, so $\sqrt{2}$ is between $\sqrt{1}$ and $\sqrt{4}$. Since $\sqrt{1}$ is 1 and $\sqrt{4}$ is 2, $\sqrt{2}$ is between 1 and 2.
$\sqrt{2}$
1 and 2

e. The number 15 is between the perfect squares 9 and 16, so $\sqrt{15}$ is between $\sqrt{9}$ and $\sqrt{16}$. Since $\sqrt{9}$ is 3 and $\sqrt{16}$ is 4, $\sqrt{15}$ is between 3 and 4.
$\sqrt{15}$
3 and 4

f. The number 40 is between the perfect squares 36 and 49, so $\sqrt{40}$ is between $\sqrt{36}$ and $\sqrt{49}$. Since $\sqrt{36}$ is 6 and $\sqrt{49}$ is 7, $\sqrt{40}$ is between 6 and 7.
$\sqrt{40}$
6 and 7

g. The number 60 is between the perfect squares 49 and 64, so $\sqrt{60}$ is between $\sqrt{49}$ and $\sqrt{64}$. Since $\sqrt{49}$ is 7 and $\sqrt{64}$ is 8, $\sqrt{60}$ is between 7 and 8.
$\sqrt{60}$
7 and 8

h. The number 70 is between the perfect squares 64 and 81, so $\sqrt{70}$ is between $\sqrt{64}$ and $\sqrt{81}$. Since $\sqrt{64}$ is 8 and $\sqrt{81}$ is 9, $\sqrt{70}$ is between 8 and 9.
$\sqrt{70}$
8 and 9

i. The number 80 is between the perfect squares 64 and 81, so $\sqrt{80}$ is between $\sqrt{64}$ and $\sqrt{81}$. Since $\sqrt{64}$ is 8 and $\sqrt{81}$ is 9, $\sqrt{80}$ is between 8 and 9.
$\sqrt{80}$
8 and 9

j. The display on the calculator will show 1.732050808. This number rounded to two decimal places is 1.73.
$\sqrt{3}$
1.73

k. The display on the calculator will show 3.162277660. This number rounded to two decimal places is 3.16.
$\sqrt{10}$
3.16

l. The display on the calculator will show 7.071067812. This number rounded to two decimal places is 7.07.
$\sqrt{50}$
7.07

LESSON 89, MIXED PRACTICE

1. $\left(\dfrac{1}{4} + \dfrac{1}{4}\right) - \left(\dfrac{1}{2} \times \dfrac{1}{2}\right)$

$$\dfrac{2}{4} \quad - \quad \dfrac{1}{4}$$

$$\dfrac{1}{4}$$

2. 4 gallons $=$ 16 quarts
16 quarts $=$ 64 cups
64 cups

3. $\dfrac{3}{\underset{2}{\cancel{4}}}$ cup $\times \dfrac{\overset{1}{\cancel{2}}}{1} = \dfrac{3}{2} = 1\dfrac{1}{2}$ **cups**

4.

	Ratio	Actual Count
Sugar	2	s
Flour	9	18

$$\dfrac{2}{9} = \dfrac{s}{18}$$

$$2 \cdot 18 = 9s$$

$$\dfrac{2 \cdot \overset{2}{\cancel{18}}}{\underset{1}{\cancel{9}}} = s$$

$$4 = s$$

4 pounds

5. C. $\sqrt{45}$

6. $7\overline{)30}$ $4\frac{2}{7}$
$\underline{28}$
2

$n = \mathbf{4\frac{2}{7}}$

7. (a) 2×4 in. = **8 in.**

(b) $C = \pi d$
$C \approx (3.14)(8\text{ in.})$
$C \approx \mathbf{25.12\text{ in.}}$

8. $A = \pi r^2$
$A \approx (3.14)(16\text{ in.})^2$
$A \approx \mathbf{50.24\text{ in}^2}$

9. $A = \dfrac{8\text{ in.} \times 5\text{ in.}}{2}$
$A = \dfrac{40\text{ in.}^2}{2}$
$A = \mathbf{20\text{ in.}^2}$

10. (a) $A = 8\text{ in.} \times 5\text{ in.}$
$A = \mathbf{40\text{ in.}^2}$

(b) 6 in. $+ 8$ in. $+ 6$ in. $+ 8$ in. $= \mathbf{28\text{ in.}}$

11. $\dfrac{5}{10} = \dfrac{1}{2}$

$3\frac{1}{4} \times \frac{1}{1} = 3\frac{1}{4}$
$- \frac{1}{2} \times \frac{2}{2} = \frac{2}{4}$
$2\frac{3}{4}$

12. $\dfrac{3}{4} \times \dfrac{25}{25} = \dfrac{75}{100}$; 0.75
0.75
$\underline{\times 0.6}$
0.450
$\mathbf{0.45}$

13. $2 \times 15 + 2 \times 12$
$30 + 24$
$\mathbf{54}$

14. $\sqrt{900}$
30
$\underline{\times\ 30}$
900
$\mathbf{30}$

15. $8\overline{)\$6.00}$ $\$0.75$
$\underline{5\ 6}$
$\ \ 40$
$\ \ \underline{40}$
$\ \ \ \ 0$
$\mathbf{\$0.75}$

16. $\dfrac{8}{2} \times \dfrac{10}{1} \times \dfrac{1}{4} = \dfrac{4}{1} = \mathbf{4}$

17. $\dfrac{75}{2} \div \dfrac{100}{1}$
$1 \div \dfrac{100}{1} = \dfrac{1}{100}$
$\dfrac{75}{2} \times \dfrac{1}{100} = \dfrac{\mathbf{3}}{\mathbf{8}}$

18. $\dfrac{3}{1} \div \dfrac{15}{2}$
$1 \div \dfrac{15}{2} = \dfrac{2}{15}$
$\dfrac{3}{1} \times \dfrac{2}{15} = \dfrac{\mathbf{2}}{\mathbf{5}}$

19. **Thousands**

20. **510.05**

21. $90 + m = 180$
$m = 180 - 90$
$m = \mathbf{90}$

22. $\dfrac{32}{1} \cdot \dfrac{1}{2} = 16$ girls
$\dfrac{16}{1} \cdot \dfrac{1}{2} = 8$ brown hair
$\dfrac{8}{1} \cdot \dfrac{1}{2} = 4$ guests

23. $4 \times 3 = \mathbf{12\text{ books}}$

24. $20 - 14 = \mathbf{6\text{ more books}}$

25. **Answers may vary. Sample answer: How many books has Pat read? 14 books**

26.
$$\frac{12}{8} = \frac{21}{m}$$
$$8 \cdot 21 = 12m$$
$$\frac{\overset{2}{\cancel{8}} \cdot \overset{7}{\cancel{21}}}{\underset{\underset{1}{\cancel{3}}}{\cancel{12}}} = m$$
$$14 = m$$

27. $\dfrac{2}{12} = \dfrac{1}{6}$

28. **A. Acute**

29. (a) 200 cm = **2 m**

(b) $\dfrac{40 \text{ in.}^2}{2} = \textbf{20 in.}^2$

30. $64 \times 1 \text{ in.}^3 = \textbf{64 in.}^3$

LESSON 90, WARM-UP

a. **10**

b. **746**

c. **$4.96**

d. **$8.00**

e. **37.5**

f. **4**

g. **2**

h. **about 3 feet**

Problem Solving

A00001 to A99999 is 99,999 license plates. Add 1 plate for A00000 to get a total of **100,000 license plates.**

LESSON 90, LESSON PRACTICE

a.

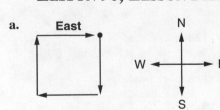

b. $2 \times 360° = \textbf{720°}$

c.

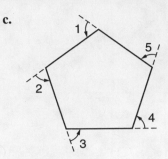

$$\frac{360°}{5} = \textbf{72°}$$

LESSON 90, MIXED PRACTICE

1.
$$\begin{array}{r} 4.2 \\ 4.8 \\ + \ 5.1 \\ \hline 14.1 \end{array} \qquad \begin{array}{r} 4.7 \\ 3\overline{)14.1} \\ 12 \\ \hline 2\ 1 \\ 2\ 1 \\ \hline 0 \end{array}$$

2.
$$\begin{array}{r} 7{:}15 \text{ p.m.} \\ + \ 2{:}00 \\ \hline \textbf{9:15 p.m.} \end{array}$$

3. $\dfrac{15}{25} \times \dfrac{4}{4} = \dfrac{60}{100} = \textbf{60\%}$

4. $\dfrac{15}{10} = \dfrac{\textbf{3}}{\textbf{2}}$

5. 9 edges and 6 vertices
3 more edges

6. (a) 12 inches ÷ 2 = **6 inches**

(b) $A = \pi r^2$
$A \approx (3.14)(6 \text{ in.})^2$
$A \approx \textbf{113.04 in.}^2$

7. **A trapezoid is a polygon with four sides. Two of the sides are parallel. The other two sides are not parallel.**

8. $-4, -2, 0, \dfrac{1}{2}, 1$

9.
$$25\overline{)70} \quad \begin{array}{r} 2\frac{20}{25} = 2\frac{4}{5} \\ 50 \\ \hline 20 \end{array}$$
$$n = 2\frac{4}{5}$$

10. $A = \dfrac{20 \text{ mm} \times 15 \text{ mm}}{2}$

$A = \dfrac{300 \text{ mm}^2}{2}$

$A = \mathbf{150 \text{ mm}^2}$

11. $25 \text{ mm} + 20 \text{ mm} + 15 \text{ mm} = \mathbf{60 \text{ mm}}$

12. $6\dfrac{25}{100} = 6\dfrac{1}{4}$

$6\dfrac{1}{4} \times \dfrac{2}{2} = 6\dfrac{\overset{10}{2}}{8}$

$- \quad \dfrac{5}{8} \times \dfrac{1}{1} = \quad \dfrac{5}{8}$

$\mathbf{5\dfrac{5}{8}}$

13.

N

180°

South

14. $\dfrac{28}{100} = \mathbf{\dfrac{7}{25}}$

15. $\dfrac{n}{12} = \dfrac{20}{30}$

$12 \cdot 20 = 30n$

$\dfrac{\overset{4}{\cancel{12}} \cdot \overset{2}{\cancel{20}}}{\underset{\underset{1}{\cancel{3}}}{\cancel{30}}} = n$

$\mathbf{8} = n$

16. $0.625 \div 10 = \mathbf{0.0625}$

17.
$$8\overline{)250.00}$$
$$31.25$$

$\begin{array}{r} 31.25 \\ 8)\overline{250.00} \\ 24 \\ \overline{10} \\ 8 \\ \overline{2\ 0} \\ 1\ 6 \\ \overline{40} \\ 40 \\ \overline{0} \end{array}$

18.

$3\dfrac{3}{8} \times \dfrac{1}{1} = 3\dfrac{3}{8}$

$+ \ 3\dfrac{3}{4} \times \dfrac{2}{2} = 3\dfrac{6}{8}$

$6\dfrac{9}{8} = \mathbf{7\dfrac{1}{8}}$

19.

$\overset{4}{\cancel{5}}\dfrac{\overset{9}{\cancel{1}}}{8}$

$- \ 1\dfrac{7}{8}$

$3\dfrac{2}{8} = \mathbf{3\dfrac{1}{4}}$

20. $\dfrac{\overset{2}{\cancel{20}}}{\underset{1}{\cancel{5}}} \times \dfrac{\overset{1}{\cancel{5}}}{\underset{1}{\cancel{10}}} \times \dfrac{4}{1} = \dfrac{8}{1} = \mathbf{8}$

21. $\dfrac{1}{\underset{1}{\cancel{3}}} \cdot \dfrac{\overset{8}{\cancel{24}}}{1} = 8$

$\overset{1}{\cancel{2}}{}^{1}4$

$- \quad 8$

$\mathbf{1\ 6 \text{ knights}}$

22.

$\begin{array}{r} 38 \text{ ounces} \\ + \ 26 \text{ ounces} \\ \hline 64 \text{ ounces} \end{array}$

$\begin{array}{r} 32 \text{ ounces} \\ 2)\overline{64 \text{ ounces}} \\ 6\phantom{4 \text{ ounces}} \\ \overline{04}\phantom{ \text{ ounces}} \\ 04\phantom{ \text{ ounces}} \\ \overline{0}\phantom{ \text{ ounces}} \end{array}$

$38 \text{ ounces} - 32 \text{ ounces} = 6$

6 ounces

23. $27 \times 1 \text{ cm}^3 = \mathbf{27 \text{ cm}^3}$

24. $0.48 \longrightarrow \mathbf{0.5}$

25. $12 - 11 = \mathbf{1}$

26. $\dfrac{\text{cats}}{\text{dogs}} = \mathbf{\dfrac{5}{6}}$

27. $10 + 10 \times 10 - 10 \div 10$

$10 + 100 - 1$

$\mathbf{109}$

28. 1 gallon = 4 quarts

4 quarts = 8 pints

$$\frac{2 \text{ pints in 2 days}}{4\overline{)8}}$$
$$\frac{8}{0}$$

1 pint per day

29. (a) 100 mm + 100 mm = **200 mm**

(b) $\frac{\overset{10}{\cancel{300}} \text{ books}}{\underset{1}{\cancel{30}} \text{ shelves}}$

10 books per shelf

30.

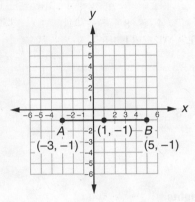

(1, −1)

INVESTIGATION 9

1.

	Freq.	Relative Frequency
Bob's Market	30	$\frac{30}{80}$ = 0.375
Corner Grocery	12	$\frac{12}{80}$ = 0.15
Express Grocery	14	$\frac{14}{80}$ = 0.175
Fine Foods	24	$\frac{24}{80}$ = 0.30

2. $\frac{14}{80}$ = **0.175**

3. $\frac{30}{80}$ = $\frac{3}{8}$

4. $\frac{24}{80}$ = 0.30

0.30 = **30%**

5. $\frac{12}{80} = \frac{a}{4000}$

$80a = 48{,}000$

$a = 600$

about 600 adults

6. See student work. The tally table should have 25 tally marks.

7. See student work. The relative frequency table should include three decimal frequencies.

8. See student work. The theoretical probability of drawing two red marbles is $\frac{2}{5}$, which is equal to 0.4. However, estimates may vary widely from the theoretical probability. Accept any answer that is based on experimental results from problems 6 and 7.

Extensions

a. See student work. The table should include relative frequencies for each of the six choices.

b. See student work. The theoretical probability of rolling a sum of 8 is $\frac{5}{36}$. The theoretical probability of rolling a sum of at least 10 is $\frac{11}{13}$. The theoretical probability of an odd sum is $\frac{1}{2}$. However, estimates may vary widely from the theoretical probability. Accept any answer that is based on experimental data.

LESSON 91, WARM-UP

a. 35

b. 125

c. $34.95

d. $250.00

e. 0.125

f. 840

g. 7

Problem Solving

The hundreds digit in the top factor must be 1.
(Digits greater than 1 would result in a four-digit
product.) We know that 2 is added to the
hundreds column from regrouping (because
$7 \times 1 + 2 = 9$), so the tens digit of the factor
must be 3 or 4. We try 3. Of the remaining
possible digits, only 6 can fill the ones place of
the factor.

$$
\begin{array}{r}
136 \\
\times 7 \\
\hline
952
\end{array}
$$

LESSON 91, LESSON PRACTICE

a. $A = lw$
$A = (8\text{ cm})(5\text{ cm})$
$A = \textbf{40 cm}^2$

b. $P = 2b + 2s$
$P = 2(10\text{ cm}) + 2(6\text{ cm})$
$P = \textbf{32 cm}$

LESSON 91, MIXED PRACTICE

1. $\dfrac{4}{8} = \dfrac{1}{2}$

2.
$$
\begin{array}{r}
\overset{17}{\cancel{8}}\text{:45 p.m.} \\
- 6\text{:15 a.m.} \\
\hline
11\text{:30}
\end{array}
$$
11 hr 30 min

3.

	Ratio	A.C.
Leapers	3	12
Duckers	2	d

$\dfrac{3}{2} = \dfrac{12}{d}$

$2 \cdot 12 = 3d$

$\dfrac{2 \cdot \overset{4}{\cancel{12}}}{\underset{1}{\cancel{3}}} = d$

8 duckers

4. 6 faces − 5 faces = 1 face
1 more face

5.

Outcomes		
Penny	**Nickel**	**Dime**
H	H	H
H	H	T
H	T	H
H	T	T
T	H	H
T	H	T
T	T	H
T	T	T

6. The minute hand of a clock makes a quarter
turn every 15 minutes. Each quarter turn is 90°.
In 45 minutes the minute hand makes three
quarter turns, so it turns $3 \times 90°$, or **270°**.

7. (a) **4 faces**

(b) **6 edges**

(c) **4 vertices**

8. (a) $P = 2b + 2s$
$P = 2(12\text{ in.}) + 2(10\text{ in.})$
$P = \textbf{44 in.}$

(b) $A = bh$
$A = (12\text{ in.})(9\text{ in.})$
$A = \textbf{108 in.}^2$

9. (a) $\dfrac{7}{20} \times \dfrac{5}{5} = \dfrac{35}{100} = \textbf{0.35}$

(b) $\dfrac{7}{20} \cdot \dfrac{5}{5} = \dfrac{35}{100} = \textbf{35\%}$

10. $P = 2l + 2w$
$P = 2(5\text{ ft}) + 2(3\text{ ft})$
$P = \textbf{16 ft}$

11.
$$
\begin{aligned}
6\tfrac{2}{3} \times \tfrac{4}{4} &= 6\tfrac{8}{12} \\
+ \ 1\tfrac{3}{4} \times \tfrac{3}{3} &= 1\tfrac{9}{12} \\
\hline
7\tfrac{17}{12} &= \mathbf{8\tfrac{5}{12}}
\end{aligned}
$$

12.
$$
\begin{aligned}
5 &= 4\tfrac{5}{5} \\
- \ 1\tfrac{2}{5} &= 1\tfrac{2}{5} \\
\hline
&\mathbf{3\tfrac{3}{5}}
\end{aligned}
$$

13.
$$4\frac{1}{4} \times \frac{2}{2} = 4\frac{\overset{10}{2}}{8}$$
$$-\ 3\frac{5}{8} \times \frac{1}{1} = 3\frac{5}{8}$$
$$\overline{\qquad\qquad\qquad \frac{5}{8}}$$

14.
$$\frac{3}{1} \times \frac{\overset{1}{\cancel{2}}}{\overset{\cancel{4}}{1}} \times \frac{\overset{2}{\cancel{8}}}{\overset{\cancel{3}}{1}} = \frac{6}{1} = 6$$

15.
$$\frac{20}{3} \div \frac{100}{1}$$
$$1 \div \frac{100}{1} = \frac{1}{100}$$
$$\frac{\overset{1}{\cancel{20}}}{3} \times \frac{1}{\underset{5}{\cancel{100}}} = \frac{1}{15}$$

16.
$$\frac{5}{2} \div \frac{15}{4}$$
$$1 \div \frac{15}{4} = \frac{4}{15}$$
$$\frac{\overset{1}{\cancel{5}}}{\underset{1}{\cancel{2}}} \times \frac{\overset{2}{\cancel{4}}}{\underset{3}{\cancel{15}}} = \frac{2}{3}$$

17.
$$\frac{9}{20} \,\text{\large\textcircled{<}}\, 50\%$$
$$\frac{9}{20} \times \frac{5}{5} \qquad \frac{50}{100}$$
$$\frac{45}{100}$$

18. (a) $\dfrac{1}{4}$

   (b) $\dfrac{1}{4} \times \dfrac{25}{25} = \dfrac{25}{100} = \mathbf{25\%}$

19.
$$\frac{5}{\underset{1}{\cancel{6}}} \cdot \frac{\overset{50}{\cancel{300}}\ \text{seeds}}{1} = \frac{250}{1} = 250$$
$$300 - 250 = \mathbf{50\ seeds}$$

20.
$$\overset{\displaystyle 1\frac{4}{6} = 1\frac{2}{3}}{6)\overline{10}}$$
$$\underline{6}$$
$$4$$

21. $20 \cdot 12 = 15w$
$$\frac{\overset{4}{\cancel{20}} \cdot \overset{4}{\cancel{12}}}{\underset{\underset{1}{\cancel{3}}}{\cancel{15}}} = w$$
$$16 = w$$

22. $A = \dfrac{1}{2} bh$

   $A = \dfrac{1}{2}\,(6\ \text{ft})\,(4\ \text{ft})$

   $A = \dfrac{1}{2}\,(24\ \text{ft}^2)$

   $A = \mathbf{12\ ft^2}$

23. $V = (12\ \text{in.})\,(12\ \text{in.})\,(12\ \text{in.})$
   $V = \mathbf{1728\ in.^3}$

24.

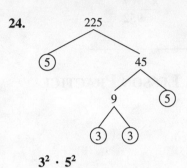

$$3^2 \cdot 5^2$$

25.
$$\begin{array}{r} 56\ \text{mm} \\ -\ 26\ \text{mm} \\ \hline \mathbf{30\ mm} \end{array}$$

26. The number 60 is between the perfect squares 49 and 64, so $\sqrt{60}$ is between $\sqrt{49}$ and $\sqrt{64}$. Since $\sqrt{49}$ is 7 and $\sqrt{64}$ is 8, $\sqrt{60}$ is between **7 and 8.**

27.
$$\begin{array}{r} 15 \\ \times\ 15 \\ \hline 75 \\ 150 \\ \hline 225 \end{array}$$
15

28.

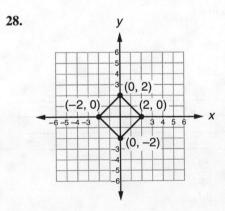

29. (a) **4 whole squares**

   (b) **8 half squares**

   (c) 8 half squares $=$ 4 whole squares
   Total area $=$ 4 whole squares
   $+$ 4 whole squares
   $= \mathbf{8\ square\ units}$

30. $A = bh$
    $A = (15 \text{ cm})(4 \text{ cm})$
    $A = 60 \text{ cm}^2$

LESSON 92, WARM-UP

a. 1500

b. 536

c. 12

d. $5.25

e. 125

f. 20

g. 3

Problem Solving
$90\% \times 5 = 450\%$
$88\% \times 4 = 352\%$
$450\% - 352\% = \textbf{98\%}$

LESSON 92, LESSON PRACTICE

a. $(2 \times 10^6) + (5 \times 10^5)$

b. **5,200,000,000**

c. $10 + 2^3 \times 3 - (7 + 2) \div \sqrt{9}$
   $10 + 2^3 \times 3 - 9 \div \sqrt{9}$
   $10 + 8 \times 3 - 9 \div 3$
   $10 + 24 - 3$
   31

d. $\dfrac{1}{2} \cdot \dfrac{1}{2} \cdot \dfrac{1}{2} = \dfrac{1}{8}$

e. 0.1
   $\times\ 0.1$
   $\overline{\ \ 0.01\ \ }$

f. $\dfrac{3}{2} \cdot \dfrac{3}{2} = \dfrac{9}{4} = 2\dfrac{1}{4}$

g. $(2 + 3)^2 - (2^2 + 3^2)$
   $(5)^2 - (4 + 9)$
   $25 - 13$
   12

LESSON 92, MIXED PRACTICE

1. It is more likely not to rain. If the chance of rain is 40%, then the chance that it will not rain is 60%. So the chance that it will not rain is greater than the chance that it will rain.

2. $\dfrac{13 \text{ hearts}}{52 \text{ cards}} = \dfrac{1}{4}$

3. $\begin{array}{r} 48 \\ 3\overline{)144} \\ 12 \\ \overline{24} \\ 24 \\ \overline{0} \end{array}$

4. True

5. 21
   $\times\ 21$
   $\overline{\ \ 21\ \ }$
   420
   $\overline{441}$

   21

6. $2 \cdot 3^2 - \sqrt{9} + (3 - 1)^3$
   $2 \cdot 3^2 - \sqrt{9} + (2)^3$
   $2 \cdot 9 - 3 + 8$
   $18 - 3 + 8$
   23

7. $P = 2l + 2w$
   $P = 2(12 \text{ in.}) + 2(6 \text{ in.})$
   $P = 24 \text{ in.} + 12 \text{ in.}$
   $P = 36 \text{ in.}$

8. **−1, 0, 0.1, 1**

9. $\dfrac{5}{\overset{1}{\cancel{6}}} \times \dfrac{\overset{5}{\cancel{30}} \text{ members}}{1} = 25 \text{ members}$

   $30 - 25 = \textbf{5 members}$

10. $\dfrac{(\overset{1}{\cancel{24}})(\overset{18}{\cancel{36}})}{\underset{\underset{1}{2}}{\cancel{48}}} = \dfrac{18}{1} = \textbf{18}$

11. $\dfrac{10}{5} = \textbf{2}$

12.

$$12\frac{5}{6} \times \frac{1}{1} = 12\frac{5}{6}$$
$$+ \ 15\frac{1}{3} \times \frac{2}{2} = 15\frac{2}{6}$$
$$27\frac{7}{6} = 28\frac{1}{6}$$

13.
$$\overset{0\ 9\ 9}{\cancel{1}\ \cancel{0}\ \cancel{0}.^1 0}$$
$$- \qquad 9.9$$
$$\mathbf{9\ 0.1}$$

14. $\frac{4}{7} \times \frac{100}{1} = \frac{400}{7} = \mathbf{57\frac{1}{7}}$

15. $\frac{5}{8} = \frac{w}{48}$

$$5 \cdot 48 = 8w$$
$$\frac{5 \cdot \overset{6}{\cancel{48}}}{\underset{1}{\cancel{8}}} = w$$
$$\mathbf{30 = w}$$

16.
$$\begin{array}{r} \$4.60 \\ \times \ 0.25 \\ \hline 2300 \\ 9200 \\ \hline \$1.1500 \end{array} \longrightarrow \mathbf{\$1.15}$$

17. $A = \pi r^2$
$$A \approx (3.14)(3\ \text{in.})^2$$
$$A \approx (3.14)(9\ \text{in.}^2)$$
$$A \approx 28.26\ \text{in.}^2$$
$$A \approx \mathbf{28\ in.^2}$$

18. $\frac{15}{4} \times \frac{25}{25} = \frac{375}{100} = 3.75$

$$\begin{array}{r} \overset{6\ \ ^{13}}{7.\cancel{4}^1 0} \\ - \ 3.7\ 5 \\ \hline \mathbf{3.6\ 5} \end{array}$$

19. A, B, C, D, E, F, G, H, I, J
$$\frac{3}{10} \times \frac{10}{10} = \frac{30}{100} = \mathbf{30\%}$$

20.

```
        W
   ┌─────────┐
   │         │
 S │         │ N
   │
   ↓
```

South

21. $7 \times 12 = \mathbf{84}$

22. $\frac{16}{10} = \frac{w}{25}$

$$16 \cdot 25 = 10w$$
$$\frac{\overset{8}{\cancel{16}} \cdot \overset{5}{\cancel{25}}}{\underset{1}{\cancel{10}}} = w$$
$$\mathbf{40 = w}$$

23. $A = \frac{1}{2}bh$

$$A = \frac{1}{2}(8\ \text{cm})(6\ \text{cm})$$
$$A = \frac{1}{2}(48\ \text{cm}^2)$$
$$A = \mathbf{24\ cm^2}$$

24. (a) **6 faces**

(b) **12 edges**

25. $P = 2l + 2w$

$$P = 2\left(\frac{3}{4}\ \text{in.}\right) + 2\left(\frac{1}{2}\ \text{in.}\right)$$
$$P = \left(\frac{6}{4}\ \text{in.}\right) + \left(\frac{2}{2}\ \text{in.}\right)$$
$$P = \left(\frac{3}{2}\ \text{in.}\right) + \left(\frac{2}{2}\ \text{in.}\right)$$
$$P = \frac{5}{2}\ \text{in.}$$
$$P = \mathbf{2\frac{1}{2}\ in.}$$

26. $A = lw$

$$A = \left(\frac{3}{4}\ \text{in.}\right)\left(\frac{1}{2}\ \text{in.}\right)$$
$$A = \mathbf{\frac{3}{8}\ in.^2}$$

27. $\frac{5}{16}, \ \frac{3}{8}, \ \frac{7}{16}, \ \frac{1}{2}$

28.

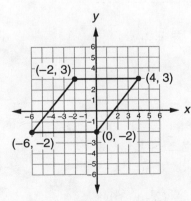

$A = bh$
$A = (6\ \text{units})(5\ \text{units})$
$A = \mathbf{30\ square\ units}$

29. (a) $(12 \text{ cm})(8 \text{ cm}) = \textbf{96 cm}^2$

(b) $\dfrac{\overset{9}{\cancel{36}}}{\underset{1}{\cancel{4}}} \dfrac{\cancel{\text{ft}} \cdot \text{ft}}{\cancel{\text{ft}}} = \textbf{9 ft}$

30. 3 gallon = 12 quarts
12 quarts = 24 pints
24 pints

___

LESSON 93, WARM-UP

a. 2400

b. 184

c. 6

d. $8.46

e. 0.012

f. 750

g. 1

Problem Solving
>; Fifty-four is the product of 6×9. Each factor on the left is greater than the respective factor on the right, so the product on the left must be greater.

___

LESSON 93, LESSON PRACTICE

a. $15 \text{ cm} \times 3 = \textbf{45 cm}$

b. True

c. False

d. Isosceles (also acute) triangle

e. False

LESSON 93, MIXED PRACTICE

1.

	Ratio	A.C.
Length	5	l
Width	2	60

$$\frac{5}{2} = \frac{l}{60}$$
$$5 \cdot 60 = 2l$$
$$\frac{5 \cdot \overset{30}{\cancel{60}}}{\underset{1}{\cancel{2}}} = l$$
$$\textbf{150 ft} = l$$

2. $4 \times 4 = \textbf{16 combinations}$

3.
$$\begin{array}{r} 36 \\ 4\overline{)144} \\ 12 \\ \hline 24 \\ 24 \\ \hline 0 \end{array}$$

4. $V = lwh$
$V = (4 \text{ cm})^3$
$V = \textbf{64 cm}^3$

5. $\dfrac{9}{25} \times \dfrac{4}{4} = \dfrac{36}{100}$
0.36, 36%

6. $\dfrac{16}{5} \times \dfrac{20}{20} = \dfrac{320}{100} = 3.20$

$$\begin{array}{r} 3.5 \\ +\ 3.20 \\ \hline 6.70 \end{array}$$
6.7

7. $\dfrac{\overset{9}{\cancel{45}}}{\underset{\underset{1}{8}}{\cancel{100}}} \times \dfrac{\overset{4}{\cancel{80}}}{1} = \dfrac{36}{1} = \textbf{36}$

8.
$$\begin{array}{r} 0.3 \\ \times\ 0.3 \\ \hline 0.09 \end{array} \qquad \begin{array}{r} 0.09 \\ \times\ 0.3 \\ \hline \textbf{0.027} \end{array}$$

9. $\dfrac{5}{2} \times \dfrac{5}{2} = \dfrac{25}{4} = \textbf{6}\dfrac{1}{4}$

10. $3 \cdot 10 = \textbf{30}$

11. $\dfrac{20}{24} = \dfrac{\textbf{5}}{\textbf{6}}$

12. $\dfrac{4}{20} = \dfrac{1}{5}$

13.
$$9\tfrac{1}{3} \times \dfrac{4}{4} = \overset{8}{\cancel{9}}\,\overset{16}{\dfrac{\cancel{4}}{12}}$$
$$-\ 4\tfrac{3}{4} \times \dfrac{3}{3} = 4\tfrac{9}{12}$$
$$\overline{\qquad\qquad\qquad} $$
$$w = \mathbf{4\tfrac{7}{12}}$$

14.
$$6 \cdot 30 = 5m$$
$$\dfrac{6 \cdot \overset{6}{\cancel{30}}}{\underset{1}{\cancel{5}}} = m$$
$$\mathbf{36} = m$$

15. **Equilateral triangle**

16. $\dfrac{3}{\underset{2}{\cancel{8}}} \times \dfrac{\overset{25}{\cancel{100}}}{1} = \dfrac{75}{2} = \mathbf{37\tfrac{1}{2}}$

17.
$$10 + 6^2 \div 3 - \sqrt{9} \times 3$$
$$10 + 36 \div 3 - 3 \times 3$$
$$10 + 12 - 9$$
$$\mathbf{13}$$

18. **5 faces**

19. $2\tfrac{1}{2}$ gallons = **10 quarts**

20. **One possibility:**

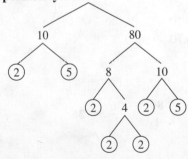

$$\mathbf{2^5 \times 5^2}$$

21. 0.125
0.1

22.
$$\begin{array}{r} \$15 \\ 8\overline{)\$120} \\ \underline{8} \\ 40 \\ \underline{40} \\ 0 \end{array}$$
$$n = \mathbf{\$15.00}$$

23.
$$A = lw$$
$$A = (26\ \text{mm})(18\ \text{mm})$$
$$A = 468\ \text{mm}^2$$
$$468\ \text{mm}^2 \div 2 = \mathbf{234\ \text{mm}^2}$$

24. $\dfrac{17}{20} \times \dfrac{5}{5} = \dfrac{85}{100} = \mathbf{85\%}$

25. **B.** $\sqrt{2}$

26. **7,250,000,000**

27. (a) $\dfrac{1}{6}$

(b) $\dfrac{5}{6}$

28.

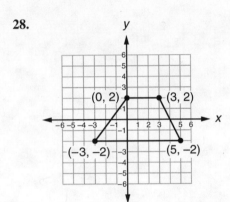

Trapezoid

29.
$$A = \dfrac{bh}{2}$$
$$A = \dfrac{(20\ \text{cm})(15\ \text{cm})}{2}$$
$$A = \mathbf{150\ \text{cm}^2}$$

30. $\dfrac{7}{16},\ \dfrac{1}{2},\ \dfrac{9}{16},\ \dfrac{5}{8}$

LESSON 94, WARM-UP

a. 3500

b. 722

c. 40

d. $3.64

e. 2

f. 32

g. 21

Problem Solving

Unknown side length: $24\text{ cm} - 16\text{ cm} = 8\text{ cm}$

Area: $\dfrac{8\text{ cm} \times 6\text{ cm}}{2} = \textbf{24 sq. cm}$

LESSON 94, LESSON PRACTICE

a. $0.5 \times 100\% = \textbf{50\%}$

b. $0.06 \times 100\% = \textbf{6\%}$

c. $0.125 \times 100\% = \textbf{12.5\%}$

d. $0.45 \times 100\% = \textbf{45\%}$

e. $1.3 \times 100\% = \textbf{130\%}$

f. $0.025 \times 100\% = \textbf{2.5\%}$

g. $0.09 \times 100\% = \textbf{9\%}$

h. $1.25 \times 100\% = \textbf{125\%}$

i. $0.625 \times 100\% = \textbf{62.5\%}$

j. $\dfrac{2}{3} \times \dfrac{100\%}{1} = \dfrac{200\%}{3} = \textbf{66}\dfrac{\textbf{2}}{\textbf{3}}\textbf{\%}$

k. $\dfrac{1}{6} \times \dfrac{100\%}{1} = \dfrac{100\%}{6} = 16\dfrac{4}{6}\% = \textbf{16}\dfrac{\textbf{2}}{\textbf{3}}\textbf{\%}$

l. $\dfrac{1}{\overset{}{\underset{2}{8}}} \times \dfrac{\overset{25}{100\%}}{1} = \dfrac{25\%}{2} = \textbf{12}\dfrac{\textbf{1}}{\textbf{2}}\textbf{\%}$

m. $\dfrac{5}{\overset{}{\underset{1}{4}}} \times \dfrac{\overset{25}{100\%}}{1} = \dfrac{125\%}{1} = \textbf{125\%}$

n. $\dfrac{14}{\overset{}{\underset{1}{5}}} \times \dfrac{\overset{20}{100\%}}{1} = \dfrac{280\%}{1} = \textbf{280\%}$

o. $\dfrac{4}{3} \times \dfrac{100\%}{1} = \dfrac{400\%}{3} = \textbf{133}\dfrac{\textbf{1}}{\textbf{3}}\textbf{\%}$

p. $\dfrac{5}{\overset{}{\underset{3}{6}}} \times \dfrac{\overset{50}{100\%}}{1} = \dfrac{250\%}{3} = \textbf{83}\dfrac{\textbf{1}}{\textbf{3}}\textbf{\%}$

q. $\dfrac{1}{3} \times \dfrac{100\%}{1} = \dfrac{100\%}{3} = \textbf{33}\dfrac{\textbf{1}}{\textbf{3}}\textbf{\%}$

LESSON 94, MIXED PRACTICE

1. $\dfrac{10}{30} = \dfrac{1}{3}$

$\dfrac{1}{3} \times 100\% = \dfrac{100\%}{3} = \textbf{33}\dfrac{\textbf{1}}{\textbf{3}}\textbf{\%}$

2. $100°\text{C} \div 2 = \textbf{50°C}$

3. $\dfrac{1}{\overset{}{\underset{1}{3}}} \times \dfrac{\overset{4}{12}}{1}\text{ cm} = \dfrac{4}{1} = \textbf{4 cm}$

$12\text{ cm} - 4\text{ cm} = \textbf{8 cm}$

4. $\dfrac{2}{\overset{}{\underset{1}{5}}} \times \dfrac{\overset{20}{100\%}}{1} = \dfrac{40\%}{1} = \textbf{40\%}$

5. $\dfrac{5}{3} \times \dfrac{100\%}{1} = \dfrac{500\%}{3} = \textbf{166}\dfrac{\textbf{2}}{\textbf{3}}\textbf{\%}$

6. $1.5 \times 100\% = \textbf{150\%}$

7. $\dfrac{25}{4} \times \dfrac{25}{25} = \dfrac{625}{100} = 6.25$

$\begin{array}{r} 6.\overset{3}{\cancel{4}}{}^{1}0 \\ -\ 6.2\ 5 \\ \hline \textbf{0.1 5} \end{array}$

8. $10{,}000 - 1000 = \textbf{9000}$

9. $\dfrac{3}{\overset{}{\underset{1}{4}}} \times \dfrac{\overset{90}{360}}{1} = \dfrac{270}{1} = \textbf{270}$

10. $C = \pi d$

$C \approx (3.14)(8\text{ cm})$

$C \approx \textbf{25.12 cm}$

SOLUTIONS

11. $A = \pi r^2$
$A \approx (3.14)(4 \text{ cm})^2$
$A \approx \mathbf{50.24 \text{ cm}^2}$

12.
$$3\frac{1}{2} \times \frac{4}{4} = 3\frac{4}{8}$$
$$1\frac{3}{4} \times \frac{2}{2} = 1\frac{6}{8}$$
$$+ \; 4\frac{5}{8} \times \frac{1}{1} = 4\frac{5}{8}$$
$$8\frac{15}{8} = \mathbf{9\frac{7}{8}}$$

13. $\dfrac{\cancel{9}^1}{\cancel{10}_2} \cdot \dfrac{\cancel{5}^1}{\cancel{6}_3} \cdot \dfrac{\cancel{8}^{\cancel{4}^2}}{\cancel{9}_1} = \mathbf{\dfrac{2}{3}}$

14. $(2 \times 10^5) + (5 \times 10^4)$

15.
$$\begin{array}{r} \$8.47 \\ \$12.00 \\ + \; \$0.95 \\ \hline \mathbf{\$21.42} \end{array}$$

16. $37.5 \div 100 = \mathbf{0.375}$

17.
$$\frac{3}{7} = \frac{21}{x}$$
$$7 \cdot 21 = 3x$$
$$\frac{7 \cdot \cancel{21}^7}{\cancel{3}_1} = x$$
$$\mathbf{49} = x$$

18.
$$\frac{100}{3} \div \frac{100}{1}$$
$$1 \div \frac{100}{1} = \frac{1}{100}$$
$$\frac{\cancel{100}^1}{3} \times \frac{1}{\cancel{100}_1} = \mathbf{\frac{1}{3}}$$

19. $100\% - 90\% = \mathbf{10\%}$

20. **120.03**

21. $-5.2, \; -2.5, \; \dfrac{2}{5}, \; \dfrac{5}{2}$

22. **8 edges**

23. $A = bh$
$A = (10 \text{ in.})(8 \text{ in.})$
$A = \mathbf{80 \text{ in.}^2}$

24. **C. Obtuse**

25.
$$\begin{array}{r} 103°F \\ + \quad 37°F \\ \hline \mathbf{140°F} \end{array}$$

26.

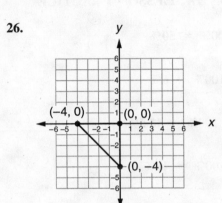

$$A = \frac{bh}{2}$$
$$A = \frac{4 \text{ units} \times 4 \text{ units}}{2} = \frac{16}{2} = \mathbf{8 \text{ sq. units}}$$

27. (a) Mode = **22**

(b) 19, 20, 21, 22, ㉒, 22, 24, 24, 25,
Mean = **22**

28. $2^3 + \sqrt{25} \times 3 - 4^2 \div \sqrt{4}$
$8 + 5 \times 3 - 16 \div 2$
$8 + 15 - 8$
15

29. 1 gallon = 4 quarts
6 gallons = 24 quarts
6 gallons

30. $\dfrac{2}{52} = \mathbf{\dfrac{1}{26}}$

LESSON 95, WARM-UP

a. **4800**

b. **287**

206   *Saxon Math 7/6—Homeschool*

c. **20**

d. **$8.27**

e. **0.175**

f. **1650**

g. **3**

Problem Solving

Add $\frac{1}{12}$ to each term to find the next term.

$$\frac{1}{12}, \frac{1}{6}, \frac{1}{4}, \frac{1}{3}, \frac{5}{12}, \frac{1}{2}, \frac{7}{12}, \frac{2}{3}, \cdots$$

LESSON 95, LESSON PRACTICE

a. $\dfrac{3 \text{ dollars}}{1 \text{ hour}} \times \dfrac{8 \text{ hours}}{1} = $ **24 dollars**

b. $\dfrac{6 \text{ baskets}}{10 \text{ shots}} \times \dfrac{100 \text{ shots}}{1} = $ **60 baskets**

c. $\dfrac{10 \text{ cents}}{1 \text{ kwh}} \times \dfrac{26.3 \text{ kwh}}{1} = $ **263 cents**

d. $\dfrac{160 \text{ km}}{2 \text{ hours}} \cdot \dfrac{10 \text{ hours}}{1} = $ **800 km**

e. $\dfrac{18 \text{ gallons}}{1} \times \dfrac{29 \text{ miles}}{1 \text{ gallon}} = $ **522 miles**

f. $\dfrac{2.3 \text{ meters}}{1} \times \dfrac{100 \text{ centimeters}}{1 \text{ meter}} = $ **230 cm**

LESSON 95, MIXED PRACTICE

1.
$$
\begin{array}{r}
\$45.79 \\
\times \quad 0.07 \\
\hline
\$3.2053
\end{array}
\longrightarrow \$3.21
$$

$$
\begin{array}{r}
\$45.79 \\
+ \quad 3.21 \\
\hline
\$49.00
\end{array}
$$

2. $\dfrac{1.67 \text{ m}}{1} \times \dfrac{100 \text{ cm}}{1 \text{ m}} = $ **167 cm**

3. $\dfrac{5}{8} \times \dfrac{40}{1} = $ **25 sprouted**

$$
\begin{array}{r}
4\,0 \\
- \ 2\,5 \\
\hline
1\,5 \text{ seeds did not sprout.}
\end{array}
$$

4. **560.73**

5. $\dfrac{1}{6} \times \dfrac{100\%}{1} = \dfrac{50\%}{3} = \mathbf{16 \dfrac{2}{3}\%}$

6. $2.5 \times 100\% = \mathbf{250\%}$

7.
$$
\begin{array}{r}
\$12.00 \\
\times \quad 0.30 \\
\hline
\$3.6000
\end{array}
\longrightarrow \$3.60
$$

8. $180°$ is a half turn

$\frac{1}{2}$ of 60 minutes = **30 minutes**

9.
$$
\begin{array}{r}
5 \text{ turns} \\
6\overline{)30} \\
\underline{30} \\
0
\end{array}
$$

10. $V = lwh$

$V = (4 \text{ ft})(3 \text{ ft})(3 \text{ ft})$

$V = \mathbf{36 \text{ ft}^3}$

11.
$$
\begin{array}{r}
\dfrac{3}{4} \times \dfrac{5}{5} = \dfrac{15}{20} \\
+ \ \dfrac{3}{5} \times \dfrac{4}{4} = \dfrac{12}{20} \\
\hline
\dfrac{27}{20} = 1\dfrac{7}{20}
\end{array}
$$

12.
$$
\begin{array}{r}
18\dfrac{1}{8} \times \dfrac{1}{1} = 18\dfrac{1}{8} \\
- 12\dfrac{1}{2} \times \dfrac{4}{4} = 12\dfrac{4}{8} \\
\hline
5\dfrac{5}{8}
\end{array}
$$

13. $\dfrac{\overset{\overset{1}{\cancel{\overset{5}{\cancel{15}}}}}{\cancel{4}}}{\underset{1}{}} \times \dfrac{\overset{\overset{1}{\cancel{2}}}{\cancel{8}}}{\underset{1}{\cancel{3}}} \times \dfrac{11}{\underset{\underset{1}{\cancel{2}}}{\cancel{10}}} = \dfrac{11}{1} = \mathbf{11}$

14. $\dfrac{\cancel{2} \cdot \cancel{2} \cdot \cancel{2} \cdot 2 \cdot 2}{\cancel{2} \cdot \cancel{2} \cdot \cancel{2}} = \dfrac{4}{1} = \mathbf{4}$

15. $\dfrac{5}{2} \div \dfrac{1}{4}$

$1 \div \dfrac{1}{4} = \dfrac{4}{1}$

$\dfrac{5}{\underset{1}{\cancel{2}}} \times \dfrac{\overset{2}{\cancel{4}}}{1} = \dfrac{10}{1} = \mathbf{10}$

16.
```
   12
  8.75
+  6.8
───────
 27.55
```

17.
```
    1.5
×   1.5
──────
     75
    150
──────
   2.25
```

18. $6\dfrac{2}{5} = 6\dfrac{4}{10} = 6.4$

$$0\,8.\overline{)64.}\;\;\;\overset{8}{}$$
```
        8
  08.)64.
       64
      ───
        0
```

19. $6 + 5 + 8 = \mathbf{19}$
Round $6\dfrac{1}{4}$ to 6, round 4.95 to 5, and round 8.21 to 8. Then add 6, 5, and 8.

20. (a) 60 inches $\div$ 2 = **30 inches**

(b) $A = \pi r^2$
$A \approx (3.14)(30 \text{ inches})^2$
$A \approx \mathbf{2826 \text{ in.}^2}$

21. $4\%, \dfrac{1}{4}, 0.4$

22.
```
    4
   8.¹0
 − 3. 4
 ──────
   1. 6
```
$y = \mathbf{1.6}$

23. $\dfrac{4}{8} = \dfrac{x}{12}$

$4 \cdot 12 = 8x$

$\dfrac{\overset{1}{\cancel{4}} \cdot \overset{6}{\cancel{12}}}{\underset{1}{\cancel{8}}} = x$

$\mathbf{6} = x$

24. (a) $A = lw$
$A = 6 \text{ cm} \times 6 \text{ cm}$
$A = \mathbf{36 \text{ cm}^2}$

(b) $V = lwh$
$V = (6 \text{ cm})(6 \text{ cm})(6 \text{ cm})$
$V = \mathbf{216 \text{ cm}^3}$

25.
```
    3
   4¹2 mm
 −  2 4 mm
 ─────────
   1 8 mm
```

26. $6^2 \div \sqrt{9} + 2 \times 2^3 - \sqrt{400}$
$36 \div 3 + 2 \times 8 - 20$
$12 + 16 - 20$
$\mathbf{8}$

27. 1 quart = 2 pints
$\dfrac{\mathbf{1}}{\mathbf{2}}$

28. $A = bh$
$A = (1.2 \text{ m})(0.9 \text{ m})$
$A = \mathbf{1.08 \text{ m}^2}$

29. $\dfrac{2.5 \text{ liters}}{1} \times \dfrac{1000 \text{ milliliters}}{1 \text{ liter}} = \mathbf{2500 \text{ milliliters}}$

30. $\dfrac{2}{4} = \dfrac{\mathbf{1}}{\mathbf{2}}$

───────────────

LESSON 96, WARM-UP

a. **6300**

b. **614**

c. **30**

d. **$4.11**

e. 1.5

f. 25

g. 9

Problem Solving
Work backward, beginning with $3 \times 3 = 9$ and $2 \times 2 = 4$.

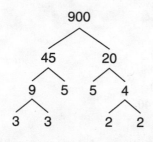

LESSON 96, LESSON PRACTICE

a. $x - 2 = y$
$10 - 2 = 8$
$y = \mathbf{8}$

b. $a + 5 = b$
$10 + 5 = 15$
$a = \mathbf{10}$

c. $3(8) + 1 = \mathbf{25}$

d. $3(4) - 1 = \mathbf{11}$

e.

Q	P
1	2
2	4
3	6
4	8

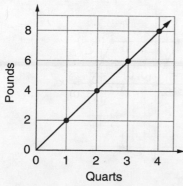

It is appropriate to use a ray, because any portion of a quart may be weighed.

LESSON 96, MIXED PRACTICE

1. $(2.0 \times 2.0) - (2.0 + 2.0)$
$4.0 - 4.0$
$\mathbf{0}$

2. $\begin{array}{r} \mathbf{10} \text{ objects} \\ 42\overline{)420} \\ \underline{42} \\ 00 \\ \underline{00} \\ 0 \end{array}$

3. $\begin{array}{r} \overset{1}{12} \\ \times\ \ 8 \\ \hline \mathbf{96} \end{array}$

4. **Trapezoid**

5. (a) $0.15 \times 100\% = \mathbf{15\%}$

(b) $1.5 \times 100\% = \mathbf{150\%}$

6. $\dfrac{5}{\cancel{6}_{3}} \times \dfrac{\overset{50}{\cancel{100}}\%}{1} = \dfrac{250\%}{3} = \mathbf{83\dfrac{1}{3}\%}$

7. **C. 0.1**

8. $11 \times 11 \times 11 = \mathbf{1331}$

9. $\dfrac{5}{\cancel{6}_{1}} \times \dfrac{\overset{60}{\cancel{360}}}{1} = \dfrac{300}{1} = \mathbf{300}$

10. The number 89 is between the perfect squares 81 and 100, so $\sqrt{89}$ is betweeen $\sqrt{81}$ and $\sqrt{100}$. Since $\sqrt{81}$ is 9 and $\sqrt{100}$ is 10, $\sqrt{89}$ is between **9 and 10.**

11. $\begin{array}{r} \mathbf{120°} \\ 3\overline{)360°} \\ \underline{3} \\ 06 \\ \underline{06} \\ 00 \\ \underline{00} \\ 0 \end{array}$

12. $2(13) - 1 = \mathbf{25}$

13. $\dfrac{(\overset{1}{\cancel{3}} \cdot \overset{1}{\cancel{3}} \cdot 5) \cdot (2 \cdot \overset{1}{\cancel{3}} \cdot \overset{1}{\cancel{3}} \cdot 3)}{\underset{1}{\cancel{3}} \cdot \underset{1}{\cancel{3}} \cdot \underset{1}{\cancel{3}} \cdot \underset{1}{\cancel{3}}} = \dfrac{30}{1} = \mathbf{30}$

14.
$$8\overline{)3000}$$
$$\begin{array}{r} 375 \\ \hline 24 \\ \hline 60 \\ 56 \\ \hline 40 \\ 40 \\ \hline 0 \end{array}$$

15. $\dfrac{50}{3} \div \dfrac{100}{1}$

$1 \div \dfrac{100}{1} = \dfrac{1}{100}$

$\dfrac{\overset{1}{\cancel{50}}}{3} \times \dfrac{1}{\cancel{100}} = \dfrac{1}{6}$
$\scriptstyle 2$

16.
$$2\tfrac{1}{2} \times \tfrac{3}{3} = 2\tfrac{3}{6}$$
$$3\tfrac{1}{3} \times \tfrac{2}{2} = 3\tfrac{2}{6}$$
$$+\ 4\tfrac{1}{6} \times \tfrac{1}{1} = 4\tfrac{1}{6}$$
$$\overline{\qquad\qquad\quad 9\tfrac{6}{6} = \mathbf{10}}$$

17. $\dfrac{6}{1} \times \dfrac{\overset{2}{\cancel{16}}}{\cancel{3}} \times \dfrac{\overset{1}{\cancel{3}}}{\cancel{8}} = \dfrac{12}{1} = \mathbf{12}$
$\scriptstyle 1 \qquad 1$

18. $\dfrac{2}{5} \times \dfrac{\$12}{1} = \dfrac{\$24}{5}$

$$5\overline{)\$24.00}$$
$$\begin{array}{r} \$4.80 \\ \hline 20 \\ \hline 4\,0 \\ 4\,0 \\ \hline 00 \\ 00 \\ \hline 0 \end{array}$$

19.
$$\begin{array}{r} \$6.50 \\ \times\ \ 0.12 \\ \hline 1300 \\ 6500 \\ \hline \$0.7800 \end{array} \longrightarrow \ \mathbf{\$0.78}$$

20. $\dfrac{15}{4} = 3.75$

$$\begin{array}{r} \overset{4}{\cancel{5}}.\overset{12}{\cancel{3}}{}^{1}0 \\ -\ 3.7\,5 \\ \hline 1.5\,5 \end{array}$$

21. $\dfrac{10}{25} = \dfrac{2}{5}$

22. $4n = 6 \cdot 14$

$n = \dfrac{\overset{3}{\cancel{6}} \cdot \overset{7}{\cancel{14}}}{\underset{\underset{1}{2}}{\cancel{4}}}$

$n = \mathbf{21}$

23.
$$3\overline{)120}$$
$$\begin{array}{r} 40 \\ \hline 12 \\ \hline 00 \\ 00 \\ \hline 0 \end{array}$$
$n = \mathbf{40}$

24.

R ———————— S ———————— T

$\dfrac{7}{4} \div \dfrac{2}{1}$

$1 \div \dfrac{2}{1} = \dfrac{1}{2}$

$\dfrac{7}{4} \times \dfrac{1}{2} = \dfrac{7}{8}$ inch

$RS = ST = \dfrac{7}{8}$ **inch**

25. $\dfrac{6}{9} = \dfrac{36}{w}$

$6w = 9 \cdot 36$

$w = \dfrac{9 \cdot \overset{6}{\cancel{36}}}{\cancel{6}}$
$\scriptstyle 1$

$w = \mathbf{54}$

26. $\dfrac{4\ \cancel{\text{hours}}}{1} \times \dfrac{6\ \text{dollars}}{1\ \cancel{\text{hour}}} = \mathbf{24\ dollars}$

27.

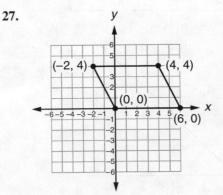

$A = bh$
$A = (6\ \text{units})(4\ \text{units})$
$A = \mathbf{24\ units^2}$

28. 1 gallon = 4 quarts
4 quarts = 8 pints
1 gallon = 8 pints
About 8 pounds

29. $3^2 + 2^3 - \sqrt{4} \times 5 + 6^2 \div \sqrt{16}$
$9 + 8 - 2 \times 5 + 36 \div 4$
$9 + 8 - 10 + 9$
16

30. $\dfrac{3}{6} = \dfrac{1}{2}$

LESSON 97, WARM-UP

a. 1000

b. 267

c. 15

d. $15.57

e. 0.001

f. 1500

g. 3

Problem Solving

Lowest: 70% + 70% + 100% = 240%
240% ÷ 3 = **80%**
Highest: 70% + 100% + 100% = 270%
270% ÷ 3 = **90%**

LESSON 97, LESSON PRACTICE

a. Line c

b. $\angle 5$

c. $\angle 4$

d. $\angle 1$

e. Angle 3, $\angle 5$, and $\angle 7$ each measure 105°.
Angle 2, $\angle 4$, $\angle 6$, and $\angle 8$ each measure 75°.

Saxon Math 7/6—Homeschool

LESSON 97, MIXED PRACTICE

1. $\dfrac{100}{1} \div \dfrac{1}{4}$
$1 \div \dfrac{1}{4} = \dfrac{4}{1}$
$\dfrac{100}{1} \cdot \dfrac{4}{1} = \dfrac{400}{1} = 400$
400 hamburgers

2.
$$\begin{array}{r} \overset{1}{2}1\ 2°F \\ -\ 3\ 2°F \\ \hline 1\ 8\ 0°F \end{array} \quad \begin{array}{r} 90°F \\ 2\overline{)180°F} \\ 18 \\ \hline 00 \\ 00 \\ \hline 0 \end{array} \quad \begin{array}{r} 32°F \\ +\ 90°F \\ \hline \mathbf{122°F} \end{array}$$

3. $1.8(30) + 32$
86

4. $\dfrac{5}{8} \,\lessdot\, 0.675$
$$\begin{array}{r} 0.625 \\ 8\overline{)5.000} \\ 4\ 8 \\ \hline 20 \\ 16 \\ \hline 40 \\ 40 \\ \hline 0 \end{array}$$

5. $\dfrac{9}{4} \times \dfrac{25}{25} = \dfrac{225}{100} = \mathbf{225\%}$

6. $\dfrac{7}{5} \times \dfrac{20}{20} = \dfrac{140}{100} = \mathbf{140\%}$

7. $\dfrac{7}{10} \times \dfrac{10}{10} = \dfrac{70}{100} = \mathbf{70\%}$

8. $\dfrac{7}{\underset{2}{8}} \times \dfrac{\overset{25}{100\%}}{1} = \dfrac{175}{2} = \mathbf{87\dfrac{1}{2}\%}$

9.
$$\begin{array}{r} 1 \\ 5\overline{)5} \\ 2\overline{)10} \\ 2\overline{)20} \\ 2\overline{)40} \\ 2\overline{)80} \\ 2\overline{)160} \\ 2\overline{)320} \end{array}$$
$2^6 \cdot 5$

10. $\dfrac{360°}{60} = \textbf{6°}$

11. $4\overline{)360°}$ with quotient $90°$

$$\begin{array}{r} 90° \\ 4\overline{)360°} \\ \underline{36} \\ 00 \\ \underline{00} \\ 0 \end{array}$$

12. $6\dfrac{3}{4} \times \dfrac{2}{2} = 6\dfrac{6}{8}$

$+\ 5\dfrac{7}{8} \times \dfrac{1}{1} = 5\dfrac{7}{8}$

$11\dfrac{13}{8} = \mathbf{12\dfrac{5}{8}}$

13. $6\dfrac{1}{3} \times \dfrac{2}{2} = 6\dfrac{2}{6}$

$-\ 2\dfrac{1}{2} \times \dfrac{3}{3} = 2\dfrac{3}{6}$

$\mathbf{3\dfrac{5}{6}}$

14. $\dfrac{5}{2} \div \dfrac{100}{1}$

$1 \div \dfrac{100}{1} = \dfrac{1}{100}$

$\dfrac{5}{2} \times \dfrac{1}{100} = \mathbf{\dfrac{1}{40}}$

15. $\begin{array}{r} 6.93 \\ 8.429 \\ +\ 12 \\ \hline \mathbf{27.359} \end{array}$

16. $(1 - 0.1)(1 \div 0.1)$

$(0.9)(10)$

$\mathbf{9}$

17. $\dfrac{7}{8} = 0.875$

$\begin{array}{r} 4.2 \\ +\ 0.875 \\ \hline \mathbf{5.075} \end{array}$

18. $3\dfrac{1}{3} \times \dfrac{2}{2} = 3\dfrac{2}{6}$

$-\ 2\dfrac{1}{2} \times \dfrac{3}{3} = 2\dfrac{3}{6}$

$\mathbf{\dfrac{5}{6}}$

19. $\dfrac{80}{100} \times \dfrac{30}{1} = 24$ moviegoers

$30 - 24 = \textbf{6 moviegoers}$

20. $\dfrac{1}{2} \div \dfrac{1}{3} \ \bigcirc\!\!\!> \ \dfrac{1}{3} \div \dfrac{1}{2}$

$1 \div \dfrac{1}{3} = \dfrac{3}{1} \qquad 1 \div \dfrac{1}{2} = \dfrac{2}{1}$

$\dfrac{1}{2} \times \dfrac{3}{1} = \dfrac{3}{2} \qquad \dfrac{1}{3} \cdot \dfrac{2}{1} = \dfrac{2}{3}$

21. Since each number in the sequence is $\frac{1}{10}$ of the previous term, the next number is $\frac{1}{10}$ of 1, which is $\frac{1}{10}$ or **0.1**.

22. $a + 130 = 180$

$\begin{array}{r} 180 \\ -\ 130 \\ \hline 50 \end{array}$

$a = \mathbf{50}$

23. $\dfrac{7}{4} = \dfrac{w}{44}$

$4w = 7 \cdot 44$

$w = \dfrac{7 \cdot 44}{4}$

$w = \mathbf{77}$

24. $b = h = \dfrac{48 \text{ in.}}{4} = 12 \text{ in.}$

$A = \dfrac{1}{2}\,bh$

$A = \dfrac{1}{2}\,(12 \text{ in.})\,(12 \text{ in.})$

$A = \mathbf{72 \text{ in.}^2}$

25. $5{:}00 \div 2 = \mathbf{2{:}30}$

26. **B. 11:00**

27. **Answers may vary. Sample answer: If Mark continued his $\frac{1}{2}$ mile record pace while running a 1-mile race, what would his 1-mile time be?**

$\begin{array}{r} 2{:}20 \\ \times \quad 2 \\ \hline \mathbf{4{:}40} \end{array}$

28. (a) $\mathbf{\angle 6}$

(b) $m\angle 5 = 180° - 78° = \mathbf{102°}$

$m\angle 8 = m\angle 2 = \mathbf{78°}$

29. $10^2 - \sqrt{49} - (10 + 8) \div 3^2$
$100 - 7 - 18 \div 9$
$100 - 7 - 2$
91

30. $\dfrac{2}{6} = \dfrac{1}{3}$

LESSON 98, WARM-UP

a. 2000

b. 743

c. 24

d. $1.28

e. 1250

f. 9

g. 2

Problem Solving

Because 20^2 equals 400 and 25^2 equals 625, $\sqrt{529}$ must be between 20 and 25. Because $22^2 = 484$ and $24^2 = 576$, $\sqrt{529}$ must be between 22 and 24. We find that $\sqrt{529} = $ **23.**

LESSON 98, LESSON PRACTICE

a. Since the sum of the interior angles of a triangle is 180°, the sum of m∠1, m∠2, and m∠3, is **180°.**

b. m∠4 + m∠5 + m∠6 = **180°**

c. (m∠1 + m∠2 + m∠3)
　　+ (m∠4 + m∠5 + m∠6)
　180° + 180°
　　360°

d. m∠P + 75° + 30° = 180°
　m∠P + 105° = 180°
　m∠P = **75°**

e. **90°**

f.

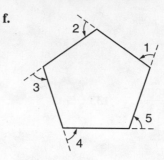

$\dfrac{360°}{5} = 72°$

LESSON 98, MIXED PRACTICE

1.
$\dfrac{1}{2} \times \dfrac{2}{2} = \dfrac{2}{4}$　　$\dfrac{1}{2} \times \dfrac{1}{4} = \dfrac{1}{8}$
$+ \dfrac{1}{4} \times \dfrac{1}{1} = \dfrac{1}{4}$
$\overline{\qquad\qquad\quad \dfrac{3}{4}}$

$\dfrac{3}{4} \div \dfrac{1}{8}$

$1 \div \dfrac{1}{8} = \dfrac{8}{1}$

$\dfrac{3}{\cancel{4}_1} \times \dfrac{\cancel{8}^2}{1} = 6$

2. $\dfrac{11}{\cancel{2}_1} \times \dfrac{\cancel{12}^6 \text{ inches}}{1} = $ **66 inches**

3. $\dfrac{4}{\cancel{5}_1} \times \dfrac{\cancel{200}^{40}}{1} = 160$

$200 - 160 = $ **40 runners**

4. (a) ∠8

(b) m∠6 = **85°**
　　m∠7 = 180° − 85° = **95°**

5. $(2 \times 10^4) + (5 \times 10^3)$ **miles**

6. $\dfrac{15}{16}$ **inch**

7. C. tire

8. 13 + 8 = **21**

9. $100\% - 20\% = 80\%$

0.8 or $\dfrac{4}{5}$

10. $\dfrac{4}{3} \times \dfrac{100\%}{1} = \dfrac{400\%}{3} = \mathbf{133\dfrac{1}{3}\%}$

11.
$$
\begin{array}{r}
\$7.50 \\
8)\overline{\$60.00} \\
56 \\
\hline
4\,0 \\
4\,0 \\
\hline
00 \\
00 \\
\hline
0
\end{array}
$$
$w = \mathbf{\$7.50}$

12. $\dfrac{0.999}{0.03} = \mathbf{33.3}$

13. $\dfrac{10}{3} \div \dfrac{100}{1}$

$1 \div \dfrac{100}{1} = \dfrac{1}{100}$

$\dfrac{\overset{1}{\cancel{10}}}{3} \times \dfrac{1}{\underset{10}{\cancel{100}}} = \mathbf{\dfrac{1}{30}}$

14. $V = lwh$
$V = (5\text{ in.})(2\text{ in.})(3\text{ in.})$
$V = \mathbf{30\text{ cubic inches}}$

15.
$$
\begin{array}{r}
6.5 \\
+ \ 4.95 \\
\hline
\mathbf{11.45}
\end{array}
$$

16. $2\dfrac{1}{6} \times \dfrac{1}{1} = \cancel{2}\overset{7}{\cancel{1}}\dfrac{1}{6}$

$\underline{- \ 1\dfrac{1}{2} \times \dfrac{3}{3} = 1\dfrac{3}{6}}$

$\dfrac{4}{6} = \mathbf{\dfrac{2}{3}}$

17.
$$
\begin{array}{r}
\$19.79 \\
\times \ \ 0.06 \\
\hline
1.1874
\end{array}
\qquad
\begin{array}{r}
\overset{1}{\$}1\overset{1}{9}.79 \\
+ \ \$1.19 \\
\hline
\mathbf{\$20.98}
\end{array}
$$

18. $\dfrac{3}{12} = \mathbf{\dfrac{1}{4}}$

19. $\dfrac{3}{100} = \mathbf{3\%}$

20. $\dfrac{5}{3} = \dfrac{45}{a}$

$5a = 3 \cdot 45$

$a = \dfrac{3 \cdot \overset{9}{\cancel{45}}}{\underset{1}{\cancel{5}}}$

$a = 27$

27 adults

21. $\mathbf{-1, \ -\dfrac{1}{2}, \ 0, \ \dfrac{1}{2}, \ 1}$

22. Trapezoid

23. Not congruent

24. (a) $40° + 110° + m\angle A = 180°$
$150° + m\angle A = 180°$
$m\angle A = \mathbf{30°}$

(b) $180° - 110° = x$
$\mathbf{70°} = x$

25. (a) $\dfrac{40}{100} = \mathbf{\dfrac{2}{5}}$

(b) $0.40 = \mathbf{0.4}$

26. $A = \pi r^2$
$A \approx (3.14)(10\text{ mm})^2$
$A \approx \mathbf{314\text{ mm}^2}$

27. $2^3 + \sqrt{81} \div 3^2 + \left(\dfrac{1}{2}\right)^2$

$8 + 9 \div 9 + \dfrac{1}{4}$

$8 + 1 + \dfrac{1}{4}$

$\mathbf{9\dfrac{1}{4}}$

28. $\dfrac{\overset{10}{\cancel{120}}\text{ in.}}{1} \times \dfrac{1\text{ ft}}{\underset{1}{\cancel{12}}\text{ in.}} = \mathbf{10\text{ ft}}$

29. (a) $\dfrac{20}{15} = \mathbf{\dfrac{4}{3}}$

(b) $\dfrac{15}{35} = \mathbf{\dfrac{3}{7}}$

30. Isosceles triangle

LESSON 99, WARM-UP

a. 3000

b. 291

c. 12

d. $17.74

e. 0.375

f. 450

g. 11

Problem Solving

$$\frac{9 \text{ cm} \times 4 \text{ cm}}{2} = \textbf{18 sq. cm}$$

$$\frac{6 \text{ cm} \times 4 \text{ cm}}{2} = \textbf{12 sq. cm}$$

LESSON 99, LESSON PRACTICE

a. $\frac{3}{5} \times \frac{20}{20} = \frac{60}{100} = \textbf{0.6}$

b. $\frac{3}{5} \times \frac{100\%}{1} = \frac{300\%}{5} = \textbf{60\%}$

c. $0.8 = \frac{8}{10} = \frac{\textbf{4}}{\textbf{5}}$

d. $\frac{80}{100} = \textbf{80\%}$

e. $\frac{20}{100} = \frac{\textbf{1}}{\textbf{5}}$

f. $20\% = 0.20 = \textbf{0.2}$

g. $\frac{3}{4} \times \frac{25}{25} = \frac{75}{100} = \textbf{0.75}$

h. $\frac{3}{4} \times 100\% = \textbf{75\%}$

i. $\frac{12}{100} = \frac{\textbf{3}}{\textbf{25}}$

j. $\frac{12}{100} = \textbf{12\%}$

k. $\frac{5}{100} = \frac{\textbf{1}}{\textbf{20}}$

l. $\frac{5}{100} = \textbf{0.05}$

LESSON 99, MIXED PRACTICE

1. $12 \div 1\frac{1}{2} \text{ lengths} = \frac{12}{1} \div \frac{3}{2}$

   $1 \div \frac{3}{2} = \frac{2}{3}$

   $\frac{\overset{4}{\cancel{12}}}{1} \times \frac{2}{\underset{1}{\cancel{3}}} = \frac{8}{1} = \textbf{8 lengths}$

2. **Cylinder**

3. $\frac{3}{8} + \frac{3}{8} = \frac{6}{8}$

   $\frac{8}{8} - \frac{6}{8} = \frac{2}{8} = \frac{\textbf{1}}{\textbf{4}}$

4. (a) $\frac{9}{12} = \frac{\textbf{3}}{\textbf{4}}$

   (b) $\frac{3}{\underset{1}{\cancel{4}}} \times \frac{\overset{25}{\cancel{100\%}}}{1} = \textbf{75\%}$

5. $V = lwh$
    $V = (4 \text{ ft})(3 \text{ ft})(4 \text{ ft})$
    $V = \textbf{48 ft}^3$

6.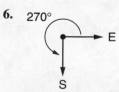

   South

7. $\frac{1}{5} \times \frac{20}{20} = \frac{20}{100} = 20\%$

   $100\% - 20\% = \textbf{80\%}$

8. $\frac{1}{7} \times \frac{100\%}{1} = \frac{100\%}{7} = 14\frac{2}{7}\%$

9. $\frac{15}{20} = \frac{24}{n}$

$15n = 20 \cdot 24$

$n = \dfrac{\overset{4}{\cancel{20}} \cdot \overset{8}{\cancel{24}}}{\underset{\underset{1}{\cancel{3}}}{\cancel{15}}}$

$n = \mathbf{32}$

10. $5 \cdot 4 \cdot 3 \cdot 2 \cdot 1 \cdot 0$

$20 \cdot 3 \cdot 2 \cdot 1 \cdot 0$

$60 \cdot 2 \cdot 1 \cdot 0$

$120 \cdot 1 \cdot 0$

$120 \cdot 0$

$\mathbf{0}$

11.
$$\begin{array}{r} 25 \\ 18\overline{)450} \\ \underline{36} \\ 90 \\ \underline{90} \\ 0 \end{array}$$

12.
$$\begin{array}{r} 40 \\ \times\ 40 \\ \hline 1600 \end{array} \qquad \sqrt{1600} = \mathbf{40}$$

13. $\sqrt{64} + 5^2 - \sqrt{25} \times (2 + 3)$

$8 + 25 - 5 \times 5$

$8 + 25 - 25$

$\mathbf{8}$

14.
$$\begin{array}{r} 6.75 \\ -\ 6.2 \\ \hline \mathbf{0.55} \end{array}$$

15. $\dfrac{25}{\underset{1}{\cancel{2}}} \times \dfrac{\overset{4}{\cancel{8}}}{\underset{1}{\cancel{5}}} \times \dfrac{\overset{1}{\cancel{5}}}{1} = \dfrac{100}{1} = \mathbf{100}$

16. $0.21 \div 7 = \mathbf{0.03}$

17.
$$\begin{array}{r} \$111.11 \\ \times\ 0.07 \\ \hline \$7.7777 \longrightarrow \ \mathbf{\$7.78} \end{array}$$

18. (a) Mode = **80%**

(b) 80%, 80%, 80%, ⟨85%⟩, 90%, 95%, 100%,
Median = **85%**

19.

$2^2 \cdot 3^2 \cdot 5^2$

20. Selection of prime numbers may vary. The GCF is 1.

21.
$$\begin{array}{r} 50\text{ cm} \\ 4\overline{)200\text{ cm}} \\ \underline{20} \\ 00 \\ \underline{00} \\ 0 \end{array}$$

22. (a) $A = \dfrac{bh}{2}$

$A = \dfrac{(\overset{4}{\cancel{8}}\text{ cm})(5\text{ cm})}{\underset{1}{\cancel{2}}}$

$A = \mathbf{20\text{ cm}^2}$

(b) **Obtuse triangle**

23. (a) $110° + 90° + 75° + m\angle B = 360°$

$m\angle B = \mathbf{85°}$

(b) $180° - 110° = 70°$
Exterior angle measure is **70°**.

24. (a) $\dfrac{6}{10} = \dfrac{3}{5}$

(b) $\dfrac{6}{10} \times \dfrac{10}{10} = \dfrac{60}{100} = \mathbf{60\%}$

25. (a) $\dfrac{15}{100} = \dfrac{3}{20}$

(b) $\dfrac{15}{100} = \mathbf{0.15}$

26. (a) $\dfrac{3}{10} = \mathbf{0.3}$

(b) $\dfrac{3}{10} \times \dfrac{10}{10} = \dfrac{30}{100} = \mathbf{30\%}$

27.

A———B———C

$1\frac{1}{4}$ inches $\div\ 2\ =\ \frac{5}{4}\ \div\ \frac{2}{1}$

$1\ \div\ \frac{2}{1}\ =\ \frac{1}{2}$

$\frac{5}{4}\ \cdot\ \frac{1}{2}\ =\ \frac{5}{8}$

$AB\ =\ BC\ =\ \frac{5}{8}$ inch

28. $\frac{13}{52}\ =\ \frac{1}{4}$

$\frac{1}{4}\ \times\ \frac{25}{25}\ =\ \frac{25}{100}\ =\ \textbf{25\%}$

29. 1 gallon $=$ 4 quarts
1 quart $<$ 1 liter
So, 4 quarts $<$ 4 liters
and 1 gallon $<$ 4 liters

30.
$$\begin{array}{r} 2.5 \\ 2\overline{)5.0} \\ 4 \\ \hline 1\,0 \\ 1\,0 \\ \hline 0 \end{array}$$

LESSON 100, WARM-UP

a. 4000

b. 930

c. 50

d. $18.11

e. 80

f. 30

g. 15

Problem Solving

First find the possibility for one letter. A00000 to
A99999 is 100,000 license plates.
100,000 $\times$ 26 letters $=$
2,600,000 license plates

LESSON 100, LESSON PRACTICE

a.

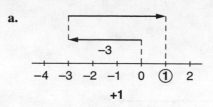

+1

b.

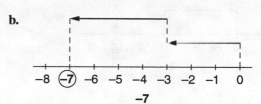

−7

c. 0

d. +1

e. −1

f. +5

g. −15

h. −5

i. +8

j. −4

k. 0

l. −3 + +4 = **+1**

m. −4 + −2 = **−6**

n. +3 + +6 = **+9**

o. −2 + +4 = **+2**

LESSON 100, MIXED PRACTICE

1.
$$\begin{array}{r} 1.2 \\ \times\ 0.6 \\ \hline 0.72 \end{array}$$

2. 50 − 12 = 38
38 − 20 = **18**

3.
$$\begin{array}{r} 3.7 \\ 4\overline{)14.8} \\ \underline{12} \\ 2\ 8 \\ \underline{2\ 8} \\ 0 \end{array}$$

4.

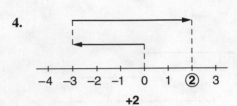

$$+2$$

5. (a) **0**

(b) **−5**

(c) **−2**

(d) **−5**

6. (a) $-2 + +5 = $ **+3**

(b) $-3 + +3 = $ **0**

(c) $+2 + +3 = $ **+5**

(d) $-2 + -3 = $ **−5**

7.
$$\begin{array}{r} 60° \\ 3\overline{)180°} \\ \underline{18} \\ 00 \\ \underline{00} \\ 0 \end{array}$$

8. $m\angle B = 180° - 70° = $ **110°**
$m\angle C = m\angle A = $ **70°**
$m\angle D = m\angle B = $ **110°**

9. (a) $\dfrac{1}{3}$

(b) $\dfrac{1}{\cancel{6}_1} \times \dfrac{\overset{5}{\cancel{30}}}{1} = \dfrac{5}{1} = 5$

about 5 times

10. $V = lwh$
$V = (7 \text{ in.})(5 \text{ in.})(6 \text{ in.})$
$V = $ **210 in.³**

11.
$$\begin{array}{r} 27 \\ \underline{-\ 12} \\ 15 \text{ girls} \end{array}$$

$$\dfrac{15}{12} = \dfrac{5}{4}$$

12. $10^2 + (5^2 - 11) \div \sqrt{49} - 3^3$
$100 + 14 \div 7 - 27$
$100 + 2 - 27$
75

13. $\dfrac{2}{3} \times \dfrac{100\%}{1} = \dfrac{200\%}{3} = \mathbf{66\dfrac{2}{3}\%}$

14. $100\% - 20\% = 80\%$
$80\% = \dfrac{80}{100} = \dfrac{4}{5}$

15. $\dfrac{16}{16} = $ **1**

16.
$$\begin{array}{r} 5\dfrac{7}{8} \times \dfrac{1}{1} = 5\dfrac{7}{8} \\ + \ 4\dfrac{3}{4} \times \dfrac{2}{2} = 4\dfrac{6}{8} \\ \hline 9\dfrac{13}{8} = \mathbf{10\dfrac{5}{8}} \end{array}$$

17. $\dfrac{3}{2} \div \dfrac{5}{2}$

$1 \div \dfrac{5}{2} = \dfrac{2}{5}$

$\dfrac{3}{\cancel{2}_1} \times \dfrac{\cancel{2}^1}{5} = \dfrac{3}{5}$

18. $5 - (3.2 + 0.4)$
$5 - 3.6$
1.4

19. $A = \pi r^2$
$A \approx (3.14)(9 \text{ ft}^2)$
$A \approx 28.26 \text{ ft}^2$

About 28 square feet

20. **Volume is a measure of space. To measure space, we use units that take up space (cubes). We do not use squares to measure volume, because squares do not take up space.**

21. $\dfrac{9}{12} = \dfrac{15}{x}$

$9x = 12 \cdot 15$

$x = \dfrac{\overset{4}{\cancel{12}} \cdot \overset{5}{\cancel{15}}}{\underset{\underset{1}{\cancel{3}}}{\cancel{9}}}$

$x = \mathbf{20}$

22. $A = \dfrac{bh}{2}$

$A = \dfrac{(8\,\text{cm})(6\,\text{cm})}{2}$

$A = \mathbf{24\ cm^2}$

23. $8\,\text{cm} + 6\,\text{cm} + 10\,\text{cm}$

24 cm

24. 21 millimeters

25. $C = \pi d$

$C \approx (3.14)(21\,\text{mm})$

$C \approx 65.94\,\text{mm}$

About 66 millimeters

26. $\dfrac{4}{12} = \dfrac{1}{3}$

27. (a) $\dfrac{9}{10} = \mathbf{0.9}$

(b) $\dfrac{9}{10} \times \dfrac{10}{10} = \dfrac{90}{100} = \mathbf{90\%}$

28. (a) $1.5 = \mathbf{1\dfrac{1}{2}}$

(b) $1.5 \times 100\% = \mathbf{150\%}$

29. (a) $\dfrac{4}{100} = \mathbf{\dfrac{1}{25}}$

(b) $\dfrac{4}{100} = \mathbf{0.04}$

30. $1\ \text{gallon} = 4\ \text{quarts} = 8\ \text{pints}$

$8\ \text{pints} - 2\ \text{pints} = 6\ \text{pints}$

$6\ \text{pints} = \mathbf{3\ quarts}$

INVESTIGATION 10

1–6.

Outcome	Probability
A, red	$\frac{1}{2} \cdot \frac{2}{3} = \frac{1}{3}$
A, white	**1.** $\frac{1}{2} \cdot \frac{2}{6} = \frac{1}{6}$
B, red	**2.** $\frac{1}{4} \cdot \frac{4}{6} = \frac{1}{6}$
B, white	**3.** $\frac{1}{4} \cdot \frac{2}{6} = \frac{1}{12}$
C, red	**4.** $\frac{1}{4} \cdot \frac{4}{6} = \frac{1}{6}$
C, white	**5.** $\frac{1}{4} \cdot \frac{2}{6} = \frac{1}{12}$
sum of probabilities	**6.** $\frac{12}{12} = 1$

7.

R, R
R, W
W, R
W, W

8.

Outcome	Probability
red, red	$\frac{4}{6} \cdot \frac{3}{5} = \frac{2}{5}$
red, white	$\frac{4}{6} \cdot \frac{2}{5} = \frac{4}{15}$
white, red	$\frac{2}{6} \cdot \frac{4}{5} = \frac{4}{15}$
white, white	$\frac{2}{6} \cdot \frac{1}{5} = \frac{1}{15}$
sum of probabilities	$\frac{15}{15} = \mathbf{1}$

9. $\dfrac{2}{6} \cdot \dfrac{1}{5} \cdot \dfrac{0}{4} = \mathbf{0}$

$\dfrac{4}{6} \cdot \dfrac{3}{5} \cdot \dfrac{2}{4} = \dfrac{1}{5}$

10.

N   Q   Outcome

H, H
H, T
T, H
T, T

11.

Outcome	Probability
H, H	$\frac{1}{2} \cdot \frac{1}{2} = \frac{1}{4}$
H, T	$\frac{1}{2} \cdot \frac{1}{2} = \frac{1}{4}$
T, H	$\frac{1}{2} \cdot \frac{1}{2} = \frac{1}{4}$
T, T	$\frac{1}{2} \cdot \frac{1}{2} = \frac{1}{4}$

12. There are two compound outcomes that satisfy the conditions, each with a probability of $\frac{1}{4}$: H, T and T, H. We add the probabilities: $\frac{1}{4} + \frac{1}{4} = \frac{1}{2}$.

13. There are three compound outcomes that satisfy the conditions, each with a probability $\frac{1}{4}$: H, H; H, T; and T, H. We add the probabilities: $\frac{1}{4} + \frac{1}{4} + \frac{1}{4} = \frac{3}{4}$.

14. There is only one outcome that satisfies the conditions: H, T. The probability of that outcome is $\frac{1}{4}$.

Extensions

a. $\frac{4}{15} + \frac{4}{15} = \frac{8}{15}$

b. $\frac{2}{5} + \frac{1}{15} = \frac{7}{15}$

c. $1 - \frac{1}{3} = \frac{2}{3}$

d. $\frac{1}{12} + \frac{1}{12} = \frac{1}{6}$

e.

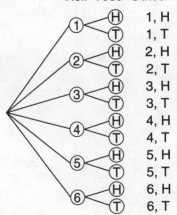

Roll Toss Outcome

1, H
1, T
2, H
2, T
3, H
3, T
4, H
4, T
5, H
5, T
6, H
6, T

f. The probability of each compound outcome is $\frac{1}{6} \times \frac{1}{2}$, or $\frac{1}{12}$.

LESSON 101, WARM-UP

a. 6000

b. 370

c. 25

d. $24.29

e. 0.0375

f. 1000

g. 10

h. 26

Problem Solving

In order to produce a four-digit product, the first digit in the bottom partial product must be 9. So the missing digits in the factors must be 9 and 1. (All other pairs of digits would result in a partial product that did not begin with 9.) Complete the multiplication.

$$
\begin{array}{r}
11 \\
\times\ 91 \\
\hline
11 \\
99 \\
\hline
1001
\end{array}
\quad \text{or} \quad
\begin{array}{r}
91 \\
\times\ 11 \\
\hline
91 \\
91 \\
\hline
1001
\end{array}
$$

LESSON 101, LESSON PRACTICE

a.

	Ratio	A.C.
Sparrows	5	s
Crows	3	c
Total	8	72

$$\frac{3}{8} = \frac{c}{72}$$
$$3 \cdot 72 = 8c$$
$$27 = c$$

27 crows

b.

	Ratio	A.C.
Raisins	2	r
Nuts	3	n
Total	5	60

$$\frac{2}{5} = \frac{r}{60}$$
$$2 \cdot 60 = 5r$$
$$24 = r$$

24 ounces

LESSON 101, MIXED PRACTICE

1.

	Ratio	A.C.
Boys	3	b
Girls	2	g
Total	5	30

$$\frac{2}{5} = \frac{g}{30}$$
$$2 \cdot 30 = 5g$$
$$12 = g$$

12 girls

2. **Rectangular prism**

3.
$$\begin{array}{r} 12 \\ \times\ \ 6 \\ \hline 72 \end{array}$$

4. $1\frac{1}{2}$ inches $\div 2 = \frac{3}{2} \div \frac{2}{1}$

$$\frac{3}{2} \times \frac{1}{2} = \frac{3}{4} \text{ inch}$$

5.
$$\begin{array}{r} \$1.65 \\ \times\ \ \ 2.6 \\ \hline 990 \\ 3300 \\ \hline \$4.290 \end{array} \longrightarrow \ \$4.29$$

6. $\dfrac{\overset{3}{\cancel{12}}}{1} \cdot \dfrac{1}{\underset{1}{\cancel{4}}} = \dfrac{3}{1} = 3 \text{ cm}$

$12 \text{ cm} - 3 \text{ cm} = \textbf{9 cm}$

7. (a) **−7**

(b) **0**

(c) **−3**

(d) **+3**

8. (a) $-3 + {^+}4 = \textbf{+1}$

(b) $+5 + {^+}5 = \textbf{+10}$

(c) $-6 + {^-}3 = \textbf{−9}$

(d) $-6 + {^+}6 = \textbf{0}$

9. There are four equally likely outcomes.

(a) **One of the four outcomes is HH, so the probability is $\frac{1}{4}$.**

(b) **Two of the four outcomes are HT and TH, so the probability is $\frac{1}{2}$.**

10. (a) $\dfrac{3}{4} \times \dfrac{25}{25} = \dfrac{75}{100} = \textbf{0.75}$

(b) $\dfrac{75}{100} = \textbf{75\%}$

11. (a) $1\dfrac{6}{10} = \mathbf{1\dfrac{3}{5}}$

(b) $\dfrac{8}{5} \times \dfrac{20}{20} = \dfrac{160}{100} = \textbf{160\%}$

12. (a) $\dfrac{5}{100} = \mathbf{\dfrac{1}{20}}$

(b) $\dfrac{5}{100} = \textbf{0.05}$

13. $\dfrac{3}{\underset{1}{\cancel{2}}} \times \dfrac{\overset{2}{\cancel{4}}}{1} = \dfrac{6}{1} = \textbf{6}$

14. $\dfrac{6}{1} \div \dfrac{3}{2}$

$$1 \div \dfrac{3}{2} = \dfrac{2}{3}$$

$$\dfrac{\overset{2}{\cancel{6}}}{1} \times \dfrac{2}{\underset{1}{\cancel{3}}} = \dfrac{4}{1} = \textbf{4}$$

15. $0.16 \div 8$

$$\begin{array}{r} 0.02 \\ 8\overline{)0.16} \\ \underline{16} \\ 00 \end{array}$$

16.
$$\dfrac{5}{1} \times \dfrac{2}{2} = \dfrac{10}{2}$$
$$-\ \dfrac{5}{2} \times \dfrac{1}{1} = \dfrac{5}{2}$$
$$\overline{}$$
$$\dfrac{5}{2} = 2\dfrac{1}{2}$$

$$x = \mathbf{2\dfrac{1}{2}}$$

17. $\dfrac{8}{5} = \dfrac{40}{x}$

$40 \cdot 5 = 8x$

$\dfrac{\overset{5}{\cancel{40}} \cdot 5}{\underset{1}{\cancel{8}}} = x$

$25 = x$

18.

$$\begin{array}{r} \$2.50 \\ 6\overline{)\$15.00} \\ \underline{12} \\ 3\,0 \\ \underline{3\,0} \\ 00 \\ \underline{00} \\ 0 \end{array}$$

$n = \$2.50$

19. $6n = 84$

$$\begin{array}{r} 14 \\ 6\overline{)84} \\ \underline{6} \\ 24 \\ \underline{24} \\ 0 \end{array}$$

$n = 14$

20. (a) $20 \times 20 = $ **400 boxes**

(b) $400 \times 8 = $ **3200 boxes**

21. $C = \pi d$

$C \approx (3.14)(2.5 \text{ in.})$

$C \approx 7.85 \text{ in.}$

C. $7\dfrac{3}{4}$ **in.**

22. $9^2 - \sqrt{9} \times 10 - 2^4 \times 2$

$81 - 3 \times 10 - 16 \times 2$

$81 - 30 - 32$

19

23. Pentagon

24. 180°

25.

$$\begin{array}{r} 15°F \\ + \ 8°F \\ \hline 23°F \end{array}$$

26. 8.8

27. The probability of rolling 1 or 4 is $\dfrac{2}{6}$ or $\dfrac{1}{3}$.

28.

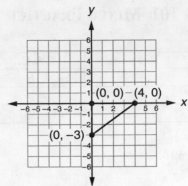

$A = \dfrac{bh}{2}$

$A = \dfrac{4 \text{ units} \cdot 3 \text{ units}}{2}$

$A = $ **6 units²**

29. $\dfrac{\overset{6}{\cancel{18}} \text{ft}}{1} \times \dfrac{1 \text{ yd}}{\underset{1}{\cancel{3}} \text{ft}} = $ **6 yd**

30. $1 \text{ gallon} = 4 \text{ quarts}$

$$\begin{array}{r} \$0.70 \\ 4\overline{)\$2.80} \\ \underline{2\,8} \\ 00 \\ \underline{00} \\ 0 \end{array}$$

$0.70 per quart

LESSON 102, WARM-UP

a. 12,000

b. 612

c. 20

d. $52.50

e. 6

f. 750

g. 50

h. 34

Problem Solving

9 inches per side $\times$ 4 sides $=$ 36 inches

36 inches $\div$ 6 sides $=$ **6 inches** per side

LESSON 102, LESSON PRACTICE

a. $1000 \text{ g} = 1 \text{ kg}$

$\frac{1}{2} \text{ kg} = \textbf{500 grams}$

b. $1 \text{ liter} = 1 \text{ kg}$

$2 \text{ liters} = 2 \text{ kg}$

$2 \text{ kg} = \textbf{2000 grams}$

c.
$$\begin{array}{r} 5 \text{ lb} \quad 10 \text{ oz} \\ + \; 1 \text{ lb} \quad \;\,9 \text{ oz} \\ \hline 6 \text{ lb} \quad 19 \text{ oz} \\ \textbf{7 lb} \quad \textbf{3 oz} \end{array}$$

d.
$$\begin{array}{r} \overset{8}{\cancel{9}} \text{ lb} \quad \overset{24}{\cancel{8}} \text{ oz} \\ - \; 6 \text{ lb} \quad 10 \text{ oz} \\ \hline \textbf{2 lb} \quad \textbf{14 oz} \end{array}$$

e. $1 \text{ ton} = 2000 \text{ pounds}$

$\frac{1}{2} \text{ ton} = \textbf{1000 pounds}$

LESSON 102, MIXED PRACTICE

1. (a) Mode = **92%**

(b) Range = $96\% - 84\% = \textbf{12\%}$

2.
$$\begin{array}{r} \overset{1}{90\%} \\ 92\% \\ 96\% \\ 92\% \\ 84\% \\ + \; 92\% \\ \hline 546\% \end{array} \qquad \begin{array}{r} 91\% \\ 6\overline{)546\%} \\ \underline{54} \\ 06 \\ \underline{06} \\ 0 \end{array}$$

3. $96 - 18 - 18 = 60$ points

30 two-point baskets

4. C. $\frac{12}{21}$

5. $\frac{4}{5} = \frac{x}{20}$

$4 \cdot 20 = 5x$

$\frac{4 \cdot \overset{4}{20}}{\underset{1}{\cancel{5}}} = x$

$16 = x$

6. $-1, -0.1, 0, 0.1, 1$

7. C. 10^5

8. (a) $\sqrt{100 \text{ mm}^2} = \textbf{10 mm}$

(b) $10 \text{ mm} \times 2 = \textbf{20 mm}$

(c) $A = \pi r^2$

$A \approx (3.14)(10 \text{ mm})^2$

$A \approx \textbf{314 mm}^2$

9. (a) $\frac{4}{25} \times \frac{4}{4} = \frac{16}{100} = \textbf{0.16}$

(b) $\frac{16}{100} = \textbf{16\%}$

10. (a) $\frac{1}{100}$

(b) $\frac{1}{100} = \textbf{1\%}$

11. (a) $\frac{90}{100} = \frac{\textbf{9}}{\textbf{10}}$

(b) $\frac{9}{10} = \textbf{0.9}$

12.
$$\begin{array}{r} 1\frac{2}{3} \times \frac{2}{2} = 1\frac{4}{6} \\ 3\frac{1}{2} \times \frac{3}{3} = 3\frac{3}{6} \\ + \; 4\frac{1}{6} \times \frac{1}{1} = 4\frac{1}{6} \\ \hline 8\frac{8}{6} = 9\frac{2}{6} = 9\frac{1}{3} \end{array}$$

13. $\dfrac{\overset{1}{\cancel{5}}}{\underset{1}{\cancel{6}}} \times \dfrac{\overset{1}{\cancel{3}}}{\underset{2}{\cancel{10}}} \times \dfrac{\overset{1}{\cancel{4}}}{1} = \dfrac{1}{1} = \textbf{1}$

14. $\frac{25}{4} \div \frac{100}{1}$

$1 \div \frac{100}{1} = \frac{1}{100}$

$\dfrac{\overset{1}{\cancel{25}}}{4} \times \dfrac{1}{\underset{4}{\cancel{100}}} = \dfrac{\textbf{1}}{\textbf{16}}$

15.
$$\begin{array}{r} 6.437 \\ 12.8 \\ + \; 7 \\ \hline \textbf{26.237} \end{array}$$

16.
$$7\overline{)1.000}$$
```
     0.142
  7)1.000
     7
     30
     28
     20
     14
      6
```
0.14

17. 8 sides − 5 sides = 3 sides
3 more sides

18. $4 \times 5^2 - 50 \div \sqrt{4} + (3^2 - 2^3)$
$\quad 4 \times 25 - 50 \div 2 + 1$
$\quad\quad 100 - 25 + 1$
$\quad\quad\quad \textbf{76}$

19. $\dfrac{1}{4} \cdot \dfrac{1}{4} = \dfrac{1}{16}$

20. $\dfrac{3}{4} \cdot 100 = 75$
About 75 times

21. 4 in. × 4 in. × 4 in. = 64 in.³
64 cubes

22. 4 × 5 = **20**

23.
```
     9        17
   10 pounds  1 ounce
 −  8 pounds  4 ounce
   1 pound   13 ounces
```

24. (a) ∠3

(b) m∠1 = m∠5 = **76°**
m∠2 = 180° − 76° = **104°**

25. 3(5) − 5 = **10**

26. 6 cm + 9 cm + 12 cm + 4 cm
+ 6 cm + 5 cm = **42 cm**

27. (a) $A = bh$
$A = (6 \text{ in.})(4 \text{ in.})$
$A = \textbf{24 in.}^2$

(b) $A = \dfrac{bh}{2}$
$A = \dfrac{(3 \text{ in.})(4 \text{ in.})}{2}$
$A = \textbf{6 in.}^2$

(c) 24 in.² + 6 in.² = **30 in.²**

28. 1000 mg = 1 g
$\dfrac{1}{2}$ g = **500 milligrams**

29.

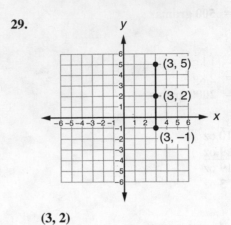

(3, 2)

30.
```
      3.1
  10)31.0
     30
     1 0
     1 0
        0
```
About 3.1 diameters

LESSON 103, WARM-UP

a. 3000

b. 293

c. 10

d. $9.64

e. 0.875

f. 25

g. 90

h. 29

Problem Solving

4D × 2 = 8N   2Q = 50¢
4D ÷ 2 = 2Q   4D = 40¢
                + 8N = 40¢
                130¢ or **$1.30**

LESSON 103, LESSON PRACTICE

a. $3\text{ cm} + x = 8\text{ cm}$
$x = 5\text{ cm}$
$5\text{ cm} + y = 12\text{ cm}$
$y = 7\text{ cm}$
$12\text{ cm} + 8\text{ cm} + 5\text{ cm} + 5\text{ cm}$
$+ 7\text{ cm} + 3\text{ cm} = \textbf{40 cm}$

b. $16\text{ mm} + x = 20\text{ mm}$
$x = 4\text{ mm}$
$7\text{ mm} + y = 15\text{ mm}$
$y = 8\text{ mm}$
$20\text{ mm} + 7\text{ mm} + 4\text{ mm} + 8\text{ mm}$
$+ 16\text{ mm} + 15\text{ mm} = \textbf{70 mm}$

LESSON 103, MIXED PRACTICE

1.
$\dfrac{1}{2} \times \dfrac{3}{3} = \dfrac{3}{6}$ $\dfrac{1}{2} \times \dfrac{1}{3} = \dfrac{1}{6}$
$+ \dfrac{1}{3} \times \dfrac{2}{2} = \dfrac{2}{6}$
$\overline{\qquad\qquad\dfrac{5}{6}}$

$\dfrac{5}{6} \div \dfrac{1}{6}$

$1 \div \dfrac{1}{6} = \dfrac{6}{1}$

$\dfrac{5}{\cancel{6}} \times \dfrac{\cancel{6}^{1}}{1} = \dfrac{5}{1} = \textbf{5}$

2. (a)
$\begin{array}{r} \overset{1}{24} \\ \times\;\; 3 \\ \hline \textbf{72 years} \end{array}$

(b) $22 + 22 + y = 72$
$72 - 44 = y$
28 years old

3. (a)
$\begin{array}{r} \textbf{9 in.} \\ 4\overline{)36} \\ \underline{36} \\ 0 \end{array}$

(b) $9\text{ in.} \times 9\text{ in.} = \textbf{81 in.}^2$

4.
$\dfrac{5}{3} = \dfrac{30}{m}$
$5m = 3 \cdot 30$
$m = \dfrac{3 \cdot \cancel{30}^{6}}{\cancel{5}_{1}}$
$m = \textbf{18}$

5.
$\begin{array}{r} \overset{2}{\cancel{3}}\!\overset{1}{}0 \\ -\;1\;4 \\ \hline 1\;6\;\text{girls} \end{array}$

$\dfrac{14}{16} = \dfrac{\textbf{7}}{\textbf{8}}$

6.

	Ratio	A.C.
Boys	4	b
Girls	7	g
Total	11	33

$\dfrac{7}{11} = \dfrac{g}{33}$
$7 \cdot 33 = 11g$
$\dfrac{7 \cdot \cancel{33}^{3}}{\cancel{11}_{1}} = g$
21 girls

7. $100 \div 10^2 + 3 \times (2^3 - \sqrt{16})$
$100 \div 100 + 3 \times 4$
$1 + 12$
13

8. (a) **2000 pounds**
(b) **Close to zero**

9. (a) $\dfrac{1}{100} = \textbf{0.01}$
(b) $\dfrac{1}{100} = \textbf{1\%}$

10. (a) $0.4 = \dfrac{4}{10} = \dfrac{\textbf{2}}{\textbf{5}}$
(b) $\dfrac{4}{10} \times \dfrac{10}{10} = \dfrac{40}{100} = \textbf{40\%}$

11. (a) $\dfrac{8}{100} = \dfrac{\textbf{2}}{\textbf{25}}$
(b) $\dfrac{8}{100} = \textbf{0.08}$

12. $\dfrac{21}{2} \div \dfrac{7}{2}$
$1 \div \dfrac{7}{2} = \dfrac{2}{7}$
$\dfrac{\cancel{21}^{3}}{\cancel{2}_{1}} \times \dfrac{\cancel{2}^{1}}{\cancel{7}_{1}} = \dfrac{3}{1} = \textbf{3}$

13. $8.4 \div 0.04$

$$
\begin{array}{r}
210 \\
4)\overline{840} \\
\underline{8} \\
04 \\
\underline{04} \\
00 \\
\underline{00} \\
0
\end{array}
$$

14.
$$7\frac{1}{2} \times \frac{2}{2} = 7\frac{2}{4}$$
$$+ \; 6\frac{3}{4} \times \frac{1}{1} = 6\frac{3}{4}$$
$$\overline{13\frac{5}{4} = 14\frac{1}{4}}$$

$$15\frac{3}{8} \times \frac{1}{1} = 15\frac{3}{8}$$
$$- \; 14\frac{1}{4} \times \frac{2}{2} = 14\frac{2}{8}$$
$$\overline{n = \; 1\frac{1}{8}}$$

15.
$$7\frac{1}{2} \times \frac{2}{2} = 7\frac{2}{4}$$
$$+ \; 1\frac{3}{4} \times \frac{1}{1} = 1\frac{3}{4}$$
$$\overline{x = 8\frac{5}{4} = 9\frac{1}{4}}$$

16. $21 \div 7 = 3$

17.
$$
\begin{array}{r}
3.18 \\
11)\overline{35.00} \\
\underline{33} \\
2\,0 \\
1\,1 \\
\overline{90} \\
\underline{88} \\
2
\end{array}
$$

About 3.2 diameters

18. $(2 \times 10^7) + (5 \times 10^5)$

19. **41, 43, 47**

20. (a) **−11**

(b) **+5**

(c) **−5**

(d) **−11**

21. $180° - 40° = 140°$

$$
\begin{array}{r}
70° \\
2)\overline{140°} \\
\underline{14} \\
00 \\
\underline{00} \\
0
\end{array}
$$

22. (a) $20\,\text{mm} + 15\,\text{mm} + 25\,\text{mm} = \textbf{60 mm}$

(b) $A = \dfrac{bh}{2}$

$$A = \dfrac{\overset{10}{\cancel{20}}\,\text{mm} \times 15\,\text{mm}}{\underset{1}{\cancel{2}}}$$

$$A = \textbf{150 mm}^2$$

23. $V = lwh$
$V = 8\,\text{ft.} \times 5\,\text{ft.} \times 3\,\text{ft.}$
$V = 120\,\text{ft}^3$
120 boxes

24. $\dfrac{\text{number of outcomes in the event}}{\text{number of possible outcomes}} = \dfrac{\textbf{1}}{\textbf{52}}$

25. (a) **−8°F**

(b) $-8°F + 12°F = \textbf{4°F}$

26. $10\,\text{mm} + x = 20\,\text{mm}$
$\phantom{10\,\text{mm} +\,} x = 10\,\text{mm}$
$30\,\text{mm} + y = 50\,\text{mm}$
$\phantom{30\,\text{mm} +\,} y = 20\,\text{mm}$
$10\,\text{mm} + 20\,\text{mm} + 10\,\text{mm} + 30\,\text{mm}$
$ + 20\,\text{mm} + 50\,\text{mm} = \textbf{140 mm}$

27. (a) $A = 10\,\text{mm} \times 8\,\text{mm} = \textbf{80 mm}^2$

(b) $A = 7\,\text{mm} \times 20\,\text{mm} = \textbf{140 mm}^2$

(c) $A = 80\,\text{mm}^2 + 140\,\text{mm}^2 = \textbf{220 mm}^2$

28.

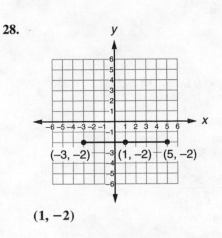

(1, −2)

29. $\frac{1}{2}$ gallon = 2 quarts

2 quarts = 4 pints

1 pint = 16 oz

16 oz = 1 pound

About 4 pounds

30. (a) $\frac{360°}{5}$ = **72°**

(b) $180° - 72°$ = **108°**

LESSON 104, ACTIVITY

Level 1

One negative

No survivors

No battle; all 7 negatives survive.

Level 2

+2

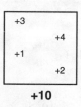

+10

0

Level 3

−4

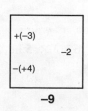

−9

−7

LESSON 104, LESSON PRACTICE

a. $-2 - 3 + 4 - 5$
$-10 + 4$
−6

b. $-3 + 2 - 5 + 6$
$-8 + 8$
0

c. $+3 - 4 - 6 + 7 + 1$
$+11 - 10$
+1

d. $+2 - 3 + 9 - 7 + 1$
$-10 + 12$
+2

e. $+3 + 5 - 4 - 2 + 8$
$+16 - 6$
+10

f. $-10 - 20 + 30 - 40$
$-70 + 30$
−40

LESSON 104, MIXED PRACTICE

1. 8 edges and 5 vertices
3 more edges

2. $\overset{11}{\cancel{12}}$ lb $\overset{22}{\cancel{6}}$ oz
$- \quad 7$ lb $\quad 8$ oz
4 lb 14 oz

3. $\frac{6}{10} = \frac{3}{5}$

4.

	Ratio	A.C.
Win	3	w
Loss	2	l
Total	5	20

$\frac{3}{5} = \frac{w}{20}$

$3 \cdot 20 = 5w$

$\frac{3 \cdot \overset{4}{\cancel{20}}}{\cancel{5}_{1}} = w$

12 games

5. $\frac{1}{2} \cdot \frac{1}{6} = \frac{1}{12}$

6. (a) 7 cm + 7 cm + 8 cm + 8 cm = **30 cm**

(b) $A = bh$
$A = 8 \text{ cm} \times 6 \text{ cm}$
$A = \textbf{48 cm}^2$

7. $180° - 59°$ = **121°**

SOLUTIONS

8. (a) **Answers may vary but should be in the vicinity of 12 to 14 sq. units.**

   (b) $A = \pi r^2$
   $A \approx (3.14)(2 \text{ units})^2$
   $A \approx$ **12.56 sq. units**

9. C. $\dfrac{4}{6}$

10. $\dfrac{6}{8} = \dfrac{a}{12}$

    $6 \cdot 12 = 8a$

    $\dfrac{\overset{3}{\cancel{6}} \cdot \overset{3}{\cancel{12}}}{\underset{\underset{1}{\cancel{4}}}{\cancel{8}}} = a$

    $9 = a$

11. $5 \text{ cm} + x = 13 \text{ cm}$
    $x = 8 \text{ cm}$
    $6 \text{ cm} + y = 12 \text{ cm}$
    $y = 6 \text{ cm}$
    $12 \text{ cm} + 5 \text{ cm} + 6 \text{ cm} + 8 \text{ cm}$
    $+ 6 \text{ cm} + 13 \text{ cm} =$ **50 cm**

12. (a) $\dfrac{3}{20} \times \dfrac{5}{5} = \dfrac{15}{100} = $ **0.15**

    (b) $\dfrac{15}{100} = $ **15%**

13. (a) $1\dfrac{2}{10} = 1\dfrac{1}{5}$

    (b) $\dfrac{12}{10} \times \dfrac{10}{10} = \dfrac{120}{100} = $ **120%**

14. (a) $\dfrac{10}{100} = \dfrac{1}{10}$

    (b) $\dfrac{1}{10} = $ **0.1**

15. $\begin{array}{r} \$6.95 \\ \times \quad 0.40 \\ \hline \$2.7800 \end{array}$

    $\begin{array}{r} \$6.\overset{8}{\cancel{9}}{}^{1}5 \\ - \ \$2.7\ 8 \\ \hline \$4.1\ 7 \end{array}$

16. The number 200 is between the perfect squares 196 and 225, so $\sqrt{200}$ is between $\sqrt{196}$ and $\sqrt{225}$. Since $\sqrt{196}$ is 14 and $\sqrt{225}$ is 15, $\sqrt{200}$ is between 14 and 15.

    14 and 15

17. $\left(\dfrac{1}{2}\right)^3 = \dfrac{1}{2} \cdot \dfrac{1}{2} \cdot \dfrac{1}{2} = \dfrac{1}{8}$

    The probability of 3 consecutive "heads" coin tosses $= \dfrac{1}{2} \cdot \dfrac{1}{2} \cdot \dfrac{1}{2} = \dfrac{1}{8}$.

    So, $\left(\dfrac{1}{2}\right)^3 = $ the probability of 3 consecutive "heads" coin tosses.

18. $\begin{array}{r} 12.4 \\ 5\overline{)62.4} \\ \underline{5} \ \ \ \\ 12 \ \ \\ \underline{10} \ \ \\ 24 \\ \underline{20} \\ 4 \end{array}$

    12

19. $20 \times 3 = $ **60**

20.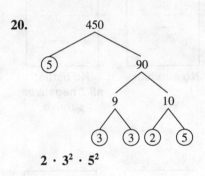

    $\mathbf{2 \cdot 3^2 \cdot 5^2}$

21. $-3 - 5 + 4 - 2$
    $-10 + 4$
    $\mathbf{-6}$

22. $81 + 25 \times 4 - 10 \times 8$
    $81 + 100 - 80$
    101

23. $V = lwh$
    $V = 12 \text{ in.} \times 6 \text{ in.} \times 5 \text{ in.}$
    $V = 360 \text{ in.}^3$
    360 blocks

24. $\dfrac{3}{\underset{1}{\cancel{4}}} \times \dfrac{\overset{15}{\cancel{60}}}{1} = 45$ athletes played

    $60 - 45 = $ **15 athletes**

25. $\dfrac{88 \text{ km}}{1 \ \cancel{\text{hr}}} \times \dfrac{4 \ \cancel{\text{hr}}}{1} = $ **352 km**

26. (a) $A = 5 \text{ cm} \times 12 \text{ cm} = $ **60 cm²**

    (b) $A = 8 \text{ cm} \times 6 \text{ cm} = $ **48 cm²**

    (c) $A = 60 \text{ cm}^2 + 48 \text{ cm}^2 = $ **108 cm²**

27.

$$\begin{array}{r} 3.12 \\ 3.2 \\ 3.15 \\ +\ 3.1 \\ \hline 12.57 \end{array}$$

$$\begin{array}{r} 3.142 \\ 4\overline{)12.570} \\ \underline{12} \\ 0\ 5 \\ \underline{4} \\ 17 \\ \underline{16} \\ 10 \\ \underline{8} \\ 2 \end{array}$$

3.14

28. There are 90 two-digit counting numbers. Since Norton was thinking of only one number, the probability of correctly guessing the number in one try is $\frac{1}{90}$.

29.

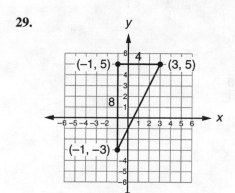

$$A = \frac{bh}{2}$$

$$A = \frac{4 \text{ units} \times \overset{4}{\cancel{8}} \text{ units}}{\underset{1}{\cancel{2}}}$$

$$A = \textbf{16 sq. units}$$

30. $\dfrac{2 \text{ g\cancel{al}}}{1} \times \dfrac{4 \text{ \cancel{qt}}}{1 \text{ g\cancel{al}}} \times \dfrac{2 \text{ pt}}{1 \text{ \cancel{qt}}} = \textbf{16 pt}$

LESSON 105, WARM-UP

a. 8000

b. 417

c. 100

d. $20.19

e. 0.075

f. 22

g. 11

h. 64

Problem Solving

Both the third row and the third column contain multiples of 3. Both the second row and the second column contain multiples of 2.
6

LESSON 105, LESSON PRACTICE

a.

	%	A.C.
Portraits	40	p
Not portraits	60	24
Total	100	t

$$\frac{60}{100} = \frac{24}{t}$$
$$60t = 100 \cdot 24$$
$$t = \frac{\overset{10}{\cancel{100}} \cdot \overset{4}{\cancel{24}}}{\underset{\underset{1}{\cancel{10}}}{\cancel{60}}}$$
$$t = \textbf{40 paintings}$$

b.

	%	A.C.
Played	70	21
Did not play	30	d
Total	100	t

$$\frac{70}{30} = \frac{21}{d}$$
$$70d = 30 \cdot 21$$
$$d = \frac{\overset{3}{\cancel{30}} \cdot \overset{3}{\cancel{21}}}{\underset{\underset{1}{\cancel{10}}}{\cancel{70}}}$$
$$d = \textbf{9 team members}$$

c. $\dfrac{70}{100} = \dfrac{21}{t}$

LESSON 105, MIXED PRACTICE

1. $\dfrac{50 \text{ mi}}{1 \text{ \cancel{hr}}} \times \dfrac{2.5 \text{ \cancel{hr}}}{1} = \textbf{125 mi}$

2. 1 inch = 2 miles
3 inches = **6 miles**

3. $\dfrac{2}{7} = \dfrac{h}{28}$

$7h = 2 \cdot 28$

$h = \dfrac{2 \cdot \overset{4}{\cancel{28}}}{\underset{1}{\cancel{7}}}$

$h = \textbf{8 humpback whales}$

4. $V = lwh$

$V = 7 \text{ in.} \times 3.5 \text{ in.} \times 5 \text{ in.}$

$V = \textbf{122.5 in.}^3$

5. (a) **0**

(b) **0**

(c) $+6 - 5 + 4$

$ 10 - 5$

$ \textbf{+5}$

6.
$$\begin{array}{r} 2.2 \text{ pounds} \\ \times 50 \\ \hline 110.0 \end{array}$$

About 110 pounds

7.

	Ratio	A.C.
Dimes	3	d
Nickels	5	n
Total	8	120

$\dfrac{3}{8} = \dfrac{d}{120}$

$8d = 3 \cdot 120$

$d = \dfrac{3 \cdot \overset{15}{\cancel{120}}}{\underset{1}{\cancel{8}}}$

$d = \textbf{45 dimes}$

8.

	%	A.C.
Discount	25	45
Not at discount	75	n
Total	100	t

$\dfrac{25}{100} = \dfrac{45}{t}$

$25t = 100 \cdot 45$

$t = \dfrac{\overset{4}{\cancel{100}} \cdot 45}{\underset{1}{\cancel{25}}}$

$t = \textbf{180 seats}$

9. (a) $\dfrac{3}{50} \times \dfrac{2}{2} = \dfrac{6}{100} = \textbf{0.06}$

(b) $\dfrac{6}{100} = \textbf{6\%}$

10. (a) $\dfrac{4}{100} = \dfrac{1}{25}$

(b) $\dfrac{4}{100} = \textbf{4\%}$

11. (a) $\dfrac{150}{100} = \dfrac{3}{2} = \textbf{1}\dfrac{\textbf{1}}{\textbf{2}}$

(b) $\dfrac{150}{100} = \textbf{1.5}$

12.
$$\begin{array}{r} 4\dfrac{1}{12} \times \dfrac{1}{1} = 4\dfrac{1}{12} \\[4pt] 5\dfrac{1}{6} \times \dfrac{2}{2} = 5\dfrac{2}{12} \\[4pt] + \; 2\dfrac{1}{4} \times \dfrac{3}{3} = 2\dfrac{3}{12} \\[4pt] \hline 11\dfrac{6}{12} = \textbf{11}\dfrac{\textbf{1}}{\textbf{2}} \end{array}$$

13. $\dfrac{4}{\underset{1}{\cancel{8}}} \times \dfrac{\overset{2}{\cancel{10}}}{\underset{1}{\cancel{5}}} \times \dfrac{\overset{1}{\cancel{2}}}{1} = \dfrac{8}{1} = \textbf{8}$

14.
$$\begin{array}{r} 0.125 \\ \times 80 \\ \hline 10.000 \end{array}$$

10

15. $1.5 \div 0.5 = \textbf{3}$

16. $\dfrac{c}{12} = \dfrac{3}{4}$

$4c = 12 \cdot 3$

$c = \dfrac{\overset{3}{\cancel{12}} \cdot 3}{\underset{1}{\cancel{4}}}$

$c = \textbf{9}$

17.
$$\begin{array}{r} \$8.75 \\ \times 0.08 \\ \hline \$0.7000 \end{array} \qquad \begin{array}{r} \overset{1}{\$8.75} \\ + \; \$0.70 \\ \hline \textbf{\$9.45} \end{array}$$

18. **105.05**

19. $m\angle B = 180° - 115° = \textbf{65°}$

$m\angle C = \textbf{90°}$

20.

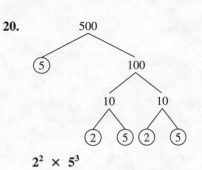

$2^2 \times 5^3$

21. 1 gallon = 4 quarts

4 liters

22. $\frac{1}{4} \times \frac{1}{4} = \frac{1}{16}$

23. 18 cm − 5 cm − 5 cm = **8 cm**

24. $A = \dfrac{bh}{2}$

$A = \dfrac{\overset{4}{\cancel{8}}\,\text{cm} \times 3\,\text{cm}}{\underset{1}{\cancel{2}}}$

$A = \mathbf{12\ cm^2}$

25. −5°F + 12°F = **7°F**

26. 100% − 30% = 70%

$70\% = \dfrac{70}{100} = 0.70 = \mathbf{0.7}$

27. 6 + x = 12

x = 6

6 + y = 12

y = 6

12 in. + 6 in. + 6 in. + 6 in.

+ 6 in. + 12 in. = **48 in.**

28. **Multiply x by 3 to find y.**

29. (a) **5 yards long and 4 yards wide.**

(b) $A = $ **5 yards × 4 yards = 20 sq. yards**

30. **The probability is $\frac{1}{6}$ because the past outcome does not affect the future outcome.**

LESSON 106, WARM-UP

a. 24,000

b. 779

c. 50

d. $0.55

e. 120

f. 720

g. 5

h. 49

Problem Solving

Since we assume the addends are reduced, the first fraction must be $\frac{1}{4}$ or $\frac{3}{4}$. Similarly, the second fraction must be $\frac{1}{6}$ or $\frac{5}{6}$. Only two of these fractions total $\frac{11}{12}$.

$$\frac{3}{4} + \frac{1}{6} = \frac{11}{12}$$

LESSON 106, LESSON PRACTICE

a. 3n + 1 = 16   check: 3(5) + 1 = 16

3n = 15   15 + 1 = 16

n = 5   16 = 16

b. 2x − 1 = 9   check: 2(5) − 1 = 9

2x = 10   10 − 1 = 9

x = 5   9 = 9

c. 3y − 2 = 22   check: 3(8) − 2 = 22

3y = 24   24 − 2 = 22

y = 8   22 = 22

d. 5m + 3 = 33   check: 5(6) + 3 = 33

5m = 30   30 + 3 = 33

m = 6   33 = 33

e. 4w − 1 = 35   check: 4(9) − 1 = 35

4w = 36   36 − 1 = 35

w = 9   35 = 35

f. 7a + 4 = 25   check: 7(3) + 4 = 25

7a = 21   21 + 4 = 25

a = 3   25 = 25

LESSON 106, MIXED PRACTICE

1.   20   60 − 28 − 15 = **17**

× 3

60

2. $\dfrac{2\frac{1}{2}\ \cancel{\text{in.}}}{1} \times \dfrac{10\ \text{mi}}{1\ \cancel{\text{in.}}} = \mathbf{25\ mi}$

3. $\dfrac{1}{\overset{}{\underset{1}{4}}} \times \dfrac{\overset{90}{\cancel{360}}}{0} = \dfrac{90}{1} = \textbf{90}$

4. $\dfrac{5}{25} \times \dfrac{4}{4} = \dfrac{20}{100} = \textbf{20\%}$

5. $\dfrac{12}{52} = \dfrac{\textbf{3}}{\textbf{13}}$

6. $8x + 1 = 25$ check: $8(3) + 1 = 25$
  $8x = 24$ $24 + 1 = 25$
  $x = \textbf{3}$ $25 = 25$

7. $3w - 5 = 25$ check: $3(10) - 5 = 25$
  $3w = 30$ $30 - 5 = 25$
  $w = \textbf{10}$ $25 = 25$

8. (a) $\textbf{+5}$

(b) $-15 - 20 = \textbf{--35}$

(c) $-3 - 2 + 1 = \textbf{--4}$

9.   1 ton = 2000 pounds
4000 pounds = **2 tons**

10. 4 quarts $-$ 1 quart = 3 quarts
      3 quarts = **6 pints**

11. $\dfrac{9}{5} = \dfrac{414}{k}$

$9k = 414 \cdot 5$

$k = \dfrac{\overset{46}{\cancel{414}} \cdot 5}{\underset{1}{\cancel{9}}}$

$k = \textbf{230 koalas}$

12. (a) $\begin{array}{r} 0.125 \\ 8\overline{)1.000} \\ \underline{8} \\ 20 \\ \underline{16} \\ 40 \\ \underline{40} \\ 0 \end{array}$

(b) $0.125 \times 100\% = \textbf{12.5\%}$

13. (a) $1\dfrac{8}{10} = 1\dfrac{\textbf{4}}{\textbf{5}}$

(b) $\dfrac{18}{10} \times \dfrac{10}{10} = \dfrac{180}{100} = \textbf{180\%}$

14. (a) $\dfrac{\textbf{3}}{\textbf{100}}$

(b) $\dfrac{3}{100} = \textbf{0.03}$

15. $\begin{aligned} 8\dfrac{1}{3} \times \dfrac{2}{2} &= 8\dfrac{\overset{8}{\cancel{2}}}{6} \\ - \ 3\dfrac{1}{2} \times \dfrac{3}{3} &= 3\dfrac{3}{6} \\ \hline &\quad 4\dfrac{5}{6} \end{aligned}$

16. $\dfrac{5}{2} \div \dfrac{100}{1}$

$\dfrac{100}{1} \div 1 = \dfrac{1}{100}$

$\dfrac{\overset{1}{\cancel{5}}}{2} \times \dfrac{1}{\underset{20}{\cancel{100}}} = \dfrac{\textbf{1}}{\textbf{40}}$

17. $\begin{array}{r} 0.028 \\ 5\overline{)0.140} \\ \underline{10} \\ 40 \\ \underline{40} \\ 0 \end{array}$

18. **60,907**

19. $\textbf{2}^{\textbf{6}} \cdot \textbf{5}^{\textbf{6}}$

20. $\begin{aligned} 12 &= 11\dfrac{4}{4} \\ - \ 5\dfrac{1}{4} &= \ 5\dfrac{1}{4} \\ \hline &\quad 6\dfrac{3}{4} \text{ inches} \end{aligned}$

21. $6 + 9(5 - 2)$
  $6 + 9(3)$
  $6 + 27$
  33

22. $V = lwh$
$V = (4 \text{ cm})(2 \text{ cm})(3 \text{ cm})$
$V = \textbf{24 cubic centimeters}$

23. $\dfrac{3}{12} = \dfrac{1}{4} \times \dfrac{25}{25} = \dfrac{25}{100} = \textbf{25\%}$

24. (a) $A = 3\text{ cm} \times 2\text{ cm} = \mathbf{6\text{ cm}^2}$

(b) $A = 4\text{ cm} \times 7\text{ cm} = \mathbf{28\text{ cm}^2}$

(c) $A = 6\text{ cm}^2 + 28\text{ cm}^2 = \mathbf{34\text{ cm}^2}$

25. $4\text{ cm} + 2\text{ cm} = x$

$6\text{ cm} = x$

$3\text{ cm} + y = 7\text{ cm}$

$y = 4\text{ cm}$

$4\text{ cm} + 7\text{ cm} + 6\text{ cm} + 3\text{ cm}$

$+ 2\text{ cm} + 4\text{ cm} = \mathbf{26\text{ cm}}$

26. $C = \pi d$

$C \approx (3.14)(2\text{ ft})$

$C \approx 6.28\text{ ft}$

B. 6 ft 3 in.

27. $A = \dfrac{bh}{2}$

$A = \dfrac{7\text{ cm} \times \overset{2}{\cancel{4}}\text{ cm}}{\underset{1}{\cancel{2}}}$

$A = \mathbf{14\text{ cm}^2}$

28. Miles: 3, 4, 5, ⑥, 7, 7, 10

6 miles

29. $7 + 3 + 6 + 10 + 5 + 4 + 7 = 42$

$$\begin{array}{r} \mathbf{6\text{ miles}} \\ 7\overline{)42\text{ miles}} \\ \underline{42} \\ 0 \end{array}$$

30. Answers may vary. Sample answer: How many more miles did Celina ride on Wednesday compared to Monday?
10 miles − 3 miles = 7 miles

LESSON 107, WARM-UP

a. 10,000

b. 226

c. 20

d. $22.88

e. 0.06

f. 6

g. 6

h. 250

Problem Solving

$(80 \times 2) + (90 \times 3) = 430$

$430 \div 5 = \mathbf{86}$

LESSON 107, LESSON PRACTICE

a.

$6\text{ in.} \times 3\text{ in.} = 18\text{ in.}^2$

$4\text{ in.} \times 8\text{ in.} = 32\text{ in.}^2$

$18\text{ in.}^2 + 32\text{ in.}^2 = \mathbf{50\text{ in.}^2}$

$10\text{ in.} \times 3\text{ in.} = 30\text{ in.}^2$

$5\text{ in.} \times 4\text{ in.} = 20\text{ in.}^2$

$30\text{ in.}^2 + 20\text{ in.}^2 = \mathbf{50\text{ in.}^2}$

b. $10\text{ cm} \times 6\text{ cm} = 60\text{ cm}^2$

$A = \dfrac{(6\text{ cm})(\overset{2}{\cancel{4}}\text{ cm})}{\underset{1}{\cancel{2}}}$

$A = 12\text{ cm}^2$

$60\text{ cm}^2 + 12\text{ cm}^2 = \mathbf{72\text{ cm}^2}$

LESSON 107, MIXED PRACTICE

1. $\dfrac{0.48}{0.8}$ $\begin{array}{r} \mathbf{0.6} \\ 8\overline{)4.8} \\ \underline{4\,8} \\ 0 \end{array}$

2. $\dfrac{1\text{ inch}}{2\text{ feet}} = \dfrac{4\text{ inches}}{t}$

$t = \mathbf{8\text{ feet}}$

3. $\dfrac{800}{600} = \dfrac{\mathbf{4}}{\mathbf{3}}$

4. $\dfrac{1}{5} \times \dfrac{20}{20} = \dfrac{20}{100} = \mathbf{20\%}$

5. $\dfrac{1}{1000}$

SOLUTIONS

6. (a) **+5**

(b) $-15 + 10 = \mathbf{-5}$

(c) $+3 - 5 + 2 - 4 = \mathbf{-4}$

7. $1000 - (100 - 10) - 1000 \div 100$

$1000 - 90 - 10$

900

8. $\dfrac{6}{n} = \dfrac{8}{12}$

$8n = 6 \cdot 12$

$n = \dfrac{\overset{3}{\cancel{6}} \cdot \overset{3}{\cancel{12}}}{\underset{1}{\cancel{8}}}$

$n = \mathbf{9}$

9. (a) **1.1**

(b) $\dfrac{11}{10} \times \dfrac{10}{10} = \dfrac{110}{100} = \mathbf{110\%}$

10. (a) $\dfrac{45}{100} = \dfrac{\mathbf{9}}{\mathbf{20}}$

(b) $\dfrac{45}{100} = \mathbf{45\%}$

11. (a) $\dfrac{80}{100} = \dfrac{\mathbf{4}}{\mathbf{5}}$

(b) $\dfrac{80}{100} = \mathbf{0.8}$

12.

$5\dfrac{3}{8} \times \dfrac{1}{1} = 5\dfrac{3}{8}$

$4\dfrac{1}{4} \times \dfrac{2}{2} = 4\dfrac{2}{8}$

$+\ 3\dfrac{1}{2} \times \dfrac{4}{4} = 3\dfrac{4}{8}$

$\overline{ 12\dfrac{9}{8} = \mathbf{13\dfrac{1}{8}}}$

13. $\dfrac{\overset{1}{\cancel{8}}}{\underset{1}{\cancel{3}}} \cdot \dfrac{\overset{1}{\cancel{3}}}{\underset{4}{\cancel{12}}} \cdot \dfrac{\overset{3}{\cancel{9}}}{\underset{2}{\cancel{10}}} = \dfrac{1}{1} = \mathbf{1}$

14.
$$\begin{array}{r} 64.8 \\ 8.42 \\ +\ 24 \\ \hline \mathbf{97.22} \end{array}$$

15. $90° - 55° = \mathbf{35°}$

16. $\dfrac{1}{2}$ pint $= 1$ cup

1 cup $= 8$ ounces

8 ounces

17. $3m + 8 = 44$ check: $3(12) + 8 = 44$

$3m = 36$ $ 36 + 8 = 44$

$m = \mathbf{12}$ $ 44 = 44$

18. $\mathbf{(1 \times 10^8) + (1 \times 10^7)}$

19. Factors of 30

1, 2, 3, 5, 6, 10, ⑮, 30

Factors of 45

1, 3, 5, 9, ⑮, 45

GCF is **15**.

20. (a) $\dfrac{1}{2}$ in. $\times \dfrac{1}{4}$ in. $= \dfrac{\mathbf{1}}{\mathbf{8}}$ **in.²**

(b) $\dfrac{\mathbf{1}}{\mathbf{8}}$

21. $3\ \text{ft} \times 3\ \text{ft} \times 3\ \text{ft} = 27\ \text{ft}^3$

27 blocks

22.
$$\begin{array}{r} \$21.30 \\ 3\overline{)\$63.90} \\ \underline{6} \\ 03 \\ \underline{3} \\ 09 \\ \underline{09} \\ 00 \\ \underline{00} \\ 0 \end{array}$$

$n = \mathbf{\$21.30}$

23. $2\ \text{cm} + x = 7\ \text{cm}$

$x = 5\ \text{cm}$

$5\ \text{cm} + y = 8\ \text{cm}$

$y = 3\ \text{cm}$

$5\ \text{cm} + 5\ \text{cm} + 3\ \text{cm} + 2\ \text{cm}$

$+\ 8\ \text{cm} + 7\ \text{cm} = \mathbf{30\ cm}$

24. $7\ \text{cm} \times 5\ \text{cm} = 35\ \text{cm}^2$

$3\ \text{cm} \times 2\ \text{cm} = 6\ \text{cm}^2$

$A = 35\ \text{cm}^2 + 6\ \text{cm}^2 = \mathbf{41\ cm^2}$

25. $A_1 = \dfrac{\cancel{6}^{3}\text{ cm} \times 10\text{ cm}}{\cancel{2}_{1}}$

$A_1 = 30\text{ cm}^2$

$A_2 = \dfrac{\cancel{14}^{7}\text{ cm} \times 6\text{ cm}}{\cancel{2}_{1}}$

$A_2 = 42\text{ cm}^2$

$A = A_1 + A_2 = 30\text{ cm}^2 + 42\text{ cm}^2 = \textbf{72 cm}^2$

26. (a) Mode = **8**

(b) 7, 7, 8, 8, ⑧, 9, 9, 10, 10
Median = **8**

27. 8 + 9 + 7 + 8 + 10 + 9
+ 8 + 7 + 10 = 76

$$\begin{array}{r} 8.44 \\ 9)\overline{76.00} \\ \underline{72} \\ 4\,0 \\ \underline{3\,6} \\ 40 \\ \underline{36} \\ 4 \end{array}$$

8.4

28. $\overset{11}{\cancel{12}}$ lb $\overset{19}{\cancel{3}}$ oz
-8 lb $\,7$ oz
$\overline{\textbf{3 lb 12 oz}}$

29. (a) 10 cm ÷ 2 = **5 cm**

(b) $C = \pi d$
$C \approx (3.14)(10\text{ cm})$
$C \approx \textbf{31.4 cm}$

30. $\dfrac{10\text{ gallons}}{1} \times \dfrac{31.5\text{ miles}}{1\text{ gallon}} = \textbf{315 miles}$

LESSON 108, WARM-UP

a. **4900**

b. **146**

c. **$5**

d. **$17.22**

e. **0.75**

f. **20**

g. **7**

h. **222**

Problem Solving
The length of the shortest side of each triangle is 12 cm − 9 cm, or 3 cm. So the area of each triangle is $\dfrac{4\text{ cm} \times 3\text{ cm}}{2}$, or **6 sq. cm.**

LESSON 108, LESSON PRACTICE

Note: Answers provided are the most likely responses; however, other answers are possible. For instance, the transformations in problems **b, d,** and **e** could each be accomplished by a reflection along a line of symmetry that either lies outside of the triangles or intersects them at only one point.

a. **Reflection**

b. **Translation**

c. **Rotation**

d. **Translation and reflection**

e. **Rotation and reflection**

LESSON 108, MIXED PRACTICE

1. 2 + 4 + 6 + 8 + 10 = **30**

2. $\dfrac{4}{3} = \dfrac{12}{l}$
$4l = 3 \cdot 12$
$l = \dfrac{3 \cdot \cancel{12}^{3}}{\cancel{4}_{1}}$
$l = \textbf{9 games}$

3.

	%	A.C.
Correct	80	c
Incorrect	20	5
Total	100	t

$$\frac{80}{20} = \frac{c}{5}$$
$$20c = 80 \cdot 5$$
$$c = \frac{\overset{4}{\cancel{80}} \cdot 5}{\underset{1}{\cancel{20}}}$$
$$c = \textbf{20 questions}$$

4. $A = \pi r^2$
$A \approx (3.14)(6 \text{ in.})^2$
$A \approx 113.04 \text{ in.}^2$
$A \approx \textbf{113 in.}^2$

5. $\dfrac{3}{\underset{1}{\cancel{8}}} \times \dfrac{\overset{6}{\cancel{48}}}{1} = \dfrac{18}{1} = 18$

18 woodwind players

6. Multiples of 6
6, 12, 18, ⓐ24, 30
Multiples of 8
8, 16, ⓐ24, 32, 40
Multiples of 12
12, ⓐ24, 36, 48, 60
LCM is **24.**

7. **Answers may vary. Sample answer: Rotate triangle I until its orientation matches triangle II's. Then translate triangle I until it is positioned on triangle II.**

8. $\dfrac{7}{20} = \dfrac{n}{100}$
$20n = 7 \cdot 100$
$n = \dfrac{7 \cdot \overset{5}{\cancel{100}}}{\underset{1}{\cancel{20}}}$
$n = \textbf{35}$

9. (a) $1\dfrac{2}{5} = 1\dfrac{4}{10} = \textbf{1.4}$

(b) $\dfrac{14}{10} \times \dfrac{10}{10} = \dfrac{140}{100} = \textbf{140\%}$

10. (a) $\dfrac{24}{100} = \dfrac{6}{25}$

(b) $\dfrac{24}{100} = \textbf{24\%}$

11. (a) $\dfrac{35}{100} = \dfrac{7}{20}$

(b) $\dfrac{35}{100} = \textbf{0.35}$

12.
$$2\dfrac{1}{4} \times \dfrac{2}{2} = 2\dfrac{\overset{10}{\cancel{2}}}{8}$$
$$-\dfrac{7}{8} \times \dfrac{1}{1} = \dfrac{7}{8}$$
$$\overline{\qquad\qquad\qquad 1\dfrac{3}{8}}$$

$$4\dfrac{3}{4} \times \dfrac{2}{2} = 4\dfrac{6}{8}$$
$$+ 1\dfrac{3}{8} \times \dfrac{1}{1} = 1\dfrac{3}{8}$$
$$\overline{\qquad\qquad\qquad 5\dfrac{9}{8} = \textbf{6}\dfrac{\textbf{1}}{\textbf{8}}}$$

13. $\dfrac{2}{1} \div \dfrac{5}{3} \qquad \dfrac{6}{5} \div \dfrac{6}{5} = \textbf{1}$

$1 \div \dfrac{5}{3} = \dfrac{3}{5}$

$\dfrac{2}{1} \times \dfrac{3}{5} = \dfrac{6}{5}$

14.
$$\begin{array}{r} \overset{8\;\;9}{\cancel{9}.\cancel{0}{}^1 0} \\ -\ 2.7\ 9 \\ \hline 6.2\ 1 \end{array} \qquad \begin{array}{r} 6.2 \\ +\ 6.21 \\ \hline \textbf{12.41} \end{array}$$

15. $-3 + 7 - 8 + 1$
$\quad -11 + 8$
$\qquad \textbf{-3}$

16.
$$\begin{array}{r} \$2.89 \\ \times\ \ 0.06 \\ \hline \$0.1734 \end{array} \longrightarrow \textbf{\$0.17}$$

17. $\dfrac{\textbf{1}}{\textbf{1000}}$

18. **0.3, 0.305, 0.31**

19. $V = lwh$
$V = 10 \text{ cm} \times 10 \text{ cm} \times 10 \text{ cm}$
$V = \textbf{1000 cm}^3$

20. $32 - 25 + 5 \times 2$
$\quad 32 - 25 + 10$
$\qquad \textbf{17}$

21. $8a - 4 = 60 \quad$ check: $\quad 8(8) - 4 = 60$
$\quad\ 8a = 64 \qquad\qquad\qquad 64 - 4 = 60$
$\qquad a = \textbf{8} \qquad\qquad\qquad\ \ 60 = 60$

22. $\dfrac{1}{\cancel{3}} \cdot \dfrac{\overset{30}{\cancel{90}}}{1} = \dfrac{30}{1} = \mathbf{30°}$

23. $20 \text{ mm} + x = 30 \text{ mm}$

$x = 10 \text{ mm}$

$15 \text{ mm} + y = 20 \text{ mm}$

$y = 5 \text{ mm}$

$20 \text{ mm} + 20 \text{ mm} + 30 \text{ mm} + 15 \text{ mm}$

$+ 10 \text{ mm} + 5 \text{ mm} = \mathbf{100 \text{ mm}}$

24. $15 \text{ mm} \times 10 \text{ mm} = 150 \text{ mm}^2$

$20 \text{ mm} \times 20 \text{ mm} = 400 \text{ mm}^2$

$A = 150 \text{ mm}^2 + 400 \text{ mm}^2 = \mathbf{550 \text{ mm}^2}$

25. 2 gallons = 8 quarts

8 quarts = 16 pint

1 pint = 1 lb

About 16 pounds

26. $A = \dfrac{bh}{2}$

$A_1 = \dfrac{\overset{10}{\cancel{20}} \text{ mm} \times 10 \text{ mm}}{\underset{1}{\cancel{2}}}$

$A_1 = 100 \text{ mm}^2$

$A_2 = \dfrac{\overset{6}{\cancel{12}} \text{ mm} \times 10 \text{ mm}}{\underset{1}{\cancel{2}}}$

$A_2 = 60 \text{ mm}^2$

$A = A_1 + A_2 = 100 \text{ mm}^2 + 60 \text{ mm}^2$

$= \mathbf{160 \text{ mm}^2}$

27. **1000 milliliters**

28. $\dfrac{6}{10} \cdot \dfrac{5}{9} = \dfrac{30}{90} = \dfrac{1}{3} \text{ m}$

29. $1 \text{ km} = 1000 \text{ m}$

$\dfrac{1}{2} \text{ km} = 500 \text{ m}$

1500 meters

30.

Obtuse angle

a. **12,000**

b. **937**

c. **$2**

d. **$3.13**

e. **50**

f. **21,000**

g. **3**

h. **190**

Problem Solving

There are 26 possible letters in the first position and 26 possible letters in the second position. Twenty-six times 26 equals 676 letter combinations. For the numerals, the range 0000 to 9999 represents 10,000 combinations. So a total of $676 \times 10,000$, or **6,760,000 license plates,** can be formed.

LESSON 109, LESSON PRACTICE

a. **True**

b. **False**

c. **True**

d. $\overline{QR}$ **or** $\overline{RQ}$

e. **I and II**

LESSON 109, MIXED PRACTICE

1. $7 \times 11 \times 13 = \mathbf{1001}$

2. $\dfrac{2 \text{ cm}}{1 \text{ km}} = \dfrac{10 \text{ cm}}{x}$

$2x = 10 \cdot 1$

$x = \dfrac{\overset{5}{\cancel{10}} \cdot 1}{\underset{1}{\cancel{2}}}$

$x = \mathbf{5 \text{ km}}$

3. $\frac{8}{52} = \mathbf{\frac{2}{13}}$

4. (a) **C.**

(b) **A.**

5. $90° \div 2 = \mathbf{45°}$

6. $\frac{8}{52} = \mathbf{\frac{2}{13}}$

7. $7w - 3 = 60$
$\quad 7w = 63$
$\quad\quad w = \mathbf{9}$

8. $\frac{8}{n} = \frac{4}{25}$

$4n = 8 \cdot 25$

$n = \dfrac{\overset{2}{\cancel{8}} \cdot 25}{\underset{1}{\cancel{4}}}$

$n = \mathbf{50}$

9. (a)
$$\begin{array}{r} 0.625 \\ 8\overline{)5.000} \\ \underline{4\,8} \\ 20 \\ \underline{16} \\ 40 \\ \underline{40} \\ 0 \end{array}$$

(b) $0.625 \times 100\% = \mathbf{62.5\%}$

10. (a) $1\frac{25}{100} = \mathbf{1\frac{1}{4}}$

(b) $\frac{125}{100} = \mathbf{125\%}$

11. (a) $\frac{70}{100} = \mathbf{\frac{7}{10}}$

(b) $\frac{7}{10} = \mathbf{0.7}$

12. (a) There are 3 numbers on the spinner that are less than 4. So the probability is $\mathbf{\frac{3}{4}}$.

(b) Two of the numbers (2 and 3) are prime. The spinner should land on a prime number about $\frac{1}{2}$ of the 100 spins.

$\frac{1}{2} \cdot 100 = \mathbf{50\ times}$

13. $\dfrac{\overset{2}{\cancel{200}}\ \cancel{cm}}{1} \cdot \dfrac{1\ m}{\underset{1}{\cancel{100}}\ \cancel{cm}} = \mathbf{2\ meters}$

14.
$$\begin{array}{r} \overset{4}{1}\overset{\,\,'1}{\cancel{8}}.\overset{1}{\cancel{2}}0 \\ -\ \ 2.79 \\ \hline \mathbf{12.41} \end{array}$$

15. $1000 \div 100 - 10$
$\quad\quad 10 - 10$
$\quad\quad\quad\ \ \mathbf{0}$

16. **Answers may vary. Sample answer:**

x	2	3	5	10
y	4	6	10	20

17.
$$\begin{array}{r} 0.666 \longrightarrow \mathbf{0.67} \\ 3\overline{)2.000} \\ \underline{1\,8} \\ 20 \\ \underline{18} \\ 20 \\ \underline{18} \\ 2 \end{array}$$

18. $V = lwh$
$V = 12\ ft \times 10\ ft \times 8\ ft$
$V = 960\ ft^2$
960 boxes

19. Similar figures have the same shape. They might or might not be the same size. Congruent figures have the same shape and are the same size. Since congruent figures have the same shape, they are similar.

20.
$$\begin{array}{r} \$35 \\ 12\overline{)\$420} \\ \underline{36} \\ 60 \\ \underline{60} \\ 0 \end{array}$$
$\mathbf{\$35.00} = m$

21. (a) $\mathbf{-1}$

(b) $\mathbf{+1}$

(c) $\mathbf{-15}$

(d) $\mathbf{+1}$

22. $A = \dfrac{4\ in. \times 3\ in.}{2}$

$A = \mathbf{6\ in.^2}$

23. Answers may vary. One possibility: rotation and translation

24. $A_1 = 7 \text{ cm} \times 6 \text{ cm}$
$A_1 = 42 \text{ cm}^2$
$A_2 = \dfrac{3 \text{ cm} \times \overset{3}{\cancel{6}} \text{ cm}}{\underset{1}{\cancel{2}}}$
$A_2 = 9 \text{ cm}^2$
$A = A_1 + A_2 = 42 \text{ cm}^2 + 9 \text{cm}^2 = \textbf{51 cm}^2$

25. (a) **34 mm**

(b) **3.4 cm**

26. **40°**

27. $\dfrac{9}{2} = \dfrac{p}{36}$
$2p = 36 \cdot 9$
$p = \dfrac{\overset{18}{\cancel{36}} \cdot 9}{\underset{1}{\cancel{2}}}$
$p = \textbf{162 peanuts}$

28. $C = \pi d$
$C \approx (3.14)(7 \text{ cm})$
$C \approx 21.98 \text{ cm}$
22 cm

29. $6\dfrac{2}{3} \div 100 = \dfrac{20}{3} \div \dfrac{100}{1}$
$1 \div \dfrac{100}{1} = \dfrac{1}{100}$
$\dfrac{\overset{1}{\cancel{20}}}{3} \cdot \dfrac{1}{\underset{5}{\cancel{100}}} = \dfrac{1}{15}$

30. $\left(\dfrac{1}{10}\right)^2 \; \ominus \; 0.01$
$\dfrac{1}{100}$
0.01

LESSON 110, WARM-UP

a. 8100

b. 476

c. $25

d. $11.60

e. 0.08

f. 7

g. 20

h. 199

Problem Solving

First think, "What number times 9 equals 99?" (11). Using only zeros or ones, we see that the first three digits of the dividend must be 1, 0, and 0, in that order. The last digit of the dividend must be 1; otherwise, the remainder would be a two-digit number. Complete the division.

$$\begin{array}{r} 91 \\ 11\overline{)1001} \\ \underline{99} \\ 11 \\ \underline{11} \\ 0 \end{array}$$

LESSON 110, LESSON PRACTICE

a.

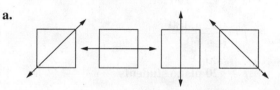

b. F

LESSON 110, MIXED PRACTICE

1.
$$\begin{array}{r} 101 \\ 99\overline{)9999} \\ \underline{99} \\ 09 \\ \underline{0} \\ 99 \\ \underline{99} \\ 0 \end{array}$$

2. $\dfrac{3}{2} = \dfrac{l}{60} \text{ mm}$
$2l = 3 \cdot 60$
$l = \dfrac{3 \cdot \overset{30}{\cancel{60}}}{\underset{1}{\cancel{2}}}$
$l = \textbf{90 mm}$

3. $V = lwh$
$V = 4 \text{ in.} \times 4 \text{ in.} \times 10 \text{ in.}$
$V = \textbf{160 in.}^3$

4.

$$\begin{array}{r} 60° \\ 6\overline{)360°} \\ \underline{36} \\ 00 \\ \underline{00} \\ 0 \end{array}$$

5. **3 lines of symmetry**

6. 6 cm $\times$ 3 = **18 cm**

7. $A = \dfrac{\overset{2}{\cancel{4}} \text{ cm} \times 3 \text{ cm}}{\underset{1}{\cancel{2}}}$

$A = $ **6 cm²**

8.

	%	A.C.
Boys	40	b
Girls	60	12
Total	100	t

$\dfrac{60}{100} = \dfrac{12}{t}$

$60t = 12 \cdot 100$

$t = \dfrac{\overset{1}{\cancel{12}} \cdot \overset{20}{\cancel{100}}}{\underset{\underset{1}{\cancel{1}}}{\cancel{60}}}$

$t = $ **20 piano students**

9. (a) $2\dfrac{3}{4} \times \dfrac{25}{25} = 2\dfrac{75}{100} = $ **2.75**

(b) $\dfrac{275}{100} = $ **275%**

10. (a) $1.1 = 1\dfrac{1}{10}$

(b) $\dfrac{11}{10} \times \dfrac{10}{10} = \dfrac{110}{100} = $ **110%**

11. (a) $\dfrac{64}{100} = \dfrac{16}{25}$

(b) $\dfrac{64}{100} = $ **0.64**

12.

$$24\dfrac{1}{6} \times \dfrac{1}{1} = 24\dfrac{1}{6}$$
$$23\dfrac{1}{3} \times \dfrac{2}{2} = 23\dfrac{2}{6}$$
$$+ \; 22\dfrac{1}{2} \times \dfrac{3}{3} = 22\dfrac{3}{6}$$
$$\overline{ 69\dfrac{6}{6} = \textbf{70}}$$

13.

$\dfrac{6}{5} \div \dfrac{2}{1}$ $\qquad$ $\dfrac{3}{5} \div \dfrac{5}{3}$

$\underset{3}{1} \div \dfrac{2}{1} = \dfrac{1}{2}$ $\qquad$ $\dfrac{5}{3} \div 1 = \dfrac{3}{5}$

$\dfrac{\cancel{6}}{5} \times \dfrac{1}{\underset{1}{\cancel{2}}} = \dfrac{3}{5}$ $\qquad$ $\dfrac{3}{5} \times \dfrac{3}{5} = \dfrac{9}{25}$

$\dfrac{9}{25}$

14. $9 - 8.99 = $ **0.01**

15.

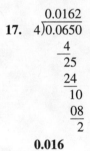

$$\begin{array}{r} \$175 \\ 36\overline{)\$6300} \\ \underline{36} \\ 270 \\ \underline{252} \\ 180 \\ \underline{180} \\ 0 \end{array}$$

\$175.00 $= m$

16.

$$\begin{array}{r} \$24.89 \\ \times 0.065 \\ \hline 12445 \\ 149340 \\ \hline \$1.61785 \end{array} \longrightarrow \textbf{\$1.62}$$

17.

$$\begin{array}{r} 0.0162 \\ 4\overline{)0.0650} \\ \underline{4} \\ 25 \\ \underline{24} \\ 10 \\ \underline{08} \\ 2 \end{array}$$

0.016

18.

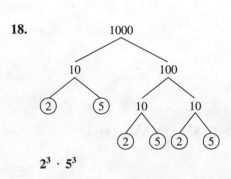

$2^3 \cdot 5^3$

19. **True**

20. $27 - 9 \div 3 - 3 \times 3$

$ 27 - 3 - 9$

$ \textbf{15}$

21. $8\,m + x = 10\,m$
$ x = \mathbf{2\,m}$
$5\,m + y = 12\,m$
$ y = 7\,m$
$8\,m + 12\,m + 10\,m + 5\,m$
$ + 2\,m + 7\,m = \mathbf{44\,m}$

22. $A_1 = 8\,m \times 12\,m \quad A_2 = 2\,m \times 5\,m$
$A_1 = 96\,m^2 A_2 = 10\,m^2$
$A = A_1 + A_2 = 96\,m^2 + 10\,m^2$
$ = \mathbf{106\,m^2}$

23. Answers may vary. One possibility: rotation, reflection, and translation

24. $24\,cm - 10\,cm - 8\,cm = \mathbf{6\,cm}$

25. $C = \pi d$
$C \approx (3.14)(2\,ft)$
$C \approx \mathbf{6.28\,ft}$

26.

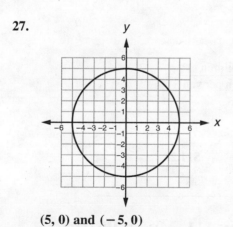

$1\dfrac{3}{4}\text{ inches} \div 2 = \dfrac{7}{4} \div \dfrac{2}{1}$

$1 \div \dfrac{2}{1} = \dfrac{1}{2}$

$\dfrac{7}{4} \times \dfrac{1}{2} = \dfrac{7}{8}\textbf{ inch}$

27.

$\mathbf{(5, 0)}$ **and** $\mathbf{(-5, 0)}$

28. $A = \pi r^2$
$A \approx (3.14)(5\,\text{units})^2$
$A \approx \mathbf{78.5\text{ sq. units}}$

29. $-3 - 4 + 5 - 7$
$ -14 + 5$
$ \mathbf{-9}$

30. $\dfrac{1}{2} \times \dfrac{1}{2} \times \dfrac{1}{2} \times \dfrac{1}{2} = \dfrac{1}{16}$

INVESTIGATION 11

1. Length: 1 inch = **10 feet**
Width: $\dfrac{1}{2}$ inch = **5 feet**

2. $\dfrac{1}{4}$ inch = 5 feet ÷ 2
$\dfrac{1}{4}$ inch = $2\dfrac{1}{2}$ **feet**

3. 1.5 ft × 8 = 12 ft
1.5 ft × 10 = 15 ft
12 ft by 15 ft
12 ft × 15 ft = **180 ft²**

4. 2 feet × 24 = **48 feet**

5. $\dfrac{7}{3} = \dfrac{14}{w}$
$7w = 42$
$w = \mathbf{6\text{ feet}}$

6. $\dfrac{7}{h} = \dfrac{14}{4}$
$14h = 28$
$h = \mathbf{2\text{ inches}}$

7. $\dfrac{3}{12} = \dfrac{F}{6}$
$12F = 18$
$F = 1\dfrac{1}{2}\textbf{ ft}$

8. $\dfrac{h}{12} = \dfrac{1}{6}$
$6h = 12$
$h = \mathbf{2\text{ cm}}$

9. (a) $7s = 28$
$s = 4$
1 cm = 4 ft

(b) Length: 2 × 4 ft = **8 ft**
Width: 1 × 4 ft = **4 ft**

10. $44s = 22$ feet
$s = \dfrac{1}{2}$, or 0.5
1 inch = 0.5 feet

11. Scale = $\dfrac{7\text{ inches}}{14\text{ feet}}$

$\phantom{\text{Scale }} = \dfrac{7\text{ inches}}{168\text{ inches}}$

$\phantom{\text{Scale }} = \dfrac{1}{24}$

Scale factor = **24**

12. 6 ft 8 in. = 80 in.

$$\frac{1}{10} = \frac{x}{80 \text{ in.}}$$

$$10x = 80 \text{ in.}$$

$$x = \textbf{8 in.}$$

13. Length: $\frac{1}{6} = \frac{3 \text{ feet}}{x}$

$$x = 18 \text{ feet}$$

Width: $\frac{1}{6} = \frac{1.5 \text{ feet}}{x}$

$$x = 9 \text{ feet}$$

18 feet by 9 feet

Extensions

a. 1 in. = 36 in.

Scale = $\frac{1}{36}$

Scale factor = **36**

b. If model is measured in inches:

$$\frac{1}{24} = \frac{5.5 \text{ in.}}{x}$$

$$x = 132 \text{ in., or 11 feet}$$

Accept answers **between 9 and 11 feet.**

If model is measured in centimeters:

$$\frac{1}{24} = \frac{14 \text{ cm}}{x}$$

$$x = 336 \text{ cm, or 3.36 m}$$

Accept answers **between 3 and 4 meters.**

c. **Answers may vary.**

d. $\frac{3}{4}$ in. = 12 in.

$$\frac{3}{4}x = 12$$

$$\frac{4}{3} \times \frac{3}{4}x = 12 \times \frac{4}{3}$$

$$x = \textbf{16}$$

LESSON 111, WARM-UP

a. **16**

b. **245**

c. **$24**

d. **$1.41**

e. **50**

f. **4**

g. **25**

h. **2000**

Problem Solving

Multiples of 2 (Sonya): 2, 4, 6, 8, 10, 12, ...
Multiples of 3 (Sid): 3, 6, 9, 12, 15, ...
Multiples of 4 (Sinéad): 4, 8, 12, 16, 20, ...
The least common multiple is 12. Twelve days after Monday is **Saturday.**

LESSON 111, LESSON PRACTICE

a.
$$\begin{array}{r} 22 \text{ R } 2 \\ 4\overline{)90} \\ 8 \\ \hline 10 \\ 8 \\ \hline 2 \end{array}$$

22, 22, 23, 23

b. $\dfrac{\overset{10}{\cancel{20 \text{ dollars}}}}{\underset{3}{\cancel{6 \text{ dollars}}} \text{ per ticket}} = 3\frac{1}{3}$

3 tickets

c. $\dfrac{\overset{14}{\cancel{28 \text{ children}}}}{\underset{3}{\cancel{6} \text{ children}} \text{ per van}} = 4\frac{2}{3}$

5 vans

d. $\dfrac{\$45.00}{4 \text{ workers}} = \textbf{\$11.25}$

LESSON 111, MIXED PRACTICE

1.
$$\begin{array}{r} 26 \text{ R } 2 \\ 3\overline{)80} \\ 6 \\ \hline 20 \\ 18 \\ \hline 2 \end{array}$$

26, 27, 27

2. 10.086
10.09

3.
$$
\begin{array}{r}
\mathbf{40\ stamps} \\
37\overline{)1480} \\
\underline{148} \\
00 \\
\underline{00} \\
0
\end{array}
$$

4. 3 cubes × 3 cubes × 3 cubes = **27 sugar cubes**

5. **5043**

6. **30 cm or 31 cm** (Accept either answer.) Twelve inches equals 30.48 cm, which is very near the midpoint of 30 cm and 31 cm.

7. (a) $\dfrac{1}{24}$

(b) **24**

8. $180° - 80° - 70° = \mathbf{30°}$

9. (a) $\dfrac{11}{20} \times \dfrac{5}{5} = \dfrac{55}{100} = \mathbf{0.55}$

(b) $\dfrac{55}{100} = \mathbf{55\%}$

10. (a) $1\dfrac{5}{10} = \mathbf{1\dfrac{1}{2}}$

(b) $\dfrac{15}{10} \times \dfrac{10}{10} = \dfrac{150}{100} = \mathbf{150\%}$

11. (a) $\dfrac{1}{100}$

(b) $\dfrac{1}{100} = \mathbf{0.01}$

12. (a) **−18**

(b) **+6**

(c) **−6**

(d) **−18**

13. $\dfrac{25}{4} \div \dfrac{100}{1}$

$1 \div \dfrac{100}{1} = \dfrac{1}{100}$

$\dfrac{\overset{1}{\cancel{25}}}{4} \times \dfrac{1}{\underset{4}{\cancel{100}}} = \mathbf{\dfrac{1}{16}}$

14.
$$
\begin{array}{r}
\mathbf{\$14.70} \\
3\overline{)\$44.10} \\
\underline{3} \\
14 \\
\underline{12} \\
2\ 1 \\
\underline{2\ 1} \\
00 \\
\underline{00} \\
0
\end{array}
$$

$m = \mathbf{\$14.70}$

15.

	%	A.C.
Kim	30	15
Team	70	t
Total	100	T

$\dfrac{30}{100} = \dfrac{15}{T}$

$30T = 15 \cdot 100$

$T = \dfrac{\overset{1}{\cancel{15}} \cdot \overset{50}{\cancel{100}}}{\underset{\underset{1}{2}}{\cancel{30}}}$

$T = \mathbf{50\ points}$

16. $6.7 + 6.6 + 6.7 + 6.7 + 6.8 = 33.5$

$33.5 \div 5 = \mathbf{6.7}$

17. Answers may vary. Sample answer: What was the range of Andrea's scores?
$7.6 - 6.5 = \mathbf{1.1}$

18. $5\,\text{m} \times 5\,\text{m} = 25\,\text{m}^2$

$\dfrac{\overset{3}{\cancel{6}}\,\text{m} \times 5\,\text{m}}{\underset{1}{\cancel{2}}} = 15\,\text{m}^2$

$25\,\text{m}^2 + 15\,\text{m}^2 = \mathbf{40\,\text{m}^2}$

19. 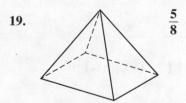 $\dfrac{5}{8}$

20. **Line** t

21. $3m + 1 = 100$ check: $3(33) + 1 = 100$

$3m = 99$ $99 + 1 = 100$

$m = \mathbf{33}$ $100 = 100$

SOLUTIONS

22.

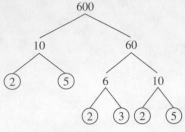

$2^3 \times 3 \times 5^2$

23. 12 grams $\times$ 10 = **120 grams**

24.
$$\begin{array}{r} \$0.89 \\ \times\ \ 0.07 \\ \hline \$0.0623 \end{array} \qquad \begin{array}{r} \$0.89 \\ +\ \$0.06 \\ \hline \mathbf{\$0.95} \end{array}$$

25. $\dfrac{999{,}999}{1{,}000{,}000}$

26. Angle B

27. Answers may vary. One possibility: rotation and translation

28. $C = \pi d$
$C \approx (3.14)(10 \text{ cm})$
$C \approx \mathbf{31.4 \text{ cm}}$

29. $\dfrac{10}{16} = \dfrac{25}{y}$

$10y = 16 \cdot 25$

$y = \dfrac{\overset{8}{\cancel{16}} \cdot \overset{5}{\cancel{25}}}{\underset{\underset{1}{2}}{\cancel{10}}}$

$y = \mathbf{40}$

30. $d = \dfrac{50 \text{ mi}}{1 \text{ hr}} \times \dfrac{5 \text{ hr}}{1}$

$d = \mathbf{250 \text{ mi}}$

LESSON 112, WARM-UP

a. 18

b. 288

c. $12

d. $7.47

e. 0.05

f. 15,000

g. 7

h. 1900

Problem Solving
$8 + 10 + 12 + 14 + 16 + 18 + 20$
$+ 22 = \mathbf{120 \text{ seats}}$

LESSON 112, LESSON PRACTICE

a. −20

b. +20

c. +20

d. −20

e. −6

f. +6

g. −6

h. +6

LESSON 112, MIXED PRACTICE

1.
$$\begin{array}{r} 2 \text{ R } 32 \\ 84\overline{)200} \\ \underline{168} \\ 32 \end{array}$$
3 buses

2. (a) $\dfrac{1}{60}$

(b) $\dfrac{12}{60} = \dfrac{10}{x}$

$12x = 60 \cdot 10$

$x = \dfrac{\overset{5}{\cancel{60}} \cdot 10}{\underset{1}{\cancel{12}}}$

$x = \mathbf{50 \text{ feet}}$

3. (a) **+12**

(b) **−3**

(c) **+1**

(d) **−12**

4. (a) **−8**

(b) **+4**

(c) **−4**

(d) **+8**

5.

	%	A.C.
Correct	90	27
Incorrect	10	i
Total	100	t

$$\frac{90}{10} = \frac{27}{i}$$
$$90i = 10 \cdot 27$$
$$i = \frac{\overset{1}{\cancel{10}} \cdot \overset{3}{\cancel{27}}}{\underset{\underset{1}{\cancel{9}}}{\cancel{90}}}$$

$i =$ **3 questions**

6. $(2 \times 10^7) + (5 \times 10^5) + (1 \times 10^4)$

7.
$$\begin{array}{r} \$3.65 \\ \times\quad 0.08 \\ \hline \$0.2920 \end{array} \longrightarrow \textbf{\$0.29}$$

8. $\frac{1}{2} \times \frac{1}{2} = \frac{1}{4}$

$\frac{1}{8} \div \frac{1}{2}$

$1 \div \frac{1}{2} = \frac{2}{1}$

$\frac{1}{\underset{4}{\cancel{8}}} \times \frac{\overset{1}{\cancel{2}}}{1} = \frac{1}{4}$

$\frac{1}{4} + \frac{1}{4} = \frac{2}{4} = \textbf{\frac{1}{2}}$

9. (a) $1\frac{4}{5} \times \frac{20}{20} = 1\frac{80}{100} = \textbf{1.8}$

(b) $\frac{180}{100} = \textbf{180\%}$

10. (a) $\frac{6}{10} = \textbf{\frac{3}{5}}$

(b) $\frac{6}{10} \times \frac{10}{10} = \frac{60}{100} = \textbf{60\%}$

11. (a) $\frac{2}{100} = \frac{1}{50}$

(b) $\frac{2}{100} = \textbf{0.02}$

12.
$$5\frac{1}{2} \times \frac{3}{3} = \overset{4}{\cancel{5}}\frac{\overset{9}{\cancel{3}}}{6}$$
$$- 2\frac{5}{6} \times \frac{1}{1} = 2\frac{5}{6}$$
$$\overline{}$$
$$m = 2\frac{4}{6} = \textbf{2\frac{2}{3}}$$

13. $\frac{6}{10} = \frac{9}{n}$

$6n = 10 \cdot 9$

$n = \frac{\overset{5}{\cancel{10}} \cdot \overset{3}{\cancel{9}}}{\underset{\underset{1}{\cancel{2}}}{\cancel{6}}}$

$n = \textbf{15}$

14. $9x - 7 = 92$

$9x = 99$

$x = \textbf{11}$

15.
$$\begin{array}{r} 160 \\ 5\overline{)800} \\ \underline{5} \\ 30 \\ \underline{30} \\ 00 \\ \underline{00} \\ 0 \end{array}$$

$w = \textbf{160}$

16. (a) 6 lb ÷ 3 = **2 lb**

(b) 2 lb × 8 = **16 lb**

17. $V = lwh$

$V = (8\text{ cm})(5\text{ cm})(2\text{ cm})$

$V = \textbf{80 cm}^3$

18. 1.2 × 1000 mm = **1200 mm**

19. 5 mm + x = 15 mm

x = 10 mm

5 mm + y = 12 mm

y = 7 mm

5 mm + 15 mm + 12 mm + 5 mm

+ 7 mm + 10 mm = **54 mm**

20. A_1 = 5 mm × 15 mm = 75 mm²

A_2 = 7 mm × 5 mm = 35 mm²

$A = A_1 + A_2 = 75$ mm² + 35 mm²

= **110 mm²**

21. Cube

22. D. 37

23. $C = \pi d$
$C \approx (3.14)(12 \text{ in.})$
$C \approx 37.68 \text{ in.}$
38 inches

24. H

25. *B*

26.

	Ratio	A.C.
Good	2	*g*
Bad	3	*b*
Total	5	30

$\dfrac{2}{5} = \dfrac{g}{30}$

$5g = 2 \cdot 30$

$g = \dfrac{2 \cdot \overset{6}{\cancel{30}}}{\underset{1}{\cancel{5}}}$

$g = $ **12 good guys**

27.

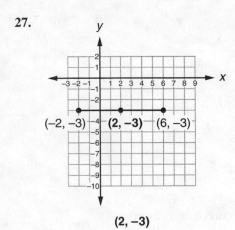

(−2, −3)   **(2, −3)**   (6, −3)

(2, −3)

28. $\dfrac{12}{51} = \dfrac{4}{17}$

29. $A_1 = \dfrac{10 \text{ in.} \times \overset{3}{\cancel{6}} \text{ in.}}{\underset{1}{\cancel{2}}}$

$A_1 = 30 \text{ in.}^2$

$A_2 = \dfrac{\overset{3}{\cancel{6}} \text{ in.} \times 8 \text{ in.}}{\underset{1}{\cancel{2}}}$

$A_2 = 24 \text{ in.}^2$
$A = A_1 + A_2 = 30 \text{ in.}^2 + 24 \text{ in.}^2 = $ **54 in.²**

30. $9 + 2 \times 25 - 50 \div 5$
$9 + 50 - 10$
    49

LESSON 113, WARM-UP

a. 12

b. 152

c. $50

d. $90.50

e. 0.012

f. 4

g. 0

h. 2004

Problem Solving
Divide 999,999 by 7.

$$\begin{array}{r} 142,857 \\ \times \qquad 7 \\ \hline 999,999 \end{array}$$

LESSON 113, LESSON PRACTICE

a.
$$\begin{array}{r} 6 \text{ ft} \quad 5 \text{ in.} \\ + \; 4 \text{ ft} \quad 8 \text{ in.} \\ \hline 10 \text{ ft} \quad 13 \text{ in.} \end{array}$$

    11 ft   1 in.

b.
$$\begin{array}{r} \overset{2}{\cancel{3}} \text{ hr} \; \overset{75}{\cancel{15}} \text{ min} \\ - \; 1 \text{ hr} \; 40 \text{ min} \\ \hline 1 \text{ hr} \; 35 \text{ min} \end{array}$$

c. 12,000

d. 1,500,000

e. 2,500,000,000

f. 250,000

LESSON 113, MIXED PRACTICE

1.
$$\begin{array}{r} \$8.75 \\ 4\overline{)\$35.00} \\ \underline{32} \\ 3\ 0 \\ \underline{2\ 8} \\ 20 \\ \underline{20} \\ 0 \end{array}$$

2. **B. 2 m**

3. $100\% - 80\% = 20\%$
$$\frac{20}{100} = \mathbf{0.2}$$

4.

	Ratio	A.C.
Win	5	w
Loss	3	l
Total	8	120

$$\frac{5}{8} = \frac{w}{120}$$
$$8w = 5 \cdot 120$$
$$w = \frac{5 \cdot \overset{15}{\cancel{120}}}{\underset{1}{\cancel{8}}}$$
$$w = \mathbf{75\ games}$$

5. **4,500,000**

6. (a) **−36**

(b) **+36**

(c) **−4**

(d) **+4**

7. (a) **−15**

(b) **−9**

(c) **−9**

(d) **+15**

8. **A variety of methods are possible. One method is to convert each fraction to a decimal number, order the decimal numbers, and then place the corresponding fractions in the same order.**

9. (a) $\frac{1}{50} \times \frac{2}{2} = \frac{2}{100} = \mathbf{0.02}$

(b) $\frac{2}{100} = \mathbf{2\%}$

10. (a) $1\frac{75}{100} = \mathbf{1\frac{3}{4}}$

(b) $\frac{175}{100} = \mathbf{175\%}$

11. (a) $\frac{25}{100} = \mathbf{\frac{1}{4}}$

(b) $\frac{25}{100} = \mathbf{0.25}$

12.
$$12\frac{1}{4}\text{ in.} \times \frac{2}{2} = \overset{11}{\cancel{12}}\frac{\overset{10}{\cancel{2}}}{8}\text{ in.}$$
$$-\ 3\frac{5}{8}\text{ in.} \times \frac{1}{1} = \ 3\frac{5}{8}\text{ in.}$$
$$\mathbf{8\frac{5}{8}\text{ in.}}$$

13. $\frac{\overset{5}{\cancel{10}}}{\underset{1}{\cancel{3}}}\text{ ft} \times \frac{\overset{3}{\cancel{9}}}{\underset{2}{\cancel{4}}}\text{ ft} = \frac{15}{2}\text{ ft}^2 = \mathbf{7\frac{1}{2}\text{ ft}^2}$

14. **27 cm³**

15. **0.3 m²**

16. $25 + 32 = \mathbf{57}$

17. $A_1 = \dfrac{3\text{ ft} \times \overset{2}{\cancel{4}}\text{ ft}}{\underset{1}{\cancel{2}}} = 6\text{ ft}^2$
$$A_2 = 7\text{ ft} \times 4\text{ ft} = 28\text{ ft}^2$$
$$A = A_1 + A_2 = 6\text{ ft}^2 + 28\text{ ft}^2 = \mathbf{34\text{ ft}^2}$$

18.
$$\begin{array}{r} \overset{1}{\cancel{2}}\text{ ft }\overset{15}{\cancel{3}}\text{ in.} \\ -\ 1\text{ ft }\ 9\text{ in.} \\ \hline \mathbf{6\text{ inches}} \end{array}$$

19. **Line g**

20. $5\text{ cm} \times 3\text{ cm} \times 2\text{ cm}$
30 cm³
30 cubes

21.
$$\begin{array}{r} \$80\text{ per day} \\ 3\overline{)\$240} \\ \underline{24} \\ 00 \\ \underline{00} \\ 0 \end{array}$$

$\$80.00 \times 10 = \mathbf{\$800.00}$

22.

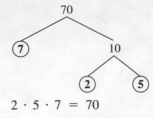

$2 \cdot 5 \cdot 7 = 70$

23. **900,000,000 miles**

24. (a) 8 in. $\times$ 4 = **32 in.**

(b) $A = bh$
$A = (8 \text{ in.})(7 \text{ in.})$
$A = \textbf{56 in.}^2$

(c) $180° - 61° = \textbf{119°}$

25.

	Ratio	A.C.
Quarters	5	120
Dimes	8	d
Total	13	t

$\dfrac{5}{8} = \dfrac{120}{d}$

$5d = 8 \cdot 120$

$d = \dfrac{8 \cdot \overset{24}{\cancel{120}}}{\underset{1}{\cancel{5}}}$

$d = \textbf{192 dimes}$

26.

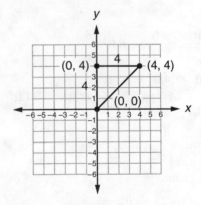

$A = \dfrac{4 \text{ units} \times \overset{2}{\cancel{4}} \text{ units}}{\underset{1}{\cancel{2}}}$

$A = \textbf{8 sq. units}$

27. 7, 8, 8, 8, 9, 9, 11, 12, 12, 16
9

28.

$$\begin{array}{r} 10 \\ 10\overline{)100} \\ \underline{10} \\ 00 \\ \underline{00} \\ 0 \end{array}$$

29. $C = \pi d$
$C \approx (3.14)(10 \text{ in.})$
$C \approx \textbf{31.4 in.}$

30. $A = (10 \text{ m})^2$
$A = \textbf{100 m}^2$

LESSON 114, WARM-UP

a. 28

b. 2880

c. $50

d. $5.64

e. 125

f. 36,000

g. 10

h. 1500

Problem Solving

There are 26 possible letters for the first position. Likewise, there are 26 possible letters for both the second and the third positions. Twenty-six times 26 times 26 equals 17,576 letter combinations. For the numerals, the range 000 to 999 represents 1000 combinations. So 17,576 $\times$ 1000, or **17,576,000 license plates,** can be formed.

LESSON 114, LESSON PRACTICE

a. $\dfrac{1 \text{ gal}}{4 \text{ qt}}, \dfrac{4 \text{ qt}}{1 \text{ gal}}$

b. $\dfrac{4 \text{ qt}}{1 \text{ gal}}$

c. $\dfrac{1 \text{ m}}{100 \text{ cm}}, \dfrac{100 \text{ cm}}{1 \text{ m}}$

d. $\dfrac{1 \text{ m}}{100 \text{ cm}}$

Saxon Math 7/6—Homeschool

e. $\dfrac{\overset{3}{\cancel{12} \, \cancel{qt}}}{1} \times \dfrac{1 \text{ gal}}{\underset{1}{\cancel{4} \, \cancel{qt}}} = \textbf{3 gal}$

f. $\dfrac{200 \, \cancel{m}}{1} \times \dfrac{100 \text{ cm}}{1 \, \cancel{m}} = \textbf{20,000 cm}$

g. $\dfrac{\overset{20}{\cancel{60} \, \cancel{ft}}}{1} \times \dfrac{1 \text{ yd}}{\underset{1}{\cancel{3} \, \cancel{ft}}} = \textbf{20 yd}$

LESSON 114, MIXED PRACTICE

1. $\begin{array}{r} 3 \text{ R } 2 \\ \$6\overline{)\$20} \\ \underline{18} \\ 2 \end{array}$

3 tickets

2. $\begin{array}{r} 2 \text{ min} \\ 3\overline{)6} \\ \underline{6} \\ 0 \end{array}$

$\begin{array}{r} 2 \text{ min} \\ \times \quad 4 \\ \hline \textbf{8 minutes} \end{array}$

3. $\dfrac{10}{25} = \dfrac{\mathbf{2}}{\mathbf{5}}$

4. $\dfrac{2}{\underset{1}{\cancel{5}}} \times \dfrac{\overset{6}{\cancel{30}}}{1} = 12 \text{ tomato plants}$

$30 - 12 = \textbf{18 plants}$

5. **6**

6. (a) $\dfrac{\textbf{1 gal}}{\textbf{4 qt}}, \dfrac{\textbf{4 qt}}{\textbf{1 gal}}$

(b) $\dfrac{\textbf{4 qt}}{\textbf{1 gal}}$

7. $\$36 + \$42 + \$27 = \textbf{\$105}$

8. $4 + 16 \div 2 - 1$
$4 + 8 - 1$
$\textbf{11}$

9. $\begin{aligned} 3\tfrac{1}{4} \times \tfrac{2}{2} &= 3\tfrac{2}{8} \\ 2\tfrac{1}{2} \times \tfrac{4}{4} &= 2\tfrac{4}{8} \\ + \; 4\tfrac{5}{8} \times \tfrac{1}{1} &= 4\tfrac{5}{8} \\ \hline 9\tfrac{11}{8} &= \mathbf{10\tfrac{3}{8} \text{ in.}} \end{aligned}$

10. (a) $\begin{array}{r} 0.125 \\ 8\overline{)1.000} \\ \underline{8} \\ 20 \\ \underline{16} \\ 40 \\ \underline{40} \\ 0 \end{array}$

(b) $0.125 \times 100\% = \textbf{12.5\%}$

11. (a) $\dfrac{\mathbf{9}}{\mathbf{10}}$

(b) $\dfrac{9}{10} \times \dfrac{10}{10} = \dfrac{90}{100} = \textbf{90\%}$

12. (a) $\dfrac{60}{100} = \dfrac{\mathbf{3}}{\mathbf{5}}$

(b) $\dfrac{60}{100} = \textbf{0.6}$

13. $3\tfrac{1}{4} \div \tfrac{2}{3}$

$\dfrac{13}{4} \div \dfrac{2}{3}$

$1 \div \dfrac{2}{3} = \dfrac{3}{2}$

$\dfrac{13}{4} \times \dfrac{3}{2} = \dfrac{39}{8} = \mathbf{4\tfrac{7}{8}}$

14. $3m - 10 = 80$
$3m = 90$
$m = \textbf{30}$

15. $\dfrac{3}{2} = \dfrac{18}{m}$

$3m = 2 \cdot 18$

$m = \dfrac{2 \cdot \overset{6}{\cancel{18}}}{\underset{1}{\cancel{3}}}$

$m = \textbf{12}$

SOLUTIONS

16. (a) **+100**

(b) **−100**

(c) **−4**

(d) **+4**

17. $\dfrac{6 \text{ hours}}{1} \times \dfrac{55 \text{ miles}}{1 \text{ hour}}$

330 miles

18.
$$9\,\text{m} + x = 15\,\text{m}$$
$$x = 6\,\text{m}$$
$$8\,\text{m} + y = 10\,\text{m}$$
$$y = 2\,\text{m}$$
$$6\,\text{m} \times 8\,\text{m} = 48\,\text{m}^2$$
$$9\,\text{m} \times 10\,\text{m} = 90\,\text{m}^2$$
$$48\,\text{m}^2 + 90\,\text{m}^2 = \mathbf{138\,m^2}$$

19. $15\,\text{m} + 10\,\text{m} + 9\,\text{m} + 2\,\text{m}$
$+ \; 6\,\text{m} + 8\,\text{m} = \mathbf{50\,m}$

20. (a) **−25**

(b) **−15**

(c) **0**

(d) **+25**

21. (a) **∠5**

(b) $180° − 45° = \mathbf{135°}$

22. $\dfrac{5}{100} = \dfrac{50}{x}$

$5x = 50 \cdot 100$

$x = \dfrac{\overset{10}{\cancel{50}} \cdot 100}{\underset{1}{\cancel{5}}}$

$x = \mathbf{1000 \text{ surveys}}$

23. **Selected pairs of prime numbers will differ. The LCM will be the product of the selected prime numbers.**

24. **1,500,000**

25. $V = lwh$
$V = (30\,\text{ft})(30\,\text{ft})(10\,\text{ft})$
$V = \mathbf{9000\,ft^3}$

26. $\dfrac{\overset{2}{\cancel{8}}\cancel{\text{ qt}}}{1} \cdot \dfrac{1 \text{ gal}}{\underset{1}{\cancel{4}}\cancel{\text{ qt}}} = \mathbf{2\,gal}$

27.

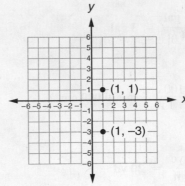

(a) **4 units**

(b) $A = \pi r^2$
$A \approx (3.14)(4\,\text{units})^2$
$A \approx \mathbf{50.24\,units^2}$

28.
$$\begin{array}{r} 95 \\ -\ 35 \\ \hline 60 \end{array}$$

29.
$$\begin{array}{r} \overset{3}{\cancel{4}}\,\text{ft}\ \overset{15}{\cancel{3}}\,\text{in.} \\ -\ 2\,\text{ft}\ \ 9\,\text{in.} \\ \hline \mathbf{1\,ft}\ \ \mathbf{6\,in.} \end{array}$$

30. **To find A, square s.**

LESSON 115, WARM-UP

a. 12

b. 4563

c. $0.25

d. $0.41

e. 0.005

f. 8

g. −1

h. 1700

Problem Solving

$7 \to 15$: $(7 \times 2) + 1 = 15$

$11 \to 23$: $(11 \times 2) + 1 = 23$

$3 \to 7$: $(3 \times 2) + 1 = 7$

Tom doubled each number and then added 1.

$5 \to 11$: $(5 \times 2) + 1 = \mathbf{11}$

LESSON 115, LESSON PRACTICE

a. $\dfrac{66\frac{2}{3}}{100}$

$\dfrac{\overset{2}{\cancel{200}}}{3} \times \dfrac{1}{\underset{1}{\cancel{100}}} = \dfrac{\mathbf{2}}{\mathbf{3}}$

b. $\dfrac{6\frac{2}{3}}{100}$

$\dfrac{\overset{1}{\cancel{20}}}{3} \times \dfrac{1}{\underset{5}{\cancel{100}}} = \dfrac{\mathbf{1}}{\mathbf{15}}$

c. $\dfrac{12\frac{1}{2}}{100}$

$\dfrac{\overset{1}{\cancel{25}}}{2} \times \dfrac{1}{\underset{4}{\cancel{100}}} = \dfrac{\mathbf{1}}{\mathbf{8}}$

d. $\dfrac{14\frac{2}{7}}{100}$

$\dfrac{\overset{1}{\cancel{100}}}{7} \times \dfrac{1}{\underset{1}{\cancel{100}}} = \dfrac{\mathbf{1}}{\mathbf{7}}$

e. $\dfrac{83\frac{1}{3}}{100}$

$\dfrac{\overset{5}{\cancel{250}}}{3} \times \dfrac{1}{\underset{2}{\cancel{100}}} = \dfrac{\mathbf{5}}{\mathbf{6}}$

LESSON 115, MIXED PRACTICE

1.
$$\begin{array}{r} \$12.60 \\ \times \quad 0.07 \\ \hline \$0.8820 \end{array} \qquad \begin{array}{r} \overset{1}{\$12.60} \\ + \ \$ 0.88 \\ \hline \mathbf{\$13.48} \end{array}$$

2. B. 6.4 lb

3.

	%	A.C.
Correct	90	c
Not Correct	10	3
Total	100	t

$\dfrac{10}{100} = \dfrac{3}{t}$

$10t = 3 \cdot 100$

$t = \dfrac{3 \cdot \overset{10}{\cancel{100}}}{\underset{1}{\cancel{10}}}$

$t = \mathbf{30 \ questions}$

4. $\dfrac{331 \text{ m}}{1 \cancel{s}} \cdot \dfrac{60 \cancel{s}}{1} = \mathbf{19{,}860 \text{ m}}$

5. 50,600

6.
$$\begin{array}{r} 0.75 \text{ meter} \\ \times \quad 2 \\ \hline 1.50 \end{array}$$
1.5 meters

7. $\dfrac{11}{3} \cdot \dfrac{8}{3} = \dfrac{88}{9} = 9\dfrac{7}{9}$

10

8. $\dfrac{3}{6} \cdot \dfrac{2}{5} = \dfrac{6}{30} = \dfrac{\mathbf{1}}{\mathbf{5}}$

9. (a) $2\dfrac{2}{5} \times \dfrac{20}{20} = 2\dfrac{40}{100} = \mathbf{2.4}$

(b) $\dfrac{240}{100} = \mathbf{240\%}$

10. (a) $\dfrac{85}{100} = \dfrac{\mathbf{17}}{\mathbf{20}}$

(b) $\dfrac{85}{100} = \mathbf{85\%}$

11. $7x - 3 = 39$

$ 7x = 42$

$ x = \mathbf{6}$

12. $\dfrac{x}{7} = \dfrac{35}{5}$

$5x = 35 \cdot 7$

$x = \dfrac{\overset{7}{\cancel{35}} \cdot 7}{\underset{1}{\cancel{5}}}$

$x = \mathbf{49}$

13. (a) **+45**

(b) **−5**

(c) **+5**

(d) **−45**

14. $-6 - 7 + 5 + 8$

0

15. $0.12 \div 30 = $ **0.004**

16.
```
      3.142
   7)22.000
     21
      1 0
        7
       30
       28
        20
        14
         6
```
3.14

17. **100**

18.
$$6 \text{ cm} + x = 10 \text{ cm}$$
$$x = 4 \text{ cm}$$
$$5 \text{ cm} + y = 8 \text{ cm}$$
$$y = 3 \text{ cm}$$
$$6 \text{ cm} \times 8 \text{ cm} = 48 \text{ cm}^2$$
$$4 \text{ cm} \times 5 \text{ cm} = 20 \text{ cm}^2$$
$$48 \text{ cm}^2 + 20 \text{ cm}^2 = \textbf{68 cm}^2$$

19. $5 \text{ cm} + 10 \text{ cm} + 8 \text{ cm} + 6 \text{ cm} + 3 \text{ cm} + 4 \text{ cm} = $ **36 cm**

20. $3 \text{ cm} \times 3 \text{ cm} \times 3 \text{ cm} = $ **27 cm³**

21. **31 days**

22.
```
   12 ounces per container
  7)84 ounces
    7
    14
    14
     0
```
12 ounces $\times$ 10 = **120 ounces**

23. **4,500,000**

24. **59,000,000**

25. *C*

26.
```
    8        23
    9̸ pounds 7̸ ounces
  − 7 pounds  9 ounces
    1 pound  14 ounces
```

27.

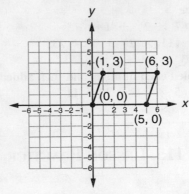

$A = bh$

$A = 5 \text{ units} (3 \text{ units})$

$A = $ **15 sq. units**

28. $\dfrac{16\frac{2}{3}}{100}$

$$\dfrac{\overset{1}{\cancel{50}}}{3} \times \dfrac{1}{\underset{2}{\cancel{100}}} = \dfrac{1}{6}$$

29.
```
    2 gal 2 qt 1 pt
  + 2 gal 2 qt 1 pt
    4 gal 4 qt 2 pt
```
5 gal 1 qt

30. **Gilbert can divide 323.4 miles by 14.2 gallons to calculate the miles per gallon.**

LESSON 116, WARM-UP

a. **6**

b. **2240**

c. **$1.25**

d. **$9.47**

e. **37.5**

f. **20,000**

g. **7**

h. **1620**

Problem Solving

Find all two-digit numbers that, when squared, equal a number in the three hundreds. Only 19^2 ends with 1 in the ones place.

$$
\begin{array}{r}
19 \\
\times\ 19 \\
\hline
171 \\
19 \\
\hline
361
\end{array}
$$

LESSON 116, LESSON PRACTICE

a. 1. $5187.48
2. $13,455.00

b. 1. $2590.06
2. $4734.73
3. $11,208.82

c. 1. $71.94
2. $452.75
3. $2246.18

Extension

See student work.

LESSON 116, MIXED PRACTICE

1.
$$
\begin{array}{r}
30\ \text{R}\ 6 \\
10\overline{)306} \\
\underline{30} \\
06 \\
\underline{0} \\
6
\end{array}
$$

30, 30, 30, 30, 31, 31, 31, 31, 31, 31
4 groups

2.
$$
\begin{array}{r}
\$1.20 \\
\times\ \ \ \ \ 2 \\
\hline
\$2.40
\end{array}
$$

3. (a) $\dfrac{6}{15} = \dfrac{2}{5}$

(b) $\dfrac{\overset{2}{\cancel{6}}}{\underset{\underset{1}{\cancel{3}}}{\cancel{15}}} \times \dfrac{\overset{20}{\cancel{100\%}}}{1} = \mathbf{40\%}$

4. $\dfrac{1}{4} \times \dfrac{25}{25} = \dfrac{25}{100} = \mathbf{25\%}$

5. $\dfrac{2}{\underset{1}{\cancel{8}}} \times \dfrac{\overset{6}{\cancel{30}}}{1} = 12$ stallions (male)

$30 - 12 = 18$ mares (female)

$\dfrac{12}{18} = \dfrac{2}{3}$

6. **1,200,000,000**

7. $c = \dfrac{\$0.65}{1\ \cancel{\text{pound}}} \times \dfrac{5\ \cancel{\text{pounds}}}{1}$

$c = \mathbf{\$3.25}$

8. **9.09, 9.9, 9.925, 9.95**

9. (a)
$$
\begin{array}{r}
0.375 \\
8\overline{)3.000} \quad 3d = \mathbf{3.375} \\
\underline{2\ 4} \\
60 \\
\underline{56} \\
40 \\
\underline{40} \\
0
\end{array}
$$

(b) $3.375 \times 100\% = \mathbf{337.5\%}$

10. (a) $\dfrac{15}{100} = \dfrac{3}{20}$

(b) $\dfrac{15}{100} = \mathbf{0.15}$

11. $9x + 17 = 80$
$9x = 63$
$x = \mathbf{7}$

12. $\dfrac{x}{3} = \dfrac{16}{12}$

$12x = 16 \cdot 3$

$x = \dfrac{\overset{4}{\cancel{16}} \cdot \overset{1}{\cancel{3}}}{\underset{\underset{1}{\cancel{4}}}{\cancel{12}}}$

$x = \mathbf{4}$

13. $-6 - 4 - 3 + 8$
$\mathbf{-5}$

14.
$$
\begin{array}{r}
6 \\
3.75 \\
+\ 4.6 \\
\hline
14.35
\end{array}
$$

15.
$$
\begin{array}{r}
210 \\
\times\ \ \ 3\ \text{feet} \\
\hline
\textbf{630 feet}
\end{array}
$$

16.

$$3\overline{)3} \quad {}^1$$
$$3\overline{)9}$$
$$3\overline{)27}$$
$$3\overline{)81}$$
$$2\overline{)162}$$
$$2\overline{)324}$$
$$2\overline{)648}$$
$$\mathbf{2^3 \cdot 3^4}$$

17.

$$\begin{array}{r} \mathbf{\$0.12} \textbf{ per ounce} \\ 32\overline{)\$3.84} \\ \underline{3\ 2} \\ 64 \\ \underline{64} \\ 0 \end{array}$$

18. $A_1 = 6\,\text{m} \times 4\,\text{m} = 24\,\text{m}^2$

$$A_2 = \frac{\overset{1}{\cancel{2}}\,\text{m} \times 4\,\text{m}}{\underset{1}{\cancel{2}}} = 4\,\text{m}^2$$

$A = A_1 + A_2 = 24\,\text{m}^2 + 4\,\text{m}^2 = \mathbf{28\,m^2}$

19. (a) $C = \pi d$

$\approx (3.14)(20\,\text{cm})$

$\approx \mathbf{62.8\ cm}$

(b) $A = \pi r^2$

$\approx (3.14)(10\,\text{cm})^2$

$\approx \mathbf{314\ cm^2}$

20. $V = 3\,\text{cm} \times 3\,\text{cm} \times 3\,\text{cm} = 27\,\text{cm}^3$

$$\frac{1}{\underset{1}{\cancel{3}}} \cdot \frac{\overset{9}{\cancel{27}}}{1} = \mathbf{9\ cm^3}$$

21.

$$\begin{array}{r} 90 \\ 6\overline{)540} \\ \underline{54} \\ 00 \\ \underline{00} \\ 0 \end{array}$$

22. (a) $\mathbf{+16}$

(b) $\mathbf{-16}$

(c) $\mathbf{-4}$

(d) $\mathbf{+4}$

23. $A = \dfrac{\overset{3}{\cancel{6}}\,\text{in.} \times 8\,\text{in.}}{\underset{1}{\cancel{2}}}$

$A = \mathbf{24\ in.^2}$

24. (a) $180° - 40° - 110° = \mathbf{30°}$

(b)

25. $\dfrac{1}{\underset{1}{\cancel{6}}} \cdot \dfrac{\overset{10}{\cancel{60}}}{1} = 10$ times

D. 10 times

26. $\sqrt{100\,\text{mm}^2} = 10\,\text{mm}$

$10\,\text{mm} \times 3 = \mathbf{30\ mm}$

27. $\dfrac{\overset{1}{\cancel{100}}}{9} \cdot \dfrac{1}{\underset{1}{\cancel{100}}} = \mathbf{\dfrac{1}{9}}$

28. Mean:

$181\,\text{cm} + 177\,\text{cm} + 189\,\text{cm} + 158\,\text{cm}$
$\quad + 195\,\text{cm} = 900\,\text{cm}$

$900\,\text{cm} \div 5 = 180\,\text{cm}$

Mean is 180 cm.

Median:

$158\,\text{cm},\ 177\,\text{cm},\ 181\,\text{cm},\ 189\,\text{cm},\ 195\,\text{cm}$

Median is 181 cm.

Range:

$195\,\text{cm} - 158\,\text{cm} = 37\,\text{cm}$

Range is 37 cm.

29.

$$\begin{array}{r} 2\ \text{lb}\ \ 12\ \text{oz} \\ + \ 3\ \text{lb}\ \ \ 8\ \text{oz} \\ \hline 5\ \text{lb}\ \ 20\ \text{oz} \\ \mathbf{6\ lb\ \ \ 4\ oz} \end{array}$$

30. C. Scalene

LESSON 117, WARM-UP

a. **14**

b. **2280**

c. **\$20**

d. **25%**

e. **24**

f. **42,000**

g. 9

h. 1900

Problem Solving

40N = $2.00 per roll
50D = $5.00 per roll
40Q = $10.00 per roll
4($10.00) + 2($5.00) + 3($2.00)
= **$56.00**

LESSON 117, LESSON PRACTICE

a.

40
8
8
8
8
8

$\frac{1}{5}$ {

b.

20
4
4
4
4
4

$\frac{2}{5}$ {

c.

12
3
3
3
3

$\frac{3}{4}$ {

d.

160
20
20
20
20
20
20
20
20

$\frac{3}{8}$ {

e.

30 bunches
6 bunches
6 bunches
6 bunches
6 bunches
6 bunches

$\frac{3}{5}$ were red grapes. {

30 bunches

LESSON 117, MIXED PRACTICE

1.

200 people
40
40
40
40
40

$\frac{3}{5}$ voted. {

2.

$$
\begin{array}{r}
32 \text{ R } 2 \\
4\overline{)130} \\
\underline{12} \\
10 \\
\underline{8} \\
2
\end{array}
$$

32, 32, 33, 33

3.

$$
\begin{array}{r}
\overset{14}{\cancel{2}}:45 \text{ p.m.} \\
-\ 11:15 \text{ a.m.} \\
\hline
3:30
\end{array}
$$

$$
\begin{array}{r}
\$1.25 \\
\times 7 \\
\hline
\$8.75
\end{array}
$$

4. $\sqrt{400 \text{ m}^2} = 20 \text{ m}$
$A = \pi r^2$
$A \approx (3.14)(20 \text{ m})^2$
$A \approx$ **1256 m²**

5. $\frac{46}{50} \times \frac{2}{2} = \frac{92}{100} =$ **92%**

6.

$A = \dfrac{5 \text{ units} \times \overset{3}{\cancel{6}} \text{ units}}{\underset{1}{\cancel{2}}}$

$A =$ **15 sq. units**

7. **0.105**

8.

$$
\begin{array}{r}
\$0.777 \longrightarrow \$0.78 \\
9\overline{)\$7.000} \\
\underline{6\ 3} \\
70 \\
\underline{63} \\
70 \\
\underline{63} \\
7
\end{array}
$$

SOLUTIONS

9. $81\% = \dfrac{81}{100} = 0.81 = 0.810$

$\dfrac{4}{5} \cdot \dfrac{20}{20} = \dfrac{80}{100} = 0.80 = 0.800$

0.800, 0.810, 0.815

$\dfrac{4}{5}$, **81%, 0.815**

10. $6x - 12 = 60$ check: $6(12) - 12 = 60$
        $6x = 72$ $72 - 12 = 60$
         $x = \mathbf{12}$ $60 = 60$

11. $\dfrac{9}{15} = \dfrac{m}{25}$ check: $\dfrac{9}{15} = \dfrac{15}{25}$

$15m = 25 \cdot 9$ $15 \cdot 15 = 25 \cdot 9$

                       $225 = 225$

$m = \dfrac{\overset{5}{\cancel{25}} \cdot \overset{3}{\cancel{9}}}{\underset{\underset{1}{\cancel{3}}}{\cancel{15}}}$

$m = \mathbf{15}$

12.

15
3
3
3
3
3

$\dfrac{2}{5}\left\{\rule{0pt}{20pt}\right.$ (bracketing bottom two rows)

13. $4\dfrac{3}{3} - 1\dfrac{2}{3} = 3\dfrac{1}{3}$

$3\dfrac{1}{3} \times \dfrac{2}{2} = \overset{2}{\cancel{3}}\dfrac{\overset{8}{\cancel{2}}}{6}$

$-\, 1\dfrac{1}{2} \times \dfrac{3}{3} = 1\dfrac{3}{6}$

                     $\mathbf{1\dfrac{5}{6}}$

14. $\dfrac{12}{5} \div \dfrac{3}{2}$

$1 \div \dfrac{3}{2} = \dfrac{2}{3}$

$\dfrac{\overset{4}{\cancel{12}}}{5} \times \dfrac{2}{\underset{1}{\cancel{3}}} = \dfrac{8}{5} = \mathbf{1\dfrac{3}{5}}$

15.
```
    0.625
  ×   2.4
    2500
  12500
  1.5000
```
1.5

16. **−15**

17.

$\mathbf{2^2 \cdot 3^2}$

18.
```
  $12.50        $12.50
×   0.06      + $ 0.75
  $0.7500       $13.25
```

19. $A_1 = 8 \text{ in.} \times 9 \text{ in.} = 72 \text{ in.}^2$

$A_2 = \dfrac{4 \text{ in.} \cdot 4 \text{ in.}}{2} = 8 \text{ in.}^2$

$A = A_1 + A_2 = 72 \text{ in.}^2 + 8 \text{ in.}^2 = \mathbf{80 \text{ in.}^2}$

20. **600,000**

21.   (a) **+5**

      (b) **−6**

      (c) **−24**

      (d) **+36**

22. $5 \text{ in.} \times 3 \text{ in.} \times 4 \text{ in.} = \mathbf{60 \text{ in.}^3}$

23. **One possibility:**

Acute triangle

24. **87**

25.   (a) **1.065**

      (b) **$1207.95**

26. $\dfrac{3}{4} \cdot \dfrac{3}{4} = \dfrac{\mathbf{9}}{\mathbf{16}}$

27. **B. Parallelogram**

28.

	Ratio	A.C.
Cattle	15	c
Horses	2	h
Total	17	1020

$\dfrac{2}{17} = \dfrac{h}{1020}$

$17h = 2 \cdot 1020$

$h = \dfrac{2 \cdot \overset{60}{\cancel{1020}}}{\underset{1}{\cancel{17}}}$

$h = \mathbf{120 \text{ horses}}$

29. $10 + 9 \times 5 - 9 \div 3$
$10 + 45 - 3$
52

30. C. ◯

LESSON 118, WARM-UP

a. 18

b. 10

c. $8

d. 25%

e. 6

f. 90,000

g. 8

h. 1969

Problem Solving

$9 \rightarrow 80$: $(9 \times 9) - 1 = 80$
$6 \rightarrow 35$: $(6 \times 6) - 1 = 35$
$7 \rightarrow 48$: $(7 \times 7) - 1 = 48$
$n^2 - 1$ (Square the number; then subtract 1.)
$10 \rightarrow 99$: $(10 \times 10) - 1 = $ **99**

LESSON 118, LESSON PRACTICE

a. 14 (or 15) square units

LESSON 118, MIXED PRACTICE

1. $\begin{array}{r} 7\,\text{R}\,3 \\ 7\overline{)52} \\ \underline{49} \\ 3 \end{array}$

3 people

2. B. 18 cm

3.
21 million
7
7
7

$\frac{1}{3}$ {

About 21 million people

4. B. $\dfrac{3}{4} \neq \dfrac{9}{16}$

5. $\begin{array}{r} \$14.49 \\ \times\ \ 0.07 \\ \hline \$1.0143 \end{array}$ $\begin{array}{r} \$14.49 \\ +\ \ \$1.01 \\ \hline \mathbf{\$15.50} \end{array}$

6. $\dfrac{48}{84} = \dfrac{4}{7}$

7. $17 + 24 + 27 + 28 = 96$
$96 \div 4 = \mathbf{24}$

8.
$6.1 = 6.10$
$\sqrt{36} = 6 = 6.00$
$6\dfrac{1}{4} = 6.25$
$6.00, 6.10, 6.25$
$\sqrt{36},\ 6.1,\ 6\dfrac{1}{4}$

9.
30 cookies
3
3
3
3
3
3
3
3
3
3

$\frac{3}{10}$ {

10. Buz can divide the circumference by π (by 3.14) to calculate the diameter.

11.
16
4
4
4
4

$\frac{3}{4}$ {

12.
$5\dfrac{1}{3} \times \dfrac{2}{2} = 5\dfrac{2}{6}$
$-\ 2\dfrac{1}{2} \times \dfrac{3}{3} = 2\dfrac{3}{6}$
$\rule{3cm}{0.4pt}$
$2\dfrac{5}{6}$

$2\dfrac{2}{3} \times \dfrac{2}{2} = 2\dfrac{4}{6}$
$+\ 2\dfrac{5}{6} \times \dfrac{1}{1} = 2\dfrac{5}{6}$
$\rule{3cm}{0.4pt}$
$4\dfrac{9}{6} = 5\dfrac{3}{6} = \mathbf{5\dfrac{1}{2}}$

13. $\dfrac{20}{3} \div \dfrac{25}{6}$

$1 \div \dfrac{25}{6} = \dfrac{6}{25}$

$\dfrac{\overset{4}{\cancel{20}}}{\underset{1}{\cancel{3}}} \times \dfrac{\overset{2}{\cancel{6}}}{\underset{5}{\cancel{25}}} = \dfrac{8}{5} = \mathbf{1\dfrac{3}{5}}$

14. $\begin{array}{r} 4.25 \\ +\ 3.2 \\ \hline \mathbf{7.45} \end{array}$

15. $1 - 0.01 = \mathbf{0.99}$

16. **21**

17. $\dfrac{16\ \text{oz}}{1\ \cancel{\text{lb}}} \times \dfrac{2.5\ \cancel{\text{lb}}}{1} = \mathbf{40\ ounces}$

18. (a) $A = bh$

$A = 6\ \text{cm} \times 4\ \text{cm}$

$A = \mathbf{24\ cm^2}$

(b) $180° - 127° = \mathbf{53°}$

19. $1.8\ \text{cm} + x = 4\ \text{cm}$

$\qquad\qquad x = 2.2\ \text{cm}$

$1.4\ \text{cm} + y = 3\ \text{cm}$

$\qquad\qquad y = 1.6\ \text{cm}$

$1.8\ \text{cm} + 1.6\ \text{cm} + 2.2\ \text{cm} + 1.4\ \text{cm}$

$\quad + 4\ \text{cm} + 3\ \text{cm} = \mathbf{14\ cm}$

20. **4 lines of symmetry**

21. $V = 4\ \text{ft} \times 4\ \text{ft} \times 4\ \text{ft}$

$V = \mathbf{64\ ft^3}$

22. $\dfrac{f}{12} = \dfrac{12}{16}$

$16f = 12 \cdot 12$

$f = \dfrac{\overset{3}{\cancel{12}} \cdot \overset{3}{\cancel{12}}}{\underset{1}{\cancel{16}}}$

$f = \mathbf{9}$

23. $y = 5x$

$40 = 5x$

$\ \ 8 = x$

24. **12,500**

25. $-5 + 2 - 3 + 4 - 1$

$\qquad\quad -9 + 6$

$\qquad\qquad\ \ \mathbf{-3}$

26. $\$4000 \times 1.075 = \$4300$

$\$4300 \times 1.075 = \mathbf{\$4622.50}$

27. $\dfrac{\dfrac{15}{2}}{100}$

$\dfrac{\overset{3}{\cancel{15}}}{2} \cdot \dfrac{1}{\underset{20}{\cancel{100}}} = \mathbf{\dfrac{3}{40}}$

28. $-3°F - 5°F = \mathbf{-8°F}$

29. $\dfrac{2}{8} = \mathbf{\dfrac{1}{4}}$

30. **8 units²**

LESSON 119, WARM-UP

a. **9**

b. **42**

c. **$35**

d. **50%**

e. **56**

f. **200**

g. **10**

h. **1776**

Problem Solving

$360° \div 12 = \mathbf{30°}$

LESSON 119, LESSON PRACTICE

a. $0.2n = 120$

$\begin{array}{r} 600 \\ 2\overline{)1200} \end{array}$

$n = \mathbf{600}$

b. $\dfrac{1}{2}n = 30$

$30 \div \dfrac{1}{2} = \dfrac{30}{1} \cdot \dfrac{2}{1} = 60$

$n = \mathbf{60}$

c. $\frac{1}{4}n = 12$

$12 \div \frac{1}{4} = \frac{12}{1} \times \frac{4}{1} = 48$

$n = \mathbf{48}$

d. $20 = 0.10n$

$$10\overline{)2000} \quad \begin{array}{c} 200 \end{array}$$

$n = \mathbf{200}$

e. $12 = 1n$

$$1\overline{)12} \quad \begin{array}{c} 12 \end{array}$$

$n = \mathbf{12}$

f. $15 = 0.15n$

$$15\overline{)1500} \quad \begin{array}{c} 100 \end{array}$$

$n = \mathbf{100}$

LESSON 119, MIXED PRACTICE

1. (a) $12\overline{)555} \quad \begin{array}{c} \mathbf{46\ R\ 3} \end{array}$

$$\begin{array}{r} \underline{48} \\ 75 \\ \underline{72} \\ 3 \end{array}$$

(b) $46\frac{3}{12} = \mathbf{46\frac{1}{4}}$

2. $9.75 + 9.8 + 9.9 + 9.9 + 9.95 = \mathbf{49.3}$

3.

	%	A.C.
Trumpet	10	6
Not Trumpet	90	n
Total	100	t

$\frac{6}{t} = \frac{10}{100}$

$10t = 6 \cdot 100$

$t = \frac{6 \cdot \overset{10}{\cancel{100}}}{\underset{1}{\cancel{10}}}$

$t = \mathbf{60\ members}$

4. $8 = \frac{2}{3}n$

$8 \div \frac{2}{3} = \frac{\overset{4}{\cancel{8}}}{1} \times \frac{3}{\underset{1}{\cancel{2}}} = \mathbf{12}$

$n = \mathbf{12}$

5. **186,000**

6. $\frac{1\ \text{inch}}{8\ \text{feet}} = \frac{2\frac{1}{2}\ \text{inches}}{x}$

$x = \mathbf{20\ feet}$

7. (a) $180° - 45° - 45° = \mathbf{90°}$

(b)

8. $\frac{\frac{25}{3}}{100}$

$\frac{\overset{1}{\cancel{25}}}{3} \times \frac{1}{\underset{4}{\cancel{100}}} = \mathbf{\frac{1}{12}}$

9. $\frac{9}{12} = \frac{3}{4}$

$\frac{3}{4} \times \frac{25}{25} = \frac{75}{100} = \mathbf{75\%}$

10. $0.20x = 12$

$$20\overline{)1200} \quad \begin{array}{c} 60 \end{array}$$

$x = \mathbf{60}$

11. $\frac{3}{10}x = 9$

$\frac{9}{1} \div \frac{3}{10} = \frac{\overset{3}{\cancel{9}}}{1} \times \frac{10}{\underset{1}{\cancel{3}}} = 30$

$x = \mathbf{30}$

12. $-5 - 6 - 7$

$\mathbf{-18}$

13. $\mathbf{+90}$

14. $\frac{60}{84} = \mathbf{\frac{5}{7}}$

15. $2\frac{1}{2} \times \frac{3}{3} = 2\frac{\overset{9}{\cancel{3}}}{6}$

$\underline{- 1\frac{2}{3} \times \frac{2}{2} = 1\frac{4}{6}}$

$\mathbf{\frac{5}{6}}$

16.

	Ratio	A.C.
Biking	5	b
Running	2	10
Total	7	t

$$\frac{2}{7} = \frac{10}{t}$$
$$2t = 10 \cdot 7$$
$$t = \frac{\overset{5}{\cancel{10}} \cdot 7}{\underset{1}{\cancel{2}}}$$
$$t = \textbf{35 kilometers}$$

17. $2.8 \text{ cm}^2 \times 2 = \textbf{5.6 cm}^2$

18. $V = lwh$
$V = 10 \text{ in.} \times 3 \text{ in.} \times 4 \text{ in.}$
$V = \textbf{120 in.}^3$

19. **2 lines of symmetry**

20. **Pyramid**

21. $3m - 5 = 25$
$ 3m = 30$
$ m = \textbf{10}$

22. $y = 4x$
$32 = 4x$
$\textbf{8} = x$

23. $1 \text{ ton} = 2000 \text{ pounds}$
$10 \text{ tons} = \textbf{20,000 pounds}$

24. **B. pentagon**

25. Area of the square $\gtrless$ area of the circle

26.

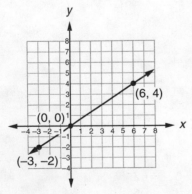

$x = \textbf{6}$

27. $\dfrac{1}{2}$

28. $\dfrac{1}{2} \times \dfrac{1}{2} \times \dfrac{1}{2} = \dfrac{1}{8}$

29. (a) $5 \text{ cm} \times 5 \text{ cm} = \textbf{25 cm}^2$

(b) $A = \pi r^2$
$A \approx (3.14)(5 \text{ cm})^2$
$A \approx \textbf{78.5 cm}^2$

30. Mode is 7; range is $10 - 2 = 8$.

LESSON 120, WARM-UP

a. 18

b. $48

c. 12

d. 100%

e. 320

f. 6000

g. 1

h. 1492

Problem Solving
Eliminate even numbers (352, 532).
Eliminate numbers divisible by 5 (235, 325).
Eliminate numbers divisible by 11 (253).
523

LESSON 120, LESSON PRACTICE

$A = \pi r^2$
$A \approx (3.14)(4 \text{ cm})^2$
$A \approx 50.24 \text{ cm}^2$
$50.24 \text{ cm}^2 \times 12 \text{ cm} = 602.88 \text{ cm}^3$
600 cm³

LESSON 120, MIXED PRACTICE

1.

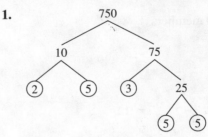

$2 \cdot 3 \cdot 5^3$

2. C. **50 mm**

3. $\dfrac{3}{24} = \dfrac{8}{x}$

$3x = 24 \cdot 8$

$x = \dfrac{\cancelto{8}{24} \cdot 8}{\cancelto{1}{3}}$

$x = \textbf{64 grams}$

4. $\dfrac{3}{24} = \dfrac{8}{w}$

$3w = 24 \cdot 8$

$w = \dfrac{\cancelto{8}{24} \cdot 8}{\cancelto{1}{3}}$

$w = \textbf{64}$

5. **7004**

6. **205,056,000**

7. (a) $17 + 23 + 25 + x = 100$

$\qquad\qquad\quad 65 + x = 100$

$\qquad\qquad\qquad\quad x = \textbf{35}$

(b) $35 - 17 = 18$

The range is 18.

8. (a) **−2**

(b) **−25**

(c) **+100**

9. $\dfrac{\frac{50}{3}}{100}$

$\dfrac{\cancelto{1}{50}}{3} \times \dfrac{1}{\cancelto{2}{100}} = \dfrac{1}{6}$

10.

30 guests

6
6
6
6
6

$\dfrac{4}{5}$ $\Big\{$

11.

$1\dfrac{1}{3} \times \dfrac{4}{4} = 1\dfrac{4}{12}$

$3\dfrac{3}{4} \times \dfrac{3}{3} = 3\dfrac{9}{12}$

$+\ 1\dfrac{1}{6} \times \dfrac{2}{2} = 1\dfrac{2}{12}$

$\rule{3cm}{0.4pt}$

$5\dfrac{15}{12} = 6\dfrac{3}{12} = \mathbf{6\dfrac{1}{4}}$

12. $\dfrac{5}{\cancelto{3}{6}} \times \dfrac{\cancelto{1}{3}}{1} \times \dfrac{\cancelto{4}{8}}{\cancelto{1}{3}} = \dfrac{20}{3} = \mathbf{6\dfrac{2}{3}}$

13.
$\overset{1}{5}.62$
0.8
$+\ 4$
$\rule{1.5cm}{0.4pt}$
10.42

14. $0.08 \div 2.5$

$\quad\ \ \textbf{0.032}$
$25\overline{)0.800}$
$\quad\ \ \underline{75}$
$\quad\ \ \ 50$
$\quad\ \ \ \underline{50}$
$\quad\ \ \ \ \ 0$

15. **−6**

16. $50 + 5$
55

17. $\quad\ \ \textbf{\$0.07 per ounce}$
$16\overline{)\$1.12}$
$\quad\ \ \underline{1\ 12}$
$\quad\quad\ \ 0$

18. $C = \pi d$

$C \approx (3.14)(10\text{ m})$

$C \approx \textbf{31.4 m}$

19. $\sqrt{36\text{ cm}^2} = 6\text{ cm}$

$6\text{ cm} \times 4 = \textbf{24 cm}$

20. $V = 4\text{ cm} \times 3\text{ cm} \times 2\text{ cm}$

$V = \textbf{24 cm}^3$

21. $.60x = 18$

$\quad\quad\ \ 30$
$60\overline{)1800}$

$x = \textbf{30 votes}$

22. $\dfrac{1}{4} \times \dfrac{1}{4} \times \dfrac{1}{4} = \dfrac{1}{64}$

23. $\dfrac{1}{\cancelto{1}{4}} \times \dfrac{\cancelto{5}{20}}{1} = \textbf{5 questions}$

24. (a) **−15**

(b) **−1**

25. $+3 - 5 + 7 - 9 + 11 - 7$

$21 - 21$

0

26.

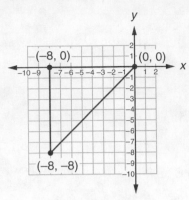

$A = \dfrac{\overset{4}{\cancel{8}} \text{ units } \times 8 \text{ units}}{\underset{1}{\cancel{2}}}$

$A = \textbf{32 sq. units}$

27. $\dfrac{1}{2} \times \dfrac{1}{2} = \dfrac{1}{4}$

28. $A = \pi r^2$

$A \approx (3.14)(4 \text{ cm})^2$

$A \approx 50.24 \text{ cm}^2$

$50.24 \text{ cm}^2 \times 7 \text{ cm} = \textbf{351.68 cm}^3$

29. **350 milliliters**

30.

	%	A.C.
Correct	90	c
Incorrect	10	4
Total	100	t

$\dfrac{90}{10} = \dfrac{c}{4}$

$10c = 90 \cdot 4$

$c = \dfrac{\overset{9}{\cancel{90}} \cdot 4}{\underset{1}{\cancel{10}}}$

$c = \textbf{36 questions}$

Extensions

a.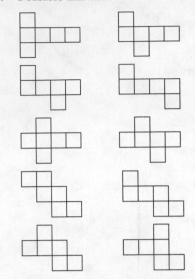

b. **Possible answers:**

c. **An icosahedron has 12 vertices and 20 faces. Each face is an equilateral triangle.**

INVESTIGATION 12

Platonic Solid	Each face is what polygon?	How many faces?	How many vertices?	How many edges?
tetrahedron	equilateral triangle	4	4	6
cube	square	6	8	12
octahedron	**triangle**	**8**	**6**	**12**
dodecahedron	**pentagon**	**12**	**20**	**30**

Solutions for

Appendix Topic

Topic A

a. 39

b. 64

c. 1919

d. 2002

Solutions for

Supplemental Practice

SUPPLEMENTAL PRACTICE, LESSON 2
SET A

1. 4608

2. $21.56

3. 4704

4. 38,754

5. $12.60

6. 2240

7. 4680

8. $44.80

9. 6300

10. 18,000

11. 20,000

12. 40,000

13. 1591

14. $45.88

15. 2322

16. 5922

17. $97.92

18. 18,759

19. 25,358

20. $257.04

21. 18,720

22. 39,480

23. $679.40

24. 16,281

SUPPLEMENTAL PRACTICE, LESSON 2
SET B

1. 71 R 2

2. $0.61

3. 127 R 3

4. 35 R 3

5. $1.03

6. 70 R 3

7. $1.04

8. 90 R 2

9. 201

10. $2.06

11. 100 R 6

12. 498 R 6

13. $0.12

14. 47 R 2

15. 59 R 8

16. 30 R 12

17. 21 R 12

18. 19 R 36

19. 57 R 2

20. 201

21. 128 R 30

22. 40 R 8

23. 20 R 17

24. 13 R 20

SUPPLEMENTAL PRACTICE, LESSON 3

1. 15
2. 25
3. 25
4. 29
5. 29
6. 46
7. 28
8. 26
9. 13
10. 60
11. 23
12. 49
13. 22
14. 62
15. 15
16. 63
17. 36
18. 69
19. 22
20. 27
21. 26
22. 52
23. 45
24. 28

SUPPLEMENTAL PRACTICE, LESSON 4

1. 6
2. 9
3. 15
4. 20
5. 25
6. 14
7. 26
8. 22
9. 12
10. 56
11. 15
12. 180
13. 70
14. 15
15. 500
16. 12
17. 15
18. 135
19. 18
20. 19
21. 16
22. 128
23. 17
24. 16

SUPPLEMENTAL PRACTICE, LESSON 12

1. 5000
2. 208
3. 1200
4. 6050
5. 943
6. 8110
7. 10,000
8. 21,000
9. 40,900
10. 1010
11. 15,021
12. 19,800
13. 100,000
14. 210,000
15. 405,000
16. 325,000
17. 1,000,000
18. 1,200,000
19. 10,150,000
20. 500,000,000
21. 2,050,000
22. 25,750,000
23. 5,000,000,000
24. 1,250,000,000
25. 21,510,000,000

26. 200,000,000,000
27. 1,000,000,000,000
28. 10,000,000,000,000
29. 2,500,000,000,000
30. 200,000,000,000,000

SUPPLEMENTAL PRACTICE, LESSON 16

1. 680
2. 80
3. 580
4. 910
5. 100
6. 1490
7. 100
8. 1320
9. 700
10. 400
11. 800
12. 1600
13. 1000
14. 1000
15. 3700
16. 5000
17. 2000
18. 2000
19. 1000

20. 4000

21. 5000

22. 2000

23. 36,000

24. 58,000

25. 376,000

SUPPLEMENTAL PRACTICE, LESSON 18

1. 18

2. 19

3. 7

4. 7

5. 250

6. 40

7. 30

8. 139

9. 96

10. 72

11. 6899

12. 86

13. 460

14. 6

15. 69

16. 72

17. 66

18. 51

19. 114

20. 289

21. 650

22. 1881

SUPPLEMENTAL PRACTICE, LESSON 20

1. 1, 2, 3, 5, 6, 10, 15, 30

2. 1, 2, 4, 5, 8, 10, 20, 40

3. 1, 2, 5, 10, 25, 50

4. 1, 2, 3, 4, 5, 6, 10, 12, 15, 20, 30, 60

5. 1, 5, 7, 35

6. 1, 2, 3, 4, 6, 9, 12, 18, 36

7. 1, 37

8. 1, 2, 19, 38

9. 1, 3, 13, 39

10. 1, 7, 49

11. 14

12. 4

13. 1

14. 5

15. 25

16. 10

17. 6

18. 3

19. 11

20. 6

21. 1

22. 2

23. 4

24. 2

25. 3

———————

SUPPLEMENTAL PRACTICE, LESSON 22

1. 21

2. $14.00

3. 28

4. $15.00

5. 45

6. $40.00

7. 12

8. $24.00

9. 12

10. $9.00

11. 20

12. $30.00

13. 32

14. $18.00

15. 25

16. $20

17. 54

18. $21

19. 9

20. 24

21. 18

22. 28

23. 8

24. 21

25. 10

———————

SUPPLEMENTAL PRACTICE, LESSON 25

1. 1

2. $1\frac{1}{4}$

3. $2\frac{1}{3}$

4. $1\frac{7}{10}$

5. 4

6. $4\frac{4}{5}$

7. 6

8. $2\frac{2}{15}$

9. 2

10. $5\frac{2}{5}$

11. $5\frac{1}{7}$

12. $4\frac{1}{6}$

13. 7

14. $2\frac{2}{5}$

15. $3\frac{1}{10}$

16. $3\frac{1}{2}$

17. 5

18. $9\frac{1}{4}$

19. 10

20. $5\frac{2}{3}$

21. $12\frac{1}{5}$

22. $5\frac{1}{10}$

23. $3\frac{1}{12}$

24. $4\frac{1}{3}$

25. $26\frac{1}{2}$

_____

SUPPLEMENTAL PRACTICE, LESSON 29
SET A

1. $\frac{1}{3}$

2. $\frac{1}{2}$

3. $\frac{2}{3}$

4. $\frac{1}{4}$

5. $\frac{1}{2}$

6. $\frac{3}{4}$

7. $\frac{1}{3}$

8. $\frac{1}{5}$

9. $\frac{2}{5}$

10. $\frac{1}{2}$

11. $\frac{4}{5}$

12. $\frac{1}{6}$

13. $\frac{1}{4}$

14. $\frac{1}{3}$

15. $\frac{1}{2}$

16. $\frac{2}{3}$

17. $\frac{3}{4}$

18. $3\frac{5}{6}$

19. $4\frac{2}{5}$

20. $1\frac{3}{4}$

21. $2\frac{5}{6}$

22. $6\frac{2}{3}$

23. $8\frac{1}{2}$

24. $9\frac{1}{3}$

25. $10\frac{5}{12}$

_____

SUPPLEMENTAL PRACTICE, LESSON 29
SET B

1. $\frac{7}{8}$

2. $\frac{3}{8}$

3. $\frac{5}{6}$

4. $\frac{1}{6}$

5. $\frac{2}{3}$

6. 0

7. $\frac{5}{9}$

8. $\frac{2}{9}$

9. $\frac{3}{4}$

10. $\frac{6}{7}$

11. $1\frac{1}{4}$

12. $\frac{1}{2}$

13. $1\frac{1}{3}$

14. $\frac{1}{4}$

15. $1\frac{3}{5}$

16. 1

17. 1

18. $\frac{1}{2}$

19. $\frac{1}{2}$

20. $\frac{4}{5}$

21. $\frac{5}{6}$

22. $\frac{1}{3}$

23. 1

24. $\frac{2}{5}$

_____

SUPPLEMENTAL PRACTICE, LESSON 30

1. 12

2. 15

3. 6

4. 12

5. 24

6. 8

7. 24

8. 8

9. 9

10. 18

11. 30

12. 20

13. 24

14. 36

15. 60

16. 10

17. 12

18. 6

19. 8

20. 12

19. 100°

20. 135°

21. 10°

22. 72°

23. 115°

24. 147°

SUPPLEMENTAL PRACTICE, INVESTIGATION 3

1. 7°

2. 45°

3. 80°

4. 125°

5. 180°

6. 33°

7. 65°

8. 108°

9. 170°

10. 15°

11. 53°

12. 95°

13. 150°

14. 30°

15. 85°

16. 127°

17. 165°

18. 55°

SUPPLEMENTAL PRACTICE, LESSON 32

1. 670

2. 5400

3. 703

4. 81

5. 9050

6. 730

7. $(5 \times 100) + (6 \times 10)$

8. $(5 \times 1000) + (6 \times 100)$

9. $(7 \times 100) + (6 \times 1)$

10. $(5 \times 1000) + (2 \times 100) + (8 \times 10)$

11. 5 hr 30 min

12. 2 hr 45 min

13. 2 hr 35 min

14. 3 hr 45 min

15. 3 hr 35 min

16. 4 hr 50 min

17. 3 hr 50 min

18. 3 hr 25 min

19. 7 hr 45 min

20. 3 hr 34 min

_____

SUPPLEMENTAL PRACTICE, LESSON 33

1. $\frac{1}{2}$

2. $\frac{1}{4}$

3. $\frac{3}{4}$

4. $\frac{1}{10}$

5. $\frac{1}{5}$

6. $\frac{9}{10}$

7. $\frac{3}{10}$

8. $\frac{3}{5}$

9. $\frac{4}{5}$

10. $\frac{1}{100}$

11. $\frac{1}{20}$

12. $\frac{99}{100}$

13. $\frac{1}{50}$

14. $\frac{7}{20}$

15. $\frac{9}{20}$

16. $\frac{1}{25}$

17. $\frac{7}{50}$

18. $\frac{6}{25}$

19. $\frac{2}{5}$

20. $\frac{7}{10}$

_____

SUPPLEMENTAL PRACTICE, LESSON 35

1. 0.5

2. 0.03

3. 0.11

4. 0.001

5. 0.025

6. 1.2

7. 10.4

8. 2.01

9. 5.12

10. 0.12

11. 0.205

12. 6.15

13. 10.1

14. 12.06

15. 10.022

16. 0.05

17. 0.012

18. 0.3

19. 0.1

20. 0.23

21. 0.124

22. 0.001

23. 0.45

24. 0.03

25. 0.052

SUPPLEMENTAL PRACTICE, LESSON 36

1. $\dfrac{1}{3}$

2. $\dfrac{2}{3}$

3. $2\dfrac{1}{4}$

4. $\dfrac{3}{5}$

5. $1\dfrac{1}{5}$

6. $3\dfrac{1}{2}$

7. $\dfrac{5}{8}$

8. $2\dfrac{3}{4}$

9. $4\dfrac{2}{3}$

10. $\dfrac{3}{10}$

11. $1\dfrac{1}{2}$

12. $5\dfrac{3}{8}$

13. $\dfrac{1}{4}$

14. $1\dfrac{5}{8}$

15. $7\dfrac{7}{10}$

SUPPLEMENTAL PRACTICE, LESSON 38

1. 6

2. 2

3. 9

4. 7

5. 11

6. 4

7. 8

8. 10

9. 5

10. 3

11. 12

12. 1

SUPPLEMENTAL PRACTICE, LESSON 40
SET A

1. 1.02

2. 0.22

3. 1.65

4. 1.05

5. 0.91

6. 0.09

7. 2.23

8. 1.01

9. 6.35

10. 0.7

11. 20.25

12. 0

13. 1.2

14. 3.75

15. 1.335

16. 0.264

17. 4.656

18. 9.3

19. 2.1

20. 0.79

———————————

SUPPLEMENTAL PRACTICE, LESSON 40
SET B

1. 1.2

2. 0.24

3. 0.06

4. 0.12

5. 1.47

6. 0.744

7. 0.144

8. 0.12

9. 9.6

10. 0.434

11. 0.144

12. 12.5

13. 4.32

14. 40.5

15. 0.00045

16. 0.675

17. 0.0375

18. 2.25

19. 0.0625

20. 1.512

21. 420

———————————

SUPPLEMENTAL PRACTICE, LESSON 41
SET A

1. a. $\dfrac{1}{10}$

    b. 0.1

2. a. $\dfrac{1}{5}$

    b. 0.2

3. a. $\dfrac{3}{10}$

    b. 0.3

4. a. $\dfrac{2}{5}$

    b. 0.4

5. a. $\dfrac{1}{2}$

    b. 0.5

6. a. $\dfrac{3}{20}$

    b. 0.15

7. a. $\dfrac{1}{4}$

    b. 0.25

8. a. $\frac{9}{20}$

   b. 0.45

9. a. $\frac{3}{4}$

   b. 0.75

10. a. $\frac{1}{100}$

    b. 0.01

11. a. $\frac{1}{50}$

    b. 0.02

12. a. $\frac{1}{25}$

    b. 0.04

13. a. $\frac{1}{20}$

    b. 0.05

14. a. $\frac{3}{50}$

    b. 0.06

15. a. $\frac{3}{25}$

    b. 0.12

16. a. $\frac{6}{25}$

    b. 0.24

17. a. $\frac{9}{10}$

    b. 0.9

18. a. $\frac{19}{20}$

    b. 0.95

19. a. $\frac{9}{25}$

    b. 0.36

20. a. $\frac{4}{5}$

    b. 0.8

SUPPLEMENTAL PRACTICE, LESSON 41
SET B

21. 25

22. 50

23. 100

24. 10

25. 5

26. 20

27. 10

28. 1

29. 100

30. 50

31. 25

32. 6

33. 60

34. 120

35. 30

36. 15

37. 75

38. 300

39. 30

40. 3

_____

SUPPLEMENTAL PRACTICE, LESSON 45

1. 0.8

2. 0.12

3. 0.06

280

4. 0.025

5. 0.24

6. 0.006

7. 0.02

8. 0.7

9. 1.9

10. 0.73

11. 4.28

12. 0.31

13. 0.038

14. 0.026

15. 1.25

16. 0.14

17. 0.1

18. 0.125

19. 0.36

20. 0.024

21. 0.015

22. 0.225

23. 0.1375

24. 0.051

SUPPLEMENTAL PRACTICE, LESSON 46

1. $(3 \times 1) + \left(5 \times \frac{1}{10}\right)$

2. $\left(2 \times \frac{1}{10}\right) + \left(6 \times \frac{1}{100}\right)$

3. $(4 \times 1) + \left(8 \times \frac{1}{100}\right)$

4. $(3 \times 1) + \left(1 \times \frac{1}{10}\right) + \left(4 \times \frac{1}{100}\right)$

5. $\left(1 \times \frac{1}{100}\right) + \left(5 \times \frac{1}{1000}\right)$

6. $\left(9 \times \frac{1}{100}\right)$

7. $(1 \times 10) + (2 \times 1) + \left(5 \times \frac{1}{10}\right)$

8. $\left(4 \times \frac{1}{10}\right) + \left(5 \times \frac{1}{1000}\right)$

9. 6.5

10. 0.75

11. 50.5

12. 0.08

13. 7.05

14. 0.039

15. 83.2

16. 70.81

SUPPLEMENTAL PRACTICE, LESSON 48

1. $\frac{4}{5}$

2. $\frac{5}{8}$

3. $1\frac{1}{2}$

4. $1\frac{2}{3}$

5. $\frac{3}{4}$

6. $1\frac{5}{8}$

SOLUTIONS

7. $\dfrac{3}{8}$

8. $4\dfrac{1}{4}$

9. $2\dfrac{7}{8}$

10. $5\dfrac{3}{5}$

11. $2\dfrac{2}{3}$

12. $2\dfrac{3}{5}$

13. $2\dfrac{7}{10}$

14. $1\dfrac{5}{8}$

15. $\dfrac{3}{4}$

16. $1\dfrac{3}{5}$

17. $2\dfrac{7}{8}$

18. $5\dfrac{3}{5}$

19. $2\dfrac{3}{4}$

20. $6\dfrac{4}{5}$

21. $2\dfrac{1}{2}$

22. $2\dfrac{3}{4}$

23. $1\dfrac{1}{3}$

24. $1\dfrac{4}{5}$

282

SUPPLEMENTAL PRACTICE, LESSON 49

1. 13
2. 0.18
3. 10.7
4. 2.5
5. 10.5
6. 90
7. 400
8. 90
9. 10
10. 0.1
11. 12.5
12. 7.2
13. 0.012
14. 0.064
15. 43.75
16. 16
17. 10
18. 200
19. 90
20. 2000
21. 20
22. 74
23. 64
24. 0.75

Saxon Math 7/6—Homeschool

SUPPLEMENTAL PRACTICE, INVESTIGATION 5

1. a. mean: 7.7
   b. median: 8
   c. mode: 8
   d. range: 6

2. a. mean: 9 s
   b. median: 9 s
   c. mode: 8.7 s
   d. range: 1.9 s

3. a. mean: 8.6 yr
   b. median: 9 yr
   c. mode: 10 yr
   d. range: 4 yr

4. a. mean: 26
   b. median: 25
   c. mode: 24
   d. range: 11

SUPPLEMENTAL PRACTICE, LESSON 51

1. 0.5
2. 0.1
3. 0.4
4. 4.3
5. 62.8
6. 0.1
7. 6.3
8. 2
9. 0.82
10. 0.67
11. 1.33
12. 4.32
13. $0.23

14. $7.68
15. $0.17
16. $3.42
17. 12
18. 5
19. 92
20. 143

SUPPLEMENTAL PRACTICE, LESSON 52

1. 42
2. 3.5
3. 1.78
4. 36.5
5. 421
6. 37.5
7. 650
8. 432.3
9. 7275
10. 6400
11. 860
12. 16.25
13. 0.42
14. 0.042
15. 4.21
16. 0.6
17. 0.875

18. 0.065

19. 0.004

20. 3.728

21. 0.1234

22. 0.0425

23. 0.0076

24. 0.004

SUPPLEMENTAL PRACTICE, LESSON 57

1. $\frac{5}{8}$

2. $\frac{3}{8}$

3. $\frac{7}{8}$

4. $\frac{5}{8}$

5. $\frac{5}{6}$

6. $\frac{1}{2}$

7. $\frac{7}{12}$

8. $\frac{1}{12}$

9. $1\frac{5}{12}$

10. $\frac{1}{12}$

11. $\frac{3}{5}$

12. $\frac{2}{5}$

13. $1\frac{1}{8}$

14. $\frac{3}{8}$

15. $1\frac{1}{6}$

16. $\frac{1}{6}$

17. $1\frac{1}{5}$

18. $\frac{1}{5}$

19. $\frac{9}{20}$

20. $\frac{1}{20}$

21. $1\frac{1}{10}$

SUPPLEMENTAL PRACTICE, LESSON 63

1. $5\frac{3}{8}$

2. $1\frac{1}{8}$

3. $2\frac{1}{2}$

4. $1\frac{1}{6}$

5. $7\frac{7}{8}$

6. $1\frac{5}{8}$

7. $6\frac{9}{10}$

8. $4\frac{3}{10}$

9. $6\frac{7}{12}$

10. $2\frac{5}{12}$

11. $8\frac{5}{6}$

12. $2\frac{1}{6}$

13. $7\frac{1}{6}$

14. $3\frac{5}{6}$

15. $2\frac{1}{4}$

16. $\frac{3}{4}$

17. $8\frac{1}{10}$

18. $5\frac{1}{10}$

19. $3\frac{9}{10}$

20. $1\frac{1}{2}$

21. $10\frac{5}{12}$

22. $6\frac{11}{12}$

23. $4\frac{5}{6}$

24. $8\frac{3}{4}$

Supplemental Practice, Lesson 65

1. 2×3

2. $2 \times 2 \times 2$

3. 3×3

4. 2×5

5. $2 \times 2 \times 3$

6. 2×7

7. 3×5

8. $2 \times 2 \times 2 \times 2$

9. $2 \times 3 \times 3$

10. $2 \times 2 \times 5$

11. 3×7

12. $2 \times 2 \times 2 \times 3$

13. $2 \times 3 \times 5$

14. $2 \times 2 \times 3 \times 3$

15. 3×13

16. $2 \times 2 \times 2 \times 5$

17. $2 \times 3 \times 7$

18. $2 \times 2 \times 2 \times 2 \times 3$

19. $2 \times 2 \times 3 \times 5$

20. $2 \times 2 \times 5 \times 5$

Supplemental Practice, Lesson 66

1. $3\frac{3}{4}$

2. $4\frac{1}{2}$

3. $1\frac{7}{8}$

4. $4\frac{1}{6}$

5. $17\frac{1}{2}$

6. $2\frac{5}{8}$

7. $5\frac{5}{9}$

8. 15

9. $\frac{24}{25}$

10. 1

11. 2

12. $2\frac{1}{2}$

13. 3

14. 11

15. 14

16. 3

17. 7

18. 6

19. $3\frac{1}{3}$

20. $2\frac{1}{3}$

21. $3\frac{1}{2}$

SUPPLEMENTAL PRACTICE, LESSON 68

1. $\frac{1}{2}$

2. 2

3. $\frac{5}{6}$

4. $1\frac{1}{5}$

5. $\frac{1}{2}$

6. 2

7. $1\frac{1}{9}$

8. $\frac{9}{10}$

9. $\frac{3}{16}$

10. $5\frac{1}{3}$

11. $\frac{24}{35}$

12. $1\frac{11}{24}$

13. 2

14. $\frac{1}{2}$

15. 4

16. $\frac{1}{4}$

17. $1\frac{1}{3}$

18. $\frac{3}{4}$

19. $1\frac{3}{7}$

20. $\frac{7}{10}$

21. $\frac{1}{3}$

SUPPLEMENTAL PRACTICE, INVESTIGATION 7

1. *y*-axis

2. *x*-axis

3. origin

4. coordinates

5. *B*

6. *R*

7. *W*

8. *X*

9. *U*

10. *F*

11. *Q*

12. *J*

13. $(-6, -1)$

14. $(1, 6)$

15. $(5, 3)$

16. $(6, 1)$

17. $(3, -5)$

18. $(-1, -6)$

19. $(-5, -3)$

20. $(0, -4)$

21. $(-6, 1)$

22. $(-3, 5)$

23. $(-2, 2)$

24. $(2, 2)$

25. $(0, 0)$

SUPPLEMENTAL PRACTICE, LESSON 73

1. $\dfrac{1}{10}$

2. $1\dfrac{1}{5}$

3. $\dfrac{2}{5}$

4. $2\dfrac{1}{2}$

5. $\dfrac{4}{5}$

6. $3\dfrac{9}{10}$

7. $\dfrac{3}{25}$

8. $4\dfrac{3}{20}$

9. $\dfrac{1}{4}$

10. $3\dfrac{3}{4}$

11. $\dfrac{1}{40}$

12. $\dfrac{1}{200}$

13. $\dfrac{5}{6}$

14. $\dfrac{2}{5}$

15. $\dfrac{13}{20}$

16. $\dfrac{1}{5}$

17. $\dfrac{1}{10}$

18. 1

SOLUTIONS

19. 3

20. 1

21. $\frac{2}{5}$

22. $\frac{1}{10}$

23. $\frac{1}{2}$

24. $\frac{1}{3}$

_____

SUPPLEMENTAL PRACTICE, LESSON 74

1. 0.5

2. 0.25

3. 0.125

4. 0.1

5. 0.75

6. 1.375

7. 2.3

8. 4.6

9. 7.625

10. 3.7

11. 4.1

12. 0.05

13. 0.24

14. 2

15. 5

16. 0.6

17. 0.9

18. 1

19. 1.5

20. 0.075

21. 0.1

22. 2.5

_____

SUPPLEMENTAL PRACTICE, LESSON 75

1. 50%

2. 60%

3. 75%

4. 100%

5. 10%

6. 20%

7. 25%

8. 50%

9. 100%

10. 75%

11. 80%

12. 12%

13. 8%

14. 80%

15. 50%

16. 100%

17. 10%

18. 20%

19. 25%

20. 50%

21. 80%

22. 100%

23. 20%

24. 30%

25. 75%

_____

SUPPLEMENTAL PRACTICE, LESSON 77

1. 5 parts

2. 5 team members

3. 2 parts

4. 10 team members

5. 3 parts

6. 15 team members

7. 6 parts

8. 50 members

9. 5 parts

10. 250 members

11. 1 part

12. 50 members

13. 10 parts

14. $600

15. 3 parts

16. $1800

17. 7 parts

18. $4200

19. 100 parts

20. 8 rooms

21. 27 parts

22. 216 rooms

23. 73 parts

24. 584 rooms

_____

SUPPLEMENTAL PRACTICE, LESSON 79

1. 24 cm^2

2. 16 cm^2

3. 10 in.^2

4. 6 in.^2

5. 60 mm^2

6. 8 ft^2

7. 24 cm^2

8. 25 in.^2

9. 20 mm^2

10. 24 mm^2

11. 39 m^2

12. 45 cm^2

_____

SUPPLEMENTAL PRACTICE, LESSON 84

1. 13

2. 5

3. 3

4. 11

5. 21

6. 13

7. 21

8. 1

9. 19

10. 21

11. 36

12. 13

13. 5

14. 50

15. 4

16. 0

17. 16

18. 122

19. 109

20. 0

SUPPLEMENTAL PRACTICE, LESSON 85

1. 15

2. 18

3. 4

4. 10

5. 15

6. 900

7. 3

8. 21

9. 20

10. 800

11. 48

12. 20

13. 12

14. 24

15. 7

16. 24

17. 54

18. 21

19. 3

20. 36

SUPPLEMENTAL PRACTICE, LESSON 86

1. 2 in.

2. 6.28 in.

3. 3.14 in.^{2}

4. 2 ft

5. 12.56 ft

6. 12.56 ft^2

7. 6 cm

8. 18.84 cm

9. 28.26 cm^2

10. 4 m

11. 25.12 m

12. 50.24 m^2

13. 10 mm

14. 31.4 mm

15. 78.5 mm²

16. 6 yd

17. 37.68 yd

18. 113.04 yd²

19. 20 km

20. 62.8 km

21. 314 km²

22. 50 mi

23. 314 mi

24. 7850 mi²

25. circumference

―――――――――――

SUPPLEMENTAL PRACTICE, LESSON 89

1. 3, 4

2. 5, 6

3. 6, 7

4. 7, 8

5. 7, 8

6. 8, 9

7. 8, 9

8. 9, 10

9. 2, 3

10. 1, 2

―――――――――――

SUPPLEMENTAL PRACTICE, LESSON 92

1. $(4 \times 10^5) + (5 \times 10^4)$

2. $(2 \times 10^7) + (5 \times 10^6)$

3. $(1 \times 10^{10}) + (6 \times 10^9)$

4. 5,000,000

5. 36,000

6. 1,500,000,000

7. $\dfrac{1}{27}$

8. $6\dfrac{1}{4}$

9. 0.001

10. $3\dfrac{3}{8}$

11. 36

12. 7

13. 12

14. 24

15. 0

―――――――――――

SUPPLEMENTAL PRACTICE, LESSON 94

1. 10%

2. 90%

3. 20%

4. 75%

5. 15%

6. 12%

7. 6%

8. 3%

9. 11%

10. 22%

11. 44%

12. 55%

13. $33\frac{1}{3}\%$

14. $66\frac{2}{3}\%$

15. $12\frac{1}{2}\%$

16. $37\frac{1}{2}\%$

17. $11\frac{1}{9}\%$

18. 125%

19. $16\frac{2}{3}\%$

20. $62\frac{1}{2}\%$

21. $14\frac{2}{7}\%$

22. 250%

23. $83\frac{1}{3}\%$

24. $77\frac{7}{9}\%$

25. $41\frac{2}{3}\%$

26. 60%

27. 340%

28. 1%

29. 120%

30. 50%

31. 100%

32. 37%

33. 450%

34. 200%

35. 10%

36. 105%

37. 60%

38. 300%

SUPPLEMENTAL PRACTICE, LESSON 99

1. a. 0.01
   b. 1%

2. a. $\frac{4}{5}$
   b. 80%

3. a. $\frac{1}{4}$
   b. 0.25

4. a. 0.75
   b. 75%

5. a. $\frac{7}{10}$
   b. 70%

6. a. $\frac{9}{10}$
   b. 0.9

7. a. 0.05
   b. 5%

8. a. $\frac{1}{2}$
   b. 50%

9. a. $\frac{1}{25}$
   b. 0.04

10. a. 0.02
    b. 2%

11. a. $\frac{9}{20}$

    b. 45%

12. a. $\frac{23}{100}$

    b. 0.23

13. a. 1.5
    b. 150%

14. a. $\frac{3}{20}$

    b. 15%

15. a. $\frac{1}{10}$

    b. 0.1

16. a. 0.125
    b. 12.5%

17. a. $\frac{1}{5}$

    b. 20%

18. a. $\frac{7}{20}$

    b. 0.35

SUPPLEMENTAL PRACTICE, LESSON 100

1. −7

2. −13

3. −2

4. −5

5. 1

6. −1

7. −1

8. 0

9. −6

10. 12

11. −24

12. 0

13. 3

14. 7

15. −7

16. 3

17. −3

18. −55

19. −5

20. −8

21. 6

22. 14

23. −32

24. 2

25. −10

26. 8

27. −8

28. 2

29. −2

30. −8

31. 11

32. 19

33. 5

34. −19

35. −19

36. −5

37. 19

38. −19

39. 0

40. 12

41. −15

42. −5

43. −15

44. 5

45. 15

46. −4

47. −20

48. −20

49. 4

50. 20

SUPPLEMENTAL PRACTICE, LESSON 107

1. 40 cm

2. 78 cm^2

3. 40 in.

4. 85 in.^{2}

5. 24 ft

6. 30 ft^2

7. 60 mm

8. 210 mm^2

9. 34 cm

10. 74 cm^2

11. 44 mm

12. 120 mm^2

13. 48 m

14. 80 m^2

15. 38 ft

16. 69 ft^2

SUPPLEMENTAL PRACTICE, LESSON 111

1. 12 R 4; $12\frac{1}{2}$; 12.5

2. 12 R 2; $12\frac{1}{2}$; 12.5

3. 5 R 6; $5\frac{3}{5}$; 5.6

4. 6 R 5; $6\frac{1}{2}$; 6.5

5. 5 R 3; $5\frac{1}{4}$; 5.25

6. 14 R 2; $14\frac{2}{5}$; 14.4

7. 12 R 1; $12\frac{1}{4}$; 12.25

8. 4 R 6; $4\frac{3}{4}$; 4.75

9. 11 R 3; $11\frac{3}{4}$; 11.75

10. 18 R 2; $18\frac{1}{4}$; 18.25

11. 19 R 10; $19\frac{1}{2}$; 19.5

12. 62 R 5; $62\frac{1}{2}$; 62.5

13. 86 R 2; $86\frac{2}{5}$; 86.4

14. 81 R 2; $81\frac{1}{4}$; 81.25

15. 16 R 5; $16\frac{1}{4}$; 16.25

16. 70 R 2; $70\frac{1}{4}$; 70.25

17. 13 R 10; $13\frac{1}{4}$; 13.25

18. 7 R 25; $7\frac{1}{2}$; 7.5

19. 2 R 40; $2\frac{2}{5}$; 2.4

20. 53 R 4; $53\frac{2}{5}$; 53.4

SUPPLEMENTAL PRACTICE, LESSON 112

1. −12

2. 96

3. −180

4. 360

5. −147

6. −136

7. 375

8. 90

9. −301

10. 288

11. −108

12. 225

13. −12

14. 8

15. −9

16. −80

17. 39

18. −12

19. −5

20. 25

SUPPLEMENTAL PRACTICE, LESSON 113

1. 6 hr 10 min

2. 4 min 25 s

3. 4 yd 1 ft 5 in.

4. 8 in.

5. 10 lb 2 oz

6. 1 lb 10 oz

7. 6 gal 1 qt

8. 3 gal 3 qt

9. 500,000

10. 25,000,000

11. 700,000

12. 1,500,000,000

13. 1,250,000,000

14. 15,000,000

15. 3,500,000

16. 500,000,000

SUPPLEMENTAL PRACTICE, LESSON 117

1. 100
2. 160
3. 150
4. 200
5. 25
6. 30
7. 40
8. 100
9. 80
10. 48
11. 16
12. 36
13. 20
14. 600
15. 300
16. 200
17. 150
18. 120
19. 100
20. 50
21. 50
22. 50
23. 40
24. 80

SUPPLEMENTAL PRACTICE, LESSON 119

1. 100
2. 50
3. 25
4. 40
5. 200
6. 500
7. 100
8. 100
9. 200
10. 2000
11. 60
12. 24
13. 600
14. 5
15. 100
16. 50
17. 500
18. 8
19. 10
20. 200
21. 40
22. 100
23. 80
24. 160
25. 80

Solutions for

Facts Practice Tests

64 Addition Facts

Add.

7 + 2 **9**	9 + 4 **13**	2 + 8 **10**	6 + 5 **11**	4 + 4 **8**	3 + 9 **12**	8 + 4 **12**	5 + 7 **12**
9 + 7 **16**	4 + 7 **11**	7 + 5 **12**	5 + 4 **9**	3 + 4 **7**	6 + 8 **14**	2 + 5 **7**	8 + 8 **16**
6 + 3 **9**	2 + 9 **11**	7 + 8 **15**	8 + 3 **11**	5 + 9 **14**	3 + 6 **9**	9 + 9 **18**	4 + 9 **13**
5 + 8 **13**	9 + 5 **14**	4 + 5 **9**	8 + 6 **14**	2 + 3 **5**	6 + 6 **12**	5 + 2 **7**	7 + 3 **10**
3 + 8 **11**	8 + 9 **17**	2 + 2 **4**	7 + 6 **13**	5 + 5 **10**	6 + 9 **15**	3 + 7 **10**	9 + 8 **17**
4 + 2 **6**	3 + 3 **6**	6 + 4 **10**	4 + 8 **12**	9 + 3 **12**	2 + 4 **6**	8 + 5 **13**	7 + 9 **16**
7 + 4 **11**	2 + 6 **8**	5 + 3 **8**	9 + 6 **15**	4 + 3 **7**	6 + 7 **13**	3 + 2 **5**	8 + 7 **15**
5 + 6 **11**	8 + 2 **10**	3 + 5 **8**	6 + 2 **8**	7 + 7 **14**	4 + 6 **10**	9 + 2 **11**	2 + 7 **9**

B 100 Addition Facts

Add.

3 + 2 **5**	8 + 3 **11**	2 + 1 **3**	5 + 6 **11**	2 + 9 **11**	4 + 8 **12**	8 + 0 **8**	3 + 9 **12**	1 + 0 **1**	6 + 3 **9**
7 + 3 **10**	1 + 6 **7**	4 + 7 **11**	0 + 3 **3**	6 + 4 **10**	5 + 5 **10**	3 + 1 **4**	7 + 2 **9**	8 + 5 **13**	2 + 5 **7**
4 + 0 **4**	5 + 7 **12**	1 + 1 **2**	5 + 4 **9**	2 + 8 **10**	7 + 1 **8**	4 + 6 **10**	0 + 2 **2**	6 + 5 **11**	4 + 9 **13**
8 + 6 **14**	0 + 4 **4**	5 + 8 **13**	7 + 4 **11**	1 + 7 **8**	6 + 6 **12**	4 + 1 **5**	8 + 2 **10**	2 + 4 **6**	6 + 0 **6**
9 + 1 **10**	8 + 8 **16**	2 + 2 **4**	4 + 5 **9**	6 + 2 **8**	0 + 0 **0**	5 + 9 **14**	3 + 3 **6**	8 + 1 **9**	2 + 7 **9**
4 + 4 **8**	7 + 5 **12**	0 + 1 **1**	8 + 7 **15**	3 + 4 **7**	7 + 9 **16**	1 + 2 **3**	6 + 7 **13**	0 + 8 **8**	9 + 2 **11**
0 + 9 **9**	8 + 9 **17**	7 + 6 **13**	1 + 3 **4**	6 + 8 **14**	2 + 0 **2**	8 + 4 **12**	3 + 5 **8**	9 + 8 **17**	5 + 0 **5**
9 + 3 **12**	2 + 6 **8**	3 + 0 **3**	6 + 1 **7**	3 + 6 **9**	5 + 2 **7**	0 + 5 **5**	6 + 9 **15**	1 + 8 **9**	9 + 6 **15**
4 + 3 **7**	9 + 9 **18**	0 + 7 **7**	9 + 4 **13**	7 + 7 **14**	1 + 4 **5**	3 + 7 **10**	7 + 0 **7**	2 + 3 **5**	5 + 1 **6**
9 + 5 **14**	1 + 5 **6**	9 + 0 **9**	3 + 8 **11**	1 + 9 **10**	5 + 3 **8**	4 + 2 **6**	9 + 7 **16**	0 + 6 **6**	7 + 8 **15**

C | 100 Subtraction Facts

Subtract.

16 − 9 **7**	7 − 1 **6**	18 − 9 **9**	11 − 3 **8**	13 − 7 **6**	8 − 2 **6**	11 − 5 **6**	5 − 0 **5**	17 − 9 **8**	6 − 1 **5**
10 − 9 **1**	6 − 2 **4**	13 − 4 **9**	4 − 0 **4**	10 − 5 **5**	5 − 1 **4**	10 − 3 **7**	12 − 6 **6**	10 − 1 **9**	6 − 4 **2**
7 − 2 **5**	14 − 7 **7**	8 − 1 **7**	11 − 6 **5**	3 − 3 **0**	16 − 7 **9**	5 − 2 **3**	12 − 4 **8**	3 − 0 **3**	11 − 7 **4**
17 − 8 **9**	6 − 0 **6**	10 − 6 **4**	4 − 1 **3**	9 − 5 **4**	9 − 0 **9**	5 − 4 **1**	12 − 5 **7**	4 − 2 **2**	9 − 3 **6**
12 − 3 **9**	16 − 8 **8**	9 − 1 **8**	15 − 6 **9**	11 − 4 **7**	13 − 5 **8**	1 − 0 **1**	8 − 5 **3**	9 − 6 **3**	11 − 2 **9**
7 − 0 **7**	10 − 8 **2**	6 − 3 **3**	14 − 5 **9**	3 − 1 **2**	8 − 6 **2**	4 − 4 **0**	11 − 8 **3**	3 − 2 **1**	15 − 9 **6**
13 − 8 **5**	7 − 4 **3**	10 − 7 **3**	0 − 0 **0**	12 − 8 **4**	5 − 5 **0**	4 − 3 **1**	8 − 7 **1**	7 − 3 **4**	7 − 6 **1**
5 − 3 **2**	7 − 5 **2**	2 − 1 **1**	6 − 6 **0**	8 − 4 **4**	2 − 2 **0**	13 − 6 **7**	15 − 8 **7**	2 − 0 **2**	13 − 9 **4**
1 − 1 **0**	11 − 9 **2**	10 − 4 **6**	9 − 2 **7**	14 − 6 **8**	8 − 0 **8**	9 − 4 **5**	10 − 2 **8**	6 − 5 **1**	8 − 3 **5**
7 − 7 **0**	14 − 8 **6**	12 − 9 **3**	9 − 8 **1**	12 − 7 **5**	9 − 9 **0**	15 − 7 **8**	8 − 8 **0**	14 − 9 **5**	9 − 7 **2**

SOLUTIONS

D | 64 Multiplication Facts

Multiply.

5 × 6 **30**	4 × 3 **12**	9 × 8 **72**	7 × 5 **35**	2 × 9 **18**	8 × 4 **32**	9 × 3 **27**	6 × 9 **54**
9 × 4 **36**	2 × 5 **10**	7 × 6 **42**	4 × 8 **32**	7 × 9 **63**	5 × 4 **20**	3 × 2 **6**	9 × 7 **63**
3 × 7 **21**	8 × 5 **40**	6 × 2 **12**	5 × 5 **25**	3 × 5 **15**	2 × 4 **8**	7 × 7 **49**	8 × 9 **72**
6 × 4 **24**	2 × 8 **16**	4 × 4 **16**	8 × 2 **16**	3 × 9 **27**	6 × 6 **36**	9 × 9 **81**	5 × 3 **15**
4 × 6 **24**	8 × 8 **64**	5 × 7 **35**	6 × 3 **18**	2 × 2 **4**	7 × 4 **28**	3 × 8 **24**	8 × 6 **48**
2 × 6 **12**	5 × 9 **45**	3 × 3 **9**	9 × 2 **18**	6 × 7 **42**	4 × 5 **20**	7 × 2 **14**	9 × 6 **54**
5 × 2 **10**	7 × 8 **56**	2 × 3 **6**	6 × 8 **48**	4 × 7 **28**	9 × 5 **45**	3 × 6 **18**	8 × 7 **56**
3 × 4 **12**	7 × 3 **21**	5 × 8 **40**	4 × 2 **8**	8 × 3 **24**	2 × 7 **14**	6 × 5 **30**	4 × 9 **36**

Saxon Math 7/6—Homeschool

E 100 Multiplication Facts

Multiply.

9 × 9 **81**	3 × 5 **15**	8 × 5 **40**	2 × 6 **12**	4 × 7 **28**	0 × 3 **0**	7 × 2 **14**	1 × 5 **5**	7 × 8 **56**	4 × 0 **0**
3 × 4 **12**	5 × 9 **45**	0 × 2 **0**	7 × 3 **21**	4 × 1 **4**	2 × 7 **14**	6 × 3 **18**	5 × 4 **20**	1 × 0 **0**	9 × 2 **18**
1 × 1 **1**	9 × 0 **0**	2 × 8 **16**	6 × 4 **24**	0 × 7 **0**	8 × 1 **8**	3 × 3 **9**	4 × 8 **32**	9 × 3 **27**	2 × 0 **0**
4 × 9 **36**	7 × 0 **0**	1 × 2 **2**	8 × 4 **32**	6 × 5 **30**	2 × 9 **18**	9 × 4 **36**	0 × 1 **0**	7 × 4 **28**	5 × 8 **40**
0 × 8 **0**	4 × 2 **8**	9 × 8 **72**	3 × 6 **18**	5 × 5 **25**	1 × 6 **6**	5 × 0 **0**	6 × 6 **36**	2 × 1 **2**	7 × 9 **63**
9 × 1 **9**	2 × 2 **4**	5 × 1 **5**	4 × 3 **12**	0 × 0 **0**	8 × 9 **72**	3 × 7 **21**	9 × 7 **63**	1 × 7 **7**	6 × 0 **0**
5 × 6 **30**	7 × 5 **35**	3 × 0 **0**	8 × 8 **64**	1 × 3 **3**	8 × 3 **24**	5 × 2 **10**	0 × 4 **0**	9 × 5 **45**	6 × 7 **42**
2 × 3 **6**	8 × 6 **48**	0 × 5 **0**	6 × 1 **6**	3 × 8 **24**	7 × 6 **42**	1 × 8 **8**	9 × 6 **54**	4 × 4 **16**	5 × 3 **15**
7 × 7 **49**	1 × 4 **4**	6 × 2 **12**	4 × 5 **20**	2 × 4 **8**	8 × 0 **0**	3 × 1 **3**	6 × 8 **48**	0 × 9 **0**	8 × 7 **56**
3 × 2 **6**	4 × 6 **24**	1 × 9 **9**	5 × 7 **35**	8 × 2 **16**	0 × 6 **0**	7 × 1 **7**	2 × 5 **10**	6 × 9 **54**	3 × 9 **27**

F 90 Division Facts

Divide.

$7\overline{)21}=3$	$2\overline{)10}=5$	$6\overline{)42}=7$	$1\overline{)3}=3$	$4\overline{)24}=6$	$3\overline{)6}=2$	$9\overline{)54}=6$	$6\overline{)18}=3$	$4\overline{)0}=0$	$5\overline{)30}=6$
$4\overline{)32}=8$	$8\overline{)56}=7$	$1\overline{)0}=0$	$6\overline{)12}=2$	$3\overline{)18}=6$	$9\overline{)72}=8$	$5\overline{)15}=3$	$2\overline{)8}=4$	$7\overline{)42}=6$	$6\overline{)36}=6$
$6\overline{)0}=0$	$5\overline{)10}=2$	$9\overline{)9}=1$	$2\overline{)6}=3$	$7\overline{)63}=9$	$4\overline{)16}=4$	$8\overline{)48}=6$	$1\overline{)2}=2$	$5\overline{)35}=7$	$3\overline{)21}=7$
$2\overline{)18}=9$	$6\overline{)6}=1$	$3\overline{)15}=5$	$8\overline{)40}=5$	$2\overline{)0}=0$	$5\overline{)20}=4$	$9\overline{)27}=3$	$1\overline{)8}=8$	$4\overline{)4}=1$	$7\overline{)35}=5$
$4\overline{)20}=5$	$9\overline{)63}=7$	$1\overline{)4}=4$	$7\overline{)14}=2$	$3\overline{)3}=1$	$8\overline{)24}=3$	$5\overline{)0}=0$	$6\overline{)24}=4$	$8\overline{)8}=1$	$2\overline{)16}=8$
$5\overline{)5}=1$	$8\overline{)64}=8$	$3\overline{)0}=0$	$4\overline{)28}=7$	$7\overline{)49}=7$	$2\overline{)4}=2$	$9\overline{)81}=9$	$3\overline{)12}=4$	$6\overline{)30}=5$	$1\overline{)5}=5$
$8\overline{)32}=4$	$1\overline{)1}=1$	$9\overline{)36}=4$	$3\overline{)27}=9$	$2\overline{)14}=7$	$5\overline{)25}=5$	$6\overline{)48}=8$	$8\overline{)0}=0$	$7\overline{)28}=4$	$4\overline{)36}=9$
$2\overline{)12}=6$	$5\overline{)45}=9$	$1\overline{)7}=7$	$4\overline{)8}=2$	$7\overline{)0}=0$	$8\overline{)16}=2$	$3\overline{)24}=8$	$9\overline{)45}=5$	$1\overline{)9}=9$	$6\overline{)54}=9$
$7\overline{)56}=8$	$9\overline{)0}=0$	$8\overline{)72}=9$	$2\overline{)2}=1$	$5\overline{)40}=8$	$3\overline{)9}=3$	$9\overline{)18}=2$	$1\overline{)6}=6$	$4\overline{)12}=3$	$7\overline{)7}=1$

Saxon Math 7/6—Homeschool

G | 30 Fractions to Reduce

Reduce each fraction to lowest terms.

$\frac{2}{8} = \frac{1}{4}$	$\frac{4}{6} = \frac{2}{3}$	$\frac{6}{10} = \frac{3}{5}$	$\frac{2}{4} = \frac{1}{2}$	$\frac{6}{16} = \frac{3}{8}$
$\frac{5}{100} = \frac{1}{20}$	$\frac{9}{12} = \frac{3}{4}$	$\frac{14}{16} = \frac{7}{8}$	$\frac{4}{10} = \frac{2}{5}$	$\frac{4}{12} = \frac{1}{3}$
$\frac{2}{10} = \frac{1}{5}$	$\frac{3}{6} = \frac{1}{2}$	$\frac{25}{100} = \frac{1}{4}$	$\frac{3}{12} = \frac{1}{4}$	$\frac{4}{16} = \frac{1}{4}$
$\frac{3}{9} = \frac{1}{3}$	$\frac{10}{16} = \frac{5}{8}$	$\frac{6}{9} = \frac{2}{3}$	$\frac{4}{8} = \frac{1}{2}$	$\frac{2}{12} = \frac{1}{6}$
$\frac{6}{12} = \frac{1}{2}$	$\frac{2}{16} = \frac{1}{8}$	$\frac{8}{10} = \frac{4}{5}$	$\frac{2}{6} = \frac{1}{3}$	$\frac{75}{100} = \frac{3}{4}$
$\frac{12}{16} = \frac{3}{4}$	$\frac{8}{12} = \frac{2}{3}$	$\frac{6}{8} = \frac{3}{4}$	$\frac{10}{12} = \frac{5}{6}$	$\frac{5}{10} = \frac{1}{2}$

H 72 Multiplication and Division Facts

Multiply or divide as indicated.

$\times \begin{array}{r}5 \\ 9 \\ \hline \mathbf{45}\end{array}$	$6\overline{)\begin{array}{c}\mathbf{6} \\ 36\end{array}}$	$\times \begin{array}{r}4 \\ 7 \\ \hline \mathbf{28}\end{array}$	$8\overline{)\begin{array}{c}\mathbf{5} \\ 40\end{array}}$	$\times \begin{array}{r}10 \\ 3 \\ \hline \mathbf{30}\end{array}$	$5\overline{)\begin{array}{c}\mathbf{4} \\ 20\end{array}}$	$\times \begin{array}{r}2 \\ 7 \\ \hline \mathbf{14}\end{array}$	$9\overline{)\begin{array}{c}\mathbf{3} \\ 27\end{array}}$
$9\overline{)\begin{array}{c}\mathbf{9} \\ 81\end{array}}$	$\times \begin{array}{r}5 \\ 7 \\ \hline \mathbf{35}\end{array}$	$6\overline{)\begin{array}{c}\mathbf{4} \\ 24\end{array}}$	$\times \begin{array}{r}2 \\ 3 \\ \hline \mathbf{6}\end{array}$	$7\overline{)\begin{array}{c}\mathbf{6} \\ 42\end{array}}$	$\times \begin{array}{r}7 \\ 9 \\ \hline \mathbf{63}\end{array}$	$5\overline{)\begin{array}{c}\mathbf{2} \\ 10\end{array}}$	$\times \begin{array}{r}4 \\ 5 \\ \hline \mathbf{20}\end{array}$
$\times \begin{array}{r}2 \\ 9 \\ \hline \mathbf{18}\end{array}$	$7\overline{)\begin{array}{c}\mathbf{4} \\ 28\end{array}}$	$\times \begin{array}{r}6 \\ 9 \\ \hline \mathbf{54}\end{array}$	$4\overline{)\begin{array}{c}\mathbf{2} \\ 8\end{array}}$	$\times \begin{array}{r}3 \\ 3 \\ \hline \mathbf{9}\end{array}$	$8\overline{)\begin{array}{c}\mathbf{3} \\ 24\end{array}}$	$\times \begin{array}{r}10 \\ 4 \\ \hline \mathbf{40}\end{array}$	$6\overline{)\begin{array}{c}\mathbf{2} \\ 12\end{array}}$
$9\overline{)\begin{array}{c}\mathbf{7} \\ 63\end{array}}$	$\times \begin{array}{r}2 \\ 5 \\ \hline \mathbf{10}\end{array}$	$8\overline{)\begin{array}{c}\mathbf{8} \\ 64\end{array}}$	$\times \begin{array}{r}7 \\ 6 \\ \hline \mathbf{42}\end{array}$	$10\overline{)\begin{array}{c}\mathbf{10} \\ 100\end{array}}$	$\times \begin{array}{r}3 \\ 5 \\ \hline \mathbf{15}\end{array}$	$9\overline{)\begin{array}{c}\mathbf{4} \\ 36\end{array}}$	$\times \begin{array}{r}10 \\ 8 \\ \hline \mathbf{80}\end{array}$
$\times \begin{array}{r}4 \\ 8 \\ \hline \mathbf{32}\end{array}$	$5\overline{)\begin{array}{c}\mathbf{5} \\ 25\end{array}}$	$\times \begin{array}{r}3 \\ 8 \\ \hline \mathbf{24}\end{array}$	$4\overline{)\begin{array}{c}\mathbf{4} \\ 16\end{array}}$	$\times \begin{array}{r}10 \\ 5 \\ \hline \mathbf{50}\end{array}$	$7\overline{)\begin{array}{c}\mathbf{2} \\ 14\end{array}}$	$\times \begin{array}{r}2 \\ 2 \\ \hline \mathbf{4}\end{array}$	$8\overline{)\begin{array}{c}\mathbf{6} \\ 48\end{array}}$
$9\overline{)\begin{array}{c}\mathbf{6} \\ 54\end{array}}$	$\times \begin{array}{r}3 \\ 6 \\ \hline \mathbf{18}\end{array}$	$5\overline{)\begin{array}{c}\mathbf{3} \\ 15\end{array}}$	$\times \begin{array}{r}2 \\ 8 \\ \hline \mathbf{16}\end{array}$	$7\overline{)\begin{array}{c}\mathbf{7} \\ 49\end{array}}$	$\times \begin{array}{r}4 \\ 6 \\ \hline \mathbf{24}\end{array}$	$6\overline{)\begin{array}{c}\mathbf{5} \\ 30\end{array}}$	$\times \begin{array}{r}5 \\ 8 \\ \hline \mathbf{40}\end{array}$
$\times \begin{array}{r}10 \\ 6 \\ \hline \mathbf{60}\end{array}$	$4\overline{)\begin{array}{c}\mathbf{3} \\ 12\end{array}}$	$\times \begin{array}{r}7 \\ 8 \\ \hline \mathbf{56}\end{array}$	$6\overline{)\begin{array}{c}\mathbf{3} \\ 18\end{array}}$	$\times \begin{array}{r}3 \\ 4 \\ \hline \mathbf{12}\end{array}$	$7\overline{)\begin{array}{c}\mathbf{5} \\ 35\end{array}}$	$\times \begin{array}{r}5 \\ 6 \\ \hline \mathbf{30}\end{array}$	$8\overline{)\begin{array}{c}\mathbf{4} \\ 32\end{array}}$
$8\overline{)\begin{array}{c}\mathbf{7} \\ 56\end{array}}$	$\times \begin{array}{r}2 \\ 4 \\ \hline \mathbf{8}\end{array}$	$8\overline{)\begin{array}{c}\mathbf{2} \\ 16\end{array}}$	$\times \begin{array}{r}6 \\ 8 \\ \hline \mathbf{48}\end{array}$	$7\overline{)\begin{array}{c}\mathbf{3} \\ 21\end{array}}$	$\times \begin{array}{r}8 \\ 9 \\ \hline \mathbf{72}\end{array}$	$3\overline{)\begin{array}{c}\mathbf{2} \\ 6\end{array}}$	$\times \begin{array}{r}3 \\ 9 \\ \hline \mathbf{27}\end{array}$
$\times \begin{array}{r}10 \\ 7 \\ \hline \mathbf{70}\end{array}$	$9\overline{)\begin{array}{c}\mathbf{2} \\ 18\end{array}}$	$\times \begin{array}{r}3 \\ 7 \\ \hline \mathbf{21}\end{array}$	$9\overline{)\begin{array}{c}\mathbf{5} \\ 45\end{array}}$	$\times \begin{array}{r}2 \\ 6 \\ \hline \mathbf{12}\end{array}$	$9\overline{)\begin{array}{c}\mathbf{8} \\ 72\end{array}}$	$\times \begin{array}{r}4 \\ 9 \\ \hline \mathbf{36}\end{array}$	$9\overline{)\begin{array}{c}\mathbf{10} \\ 90\end{array}}$

I | **28 Improper Fractions to Simplify**

Write each improper fraction as a mixed number or a whole number.

$\frac{5}{4} = 1\frac{1}{4}$	$\frac{16}{12} = 1\frac{1}{3}$	$\frac{12}{8} = 1\frac{1}{2}$	$\frac{8}{6} = 1\frac{1}{3}$
$\frac{12}{6} = 2$	$\frac{12}{10} = 1\frac{1}{5}$	$\frac{6}{4} = 1\frac{1}{2}$	$\frac{20}{12} = 1\frac{2}{3}$
$\frac{5}{3} = 1\frac{2}{3}$	$\frac{10}{8} = 1\frac{1}{4}$	$\frac{25}{10} = 2\frac{1}{2}$	$\frac{10}{3} = 3\frac{1}{3}$
$\frac{15}{10} = 1\frac{1}{2}$	$\frac{3}{2} = 1\frac{1}{2}$	$\frac{9}{6} = 1\frac{1}{2}$	$\frac{7}{4} = 1\frac{3}{4}$
$\frac{18}{12} = 1\frac{1}{2}$	$\frac{8}{3} = 2\frac{2}{3}$	$\frac{15}{6} = 2\frac{1}{2}$	$\frac{14}{4} = 3\frac{1}{2}$
$\frac{8}{4} = 2$	$\frac{10}{6} = 1\frac{2}{3}$	$\frac{5}{2} = 2\frac{1}{2}$	$\frac{21}{12} = 1\frac{3}{4}$
$\frac{15}{12} = 1\frac{1}{4}$	$\frac{10}{4} = 2\frac{1}{2}$	$\frac{15}{8} = 1\frac{7}{8}$	$\frac{4}{3} = 1\frac{1}{3}$

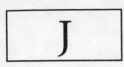

24 Mixed Numbers to Write as Improper Fractions

Write each mixed number as an improper fraction.

$2\frac{1}{2} = \frac{5}{2}$	$2\frac{2}{5} = \frac{12}{5}$	$1\frac{3}{4} = \frac{7}{4}$	$2\frac{3}{4} = \frac{11}{4}$
$2\frac{1}{8} = \frac{17}{8}$	$1\frac{2}{3} = \frac{5}{3}$	$10\frac{1}{2} = \frac{21}{2}$	$2\frac{1}{3} = \frac{7}{3}$
$3\frac{1}{2} = \frac{7}{2}$	$1\frac{5}{6} = \frac{11}{6}$	$2\frac{1}{4} = \frac{9}{4}$	$1\frac{1}{8} = \frac{9}{8}$
$5\frac{1}{2} = \frac{11}{2}$	$1\frac{3}{8} = \frac{11}{8}$	$5\frac{1}{3} = \frac{16}{3}$	$3\frac{1}{4} = \frac{13}{4}$
$4\frac{1}{2} = \frac{9}{2}$	$1\frac{7}{8} = \frac{15}{8}$	$2\frac{2}{3} = \frac{8}{3}$	$3\frac{3}{10} = \frac{33}{10}$
$1\frac{5}{8} = \frac{13}{8}$	$3\frac{3}{4} = \frac{15}{4}$	$2\frac{3}{8} = \frac{19}{8}$	$7\frac{1}{2} = \frac{15}{2}$

K   Linear Measurement

Write the abbreviation for each of the following units.

Metric Units

1. millimeter ___ **mm** ___
2. centimeter ___ **cm** ___
3. meter ___ **m** ___
4. kilometer ___ **km** ___

U.S. Customary Units

5. inch ___ **in.** ___
6. foot ___ **ft** ___
7. yard ___ **yd** ___
8. mile ___ **mi** ___

Complete each unit conversion.

Metric Conversions

9. 1 centimeter = ___ **10** ___ millimeters
10. 1 meter = ___ **1000** ___ millimeters
11. 1 meter = ___ **100** ___ centimeters
12. 1 kilometer = ___ **1000** ___ meters

U.S. Customary Conversions

13. 1 foot = ___ **12** ___ inches
14. 1 yard = ___ **36** ___ inches
15. 1 yard = ___ **3** ___ feet
16. 1 mile = ___ **5280** ___ feet
17. 1 mile = ___ **1760** ___ yards

Conversions between systems

18. 1 inch = ___ **2.54** ___ centimeters
19. 1 mile ≈ ___ **1609** ___ meters

20. 1 meter ≈ ___ **39** ___ inches
21. 1 kilometer ≈ ___ $\frac{5}{8}$ ___ mile

Write an appropriate unit for each physical reference.

Metric Units

22. The thickness of a dime: ___ **1 mm** ___
23. The width of a little finger: ___ **1 cm** ___
24. The length of one BIG step: ___ **1 m** ___

U.S. Customary Units

25. The width of two fingers: ___ **1 in.** ___
26. The length of a man's shoe: ___ **1 ft** ___
27. The length of one big step: ___ **1 yd** ___

Arrange each set of units in order from shortest to longest.

28. m, cm, mm, km ___ **mm** ___, ___ **cm** ___, ___ **m** ___, ___ **km** ___
29. ft, mi, in., yd ___ **in.** ___, ___ **ft** ___, ___ **yd** ___, ___ **mi** ___

Find each equivalent measure.

30. 10 cm = ___ **100** ___ mm
31. 2 m = ___ **200** ___ cm or ___ **2000** ___ mm
32. 5 km = ___ **5000** ___ m
33. 2.5 cm = ___ **25** ___ mm
34. 1.5 m = ___ **150** ___ cm or ___ **1500** ___ mm
35. 7.5 km = ___ **7500** ___ m

36. $\frac{1}{2}$ ft = ___ **6** ___ in.
37. 2 ft = ___ **24** ___ in.
38. 3 ft = ___ **36** ___ in.
39. 2 yd = ___ **6** ___ ft
40. 10 yd = ___ **30** ___ ft
41. 100 yd = ___ **300** ___ ft

© Saxon Publishers, Inc., and Stephen Hake. Reproduction prohibited.

L Liquid Measurement

Write the abbreviation for each of the following units.

Metric Units

1. liter ____L____
2. milliliter ____mL____

U.S. Customary Units

3. ounce ____oz____
4. pint ____pt____
5. quart ____qt____
6. gallon ____gal____

Complete each unit conversion.

Metric Conversions

7. 1 liter = ____1000____ milliliters

U.S. Customary Conversions

8. 1 cup = ____8____ ounces
9. 1 pint = ____16____ ounces
10. 1 quart = ____2____ pints
11. 1 gallon = ____4____ quarts

Conversions between systems

12. 1 liter ≈ ____1.06____ quarts
13. 1 gallon ≈ ____3.78____ liters

Complete each statement.

14. One fourth of a dollar is a quarter, and one fourth of a gallon is a ____quart____.

15. A two-liter bottle of soda is a little more than 2 ____quarts____ or $\frac{1}{2}$ of a ____gallon____.

16. "A pint's a pound the world around" means a ____pint____ of water weighs about 1 pound.

17. In the United States, gasoline is sold by the ____gallon____. In other countries, gasoline is sold by the ____liter____. A gallon of gasoline is more than ____3____ liters but a little less than ____4____ liters.

18. 2 cups make a ____pint____.

19. 2 pints make a ____quart____.

20. 2 quarts make a ____half gallon____.

21. 2 half gallons make a ____gallon____.

22. A gallon of milk will fill ____16____ cups.

23. If you drink 8 cups of water each day, you drink a ____half gallon____ of water.

Find each equivalent measure.

24. 2 liters = ____2000____ milliliters
25. 2 liters ≈ ____2 (or 2.12)____ quarts
26. 3.78 liters = ____3780____ milliliters
27. 0.5 liter = ____500____ milliliters

28. $\frac{1}{2}$ pint (1 cup) = ____8____ ounces
29. 1 quart (2 pints) = ____32____ ounces
30. $\frac{1}{2}$ gallon = ____2____ quarts
31. 1 gallon = ____8____ pints
32. 2 gallons = ____8____ quarts

M 24 Percent-Fraction-Decimal Equivalents

Write each percent as a reduced fraction and a decimal.

Percent	Fraction	Decimal
40%	$\frac{2}{5}$	0.4
5%	$\frac{1}{20}$	0.05
80%	$\frac{4}{5}$	0.8
2%	$\frac{1}{50}$	0.02
3%	$\frac{3}{100}$	0.03
20%	$\frac{1}{5}$	0.2
25%	$\frac{1}{4}$	0.25
60%	$\frac{3}{5}$	0.6
1%	$\frac{1}{100}$	0.01
90%	$\frac{9}{10}$	0.9
75%	$\frac{3}{4}$	0.75
10%	$\frac{1}{10}$	0.1

Percent	Fraction	Decimal
70%	$\frac{7}{10}$	0.7
4%	$\frac{1}{25}$	0.04
100%	1	1
30%	$\frac{3}{10}$	0.3
50%	$\frac{1}{2}$	0.5
$12\frac{1}{2}$%	$\frac{1}{8}$	0.125
$37\frac{1}{2}$%	$\frac{3}{8}$	0.375
$62\frac{1}{2}$%	$\frac{5}{8}$	0.625
$87\frac{1}{2}$%	$\frac{7}{8}$	0.875
$33\frac{1}{3}$%	$\frac{1}{3}$	Rounds to 0.333
$66\frac{2}{3}$%	$\frac{2}{3}$	Rounds to 0.667
$16\frac{2}{3}$%	$\frac{1}{6}$	Rounds to 0.167

N Measurement Facts

1. Draw a segment 1 cm long. **See student work.**

2. Draw a segment 1 in. long. **See student work.**

3. One inch is how many centimeters? **2.54 cm**

4. Which is longer, 1 km or 1 mi? **1 mi**

5. Which is longer, 1 km or $\frac{1}{2}$ mi? **1 km**

6. How many ounces are in a pound? **16 oz**

7. How many pounds are in a ton? **2000 lb**

8. A dollar bill has a mass of about one **gram**.

9. Your math book has a mass of about one **kilogram**.

10. On earth a kilogram mass weighs about **2.2** pounds.

11. A metric ton is **1000** kilograms.

12. On earth a metric ton weighs about **2200** pounds.

13. The earth rotates on its axis once in a **day**.

14. The earth revolves around the sun once in a **year**.

Abbreviations:

15. milligram **mg**

16. gram **g**

17. kilogram **kg**

18. **Celsius** C

19. ounce **oz**

20. pound **lb**

21. ton **tn**

22. **Fahrenheit** F

Equivalents:

23. 1 gram = **1000** milligrams

24. 1 kilogram = **1000** grams

25. $\frac{1}{2}$ ton = **1000** pounds

26. **365** days = a common year

27. **366** days = a leap year

28. **52** weeks ≈ a year

29. **10** years = a decade

30. **100** years = a century

31. **1000** years = a millennium

How many days are in

32. Jan. **31**

33. Feb. **28** or **29**

34. Mar. **31**

35. Apr. **30**

36. May **31**

37. June **30**

38. July **31**

39. Aug. **31**

40. Sept. **30**

41. Oct. **31**

42. Nov. **30**

43. Dec. **31**

Write the indicated temperatures.

44. Water boils **212** °F

45. **100** °C

46. Normal body temperature **98.6** °F

47. **37** °C

Cool room temperature **68** °F

48. **20** °C

49. Water freezes **32** °F

50. **0** °C

51. A cubic container 1 cm on each edge has a volume of one **cubic centimeter** and can hold one **milliliter** of water, which has a mass of one **gram**.

52. A cubic container 10 cm on each edge has a volume of **1000** cubic centimeters and can hold one **liter** of water which has a mass of one **kilogram**.

Solutions for

Tests

TEST 1

1. $8 + 15 = 23$ $15 + 8 = 23$
    $23 - 15 = 8$ $23 - 8 = 15$

2. $18 \times 4 = 72$ $4 \times 18 = 72$
    $72 \div 4 = 18$ $72 \div 18 = 4$

3.
```
  4525
   545
+ 2608
 7678
```

4.
```
  $5.00
- $3.15
  $1.85
```

5.
```
  $1.95
×     5
  $9.75
```

6. 5×10 dimes = **50 dimes**

7. $275 \div 25 =$ **11 groups**

8. Four dozen (4×12) is 48.
    $48 - 5 =$ **43 pens**

9.
```
   $4.25
   $0.85
+ $15.00
  $20.10
```

10. $32 + 32 + 32 + 32 + 32 + 32$
```
      32
×      6
     192
```

11.
```
  4636
-  364
  4272
```

12.
```
    467
×    39
   4203
   1401
 18,213
```

13.
```
    506
×    57
   3542
   2530
 28,842
```

14.
```
   1603
6)9618
  6
  36
  36
  018
   18
    0
```

15.
```
    164
25)4100
   25
   160
   150
   100
   100
     0
```

16. $30 - (15 - 10) = 30 - 5 = \mathbf{25}$

17. $40 \div (8 \div 2) = 40 \div 4 = \mathbf{10}$

18. $64 + m = 100$
```
  100
-  64
   36
```

19. $1000 - n = 456$
```
   1000
-   456
    544
```

20. $6x = 102$
```
     17
6)102
   6
   42
   42
    0
```
$x = \dfrac{102}{6}$

TEST 2

1.
```
   $0.58
   $1.17
+  $0.93
   $2.68
```

2. $10 \times \$0.55 = \mathbf{\$5.50}$

3.
```
   4396
    866
+  1207
   6469
```

4.
$$\begin{array}{r} 5307 \\ -\ 1629 \\ \hline \mathbf{3678} \end{array}$$

5.
$$\begin{array}{r} 57 \\ \times\ 26 \\ \hline 342 \\ 114 \\ \hline \mathbf{1482} \end{array}$$

6.
$$\begin{array}{r} 16 \\ 15\overline{)240} \\ \underline{15} \\ 90 \\ \underline{90} \\ 0 \end{array}$$

7.
$$\begin{array}{r} \$7.00 \\ \$2.46 \\ +\ \$0.35 \\ \hline \mathbf{\$9.81} \end{array}$$

8. $36 - (15 - 4) = 36 - 11 = \mathbf{25}$

9.
$$\begin{array}{r} 605 \\ 8\overline{)4840} \\ \underline{48} \\ 040 \\ \underline{40} \\ 0 \end{array}$$

10. $76 + n = 150$
$$\begin{array}{r} 150 \\ -\ 76 \\ \hline \mathbf{74} \end{array}$$

11. $9m = 378$
$m = \dfrac{378}{9}$
$$\begin{array}{r} 42 \\ 9\overline{)378} \\ \underline{36} \\ 18 \\ \underline{18} \\ 0 \end{array}$$

12. Four out of seven is $\frac{4}{7}$.

13. $90 \div 2 = \mathbf{45}$

14. $1\frac{1}{2}$ **in.**

15. $6 \times 6 \;\text{\large ⊙}\!\!>\; 40 - 10$ because $36 > 30$

16. Perimeter = 15 mm + 25 mm +
15 mm + 25 mm = **80 mm**

17. 1, 6, 11, 16, **21**
Each number is 5 more than the preceding number.

18. **A. 999**   Note that it ends with an odd digit.

19. $3\overline{)21}$ $21 \div 3$ $\dfrac{21}{3}$

20. $20 + 30 = 50$ $30 + 20 = 50$
$50 - 30 = 20$ $50 - 20 = 30$

TEST 3

1. 2000 dimes ÷ 50 dimes in each roll = **40 rolls**

2.
$$\begin{array}{r} 1453 \\ -\ 1337 \\ \hline \mathbf{116\ years} \end{array}$$

3. **thousands**

4. **twenty-five million**

5. $(9 + 3) \div (9 - 3) = 12 \div 6 = \mathbf{2}$

6.
$$\begin{array}{r} 6010 \\ \times\ \ \ 98 \\ \hline 48080 \\ 54090 \\ \hline \mathbf{588,980} \end{array}$$

7.
$$\begin{array}{r} 354 \\ -\ 187 \\ \hline \mathbf{167\ steps} \end{array}$$

8. **18°F**   Note that the tick marks are spaced 2° apart.

9. Three out of 7 is $\frac{3}{7}$.

10. $-3 \;\text{\large ⊙}\!\!>\; -6$

11.
$$
\begin{array}{r}
4{,}763 \\
32{,}867 \\
+\;\;\;\;984 \\
\hline
\mathbf{38{,}614}
\end{array}
$$

12.
$$
\begin{array}{r}
84 \\
\times\;\;36 \\
\hline
504 \\
252\;\;\; \\
\hline
\mathbf{3024}
\end{array}
$$

13.
$$
\begin{array}{r}
\$1.55 \\
\times\;\;\;\;\;8 \\
\hline
\mathbf{\$12.40}
\end{array}
$$

14.
$$
\begin{array}{r}
\mathbf{605}\;\;\; \\
7\overline{)4235} \\
42\;\;\;\;\; \\
\hline
035 \\
35 \\
\hline
0
\end{array}
$$

15.
$$
\begin{array}{r}
\mathbf{29}\;\; \\
29\overline{)841} \\
58\;\;\; \\
\hline
261 \\
261 \\
\hline
0
\end{array}
$$

16. $20 \div (10 \div 5) = 20 \div 2 = \mathbf{10}$

17. $2000 - m = 635$
$$
\begin{array}{r}
2000 \\
-\;\;635 \\
\hline
\mathbf{1365}
\end{array}
$$

18. $7n = 77$
$$
\begin{array}{r}
\mathbf{11} \\
7\overline{)77}
\end{array}
$$
$$n = \frac{77}{7}$$

19. Perimeter $= 4 \times 15$ cm $= \mathbf{60\ cm}$

20.
$7 \times 18 = 126$	$18 \times 7 = 126$
$126 \div 18 = 7$	$126 \div 7 = 18$

TEST 4

1. $(6 + 2) \div (6 - 2) = 8 \div 4 = \mathbf{2}$

2. **4**

3.
$$
\begin{array}{r}
213 \\
-\;\;97 \\
\hline
\mathbf{116}\ \textbf{pages}
\end{array}
$$

4.
$$
\begin{array}{l}
8° \\
0° \quad 7 + 8 = 15 \\
-7°
\end{array}
$$

From $-7°$ to $8°C$ is **15°C**.

5. $Q - 50 - 35$
$$
\begin{array}{r}
35 \\
+\;\;50 \\
\hline
\mathbf{85}
\end{array}
$$

6. $4n = 112$
$$n = \frac{112}{4}$$
$$
\begin{array}{r}
\mathbf{28} \\
4\overline{)112} \\
8\;\; \\
\hline
32 \\
32 \\
\hline
0
\end{array}
$$

7. The number 38,425 is between 38,000 and 39,000. The number 38,425 is closer to **38,000**.

8. $59 \times 42 \approx 60 \times 40 = \mathbf{2400}$

9.
$$
\begin{array}{r}
\mathbf{47} \\
3\overline{)141} \\
12\;\; \\
\hline
21 \\
21 \\
\hline
0
\end{array}
$$

10.
$$
\begin{array}{r}
7 \quad\quad \mathbf{10} \\
\quad 5\overline{)50} \\
10 \\
10 \\
12 \\
+\;11 \\
\hline
50
\end{array}
$$

11. $\dfrac{16 + 90}{2} = \dfrac{106}{2} = \mathbf{53}$

12. The factors of 20 are **1, 2, 4, 5, 10,** and **20.**

13.
$$
\begin{array}{r}
\$0.37 \\
\times\;\;\;25 \\
\hline
185 \\
74\;\; \\
\hline
\mathbf{\$9.25}
\end{array}
$$

14. Factors of 14: 1, 2, 7, 14
Factors of 35: 1, 5, 7, 35
The GCF of 14 and 35 is **7.**

15. Monday

16. 9 answers

17. $4.50 ÷ 2 = **$2.25**

18. Perimeter = 8 cm + 21 cm +
8 cm + 21 cm = **58 cm**

19. 12, 24, 36, 48, **60**
Each number is 12 more than the preceding number.

20. $2\frac{3}{8}$ **in.**

TEST 5

1. $(15 \times 10) - (15 + 10)$
$= 150 - 25 = $ **125**

2. $\begin{array}{r} 27 \\ -13 \\ \hline 14 \end{array}$ $\dfrac{\text{boys}}{\text{girls}} = \dfrac{13}{14}$

3. 18,306,000

4. $91 \times 18 \approx 90 \times 20 = $ **1800**

5. One solution is to add the digits in each number.
1356: $1 + 3 + 5 + 6 = 15$
3215: $3 + 2 + 1 + 5 = 11$
2641: $2 + 6 + 4 + 1 = 13$
A. 1356

6. $\begin{array}{r} 3 \\ 4\overline{)15} \\ 12 \\ \hline 3 \end{array}$ $\quad \dfrac{15}{4} = 3\dfrac{3}{4}$

7. $\begin{array}{r} 46 \\ 75 \\ 89 \\ +70 \\ \hline 280 \end{array}$ $\quad \begin{array}{r} 70 \\ 4\overline{)280} \end{array}$

8. 16

9. Factors of 16: 1, 2, 4, 8, 16
Factors of 48: 1, 2, 3, 4, 6, 8, 12, 16, 24, 48
The GCF of 16 and 48 is **16.**

10. $\dfrac{7}{15}$

11. $\begin{array}{r} 16 \\ 5\overline{)80} \\ 5 \\ \hline 30 \\ 30 \\ \hline 0 \end{array}$

12. $\begin{array}{r} 4 \\ 4\overline{)16} \end{array}$ $\quad 3 \times 4 = $ **12**

13. $\dfrac{3}{8} + \dfrac{4}{8} = \dfrac{7}{8}$

14. $\dfrac{13}{14} - \dfrac{4}{14} = \dfrac{9}{14}$

15. $1 - \dfrac{7}{8} = \dfrac{8}{8} - \dfrac{7}{8} = \dfrac{1}{8}$

16. $\dfrac{2}{6} + \dfrac{4}{6} = \dfrac{6}{6} = $ **1**

17. $\dfrac{1}{4}$ of a circle is **25%** of a circle.

18. $\dfrac{1}{5} + \dfrac{1}{5} + \dfrac{1}{5} = \dfrac{3}{5}$

19. $33.50 ÷ 10 = **$3.35**

20. $\dfrac{1}{4}$ ⊖ $\dfrac{1}{8}$

TEST 6

1. $\begin{array}{r} 6 \\ 4\overline{)25} \\ 24 \\ \hline 1 \end{array}$ $\quad 6\dfrac{1}{4}$ **inches**

2. $\begin{array}{r} 30 \\ -20 \\ \hline 10 \end{array}$ $\dfrac{\text{cats}}{\text{dogs}} = \dfrac{10}{20} = \dfrac{1}{2}$

3. $78 \times 32 \approx 80 \times 30 = \mathbf{2400}$

4.
$$\begin{array}{r} 64 \\ 8)\overline{512} \\ \underline{48} \\ 32 \\ \underline{32} \\ 0 \end{array}$$

5.
$$\begin{array}{r} 75 \\ 91 \\ +\ 89 \\ \hline 255 \end{array} \qquad \begin{array}{r} 85 \\ 3)\overline{255} \\ \underline{24} \\ 15 \\ \underline{15} \\ 0 \end{array}$$

6. The factors of 24 are **1, 2, 3, 4, 6, 8, 12, 24.**

7. Factors of 27: 1, 3, 9, 27
Factors of 18: 1, 2, 3, 6, 9, 18
The GCF of 27 and 18 is **9.**

8. **C. diameter**

9. $3)\overline{18}$ (with 6 above) $\quad 2 \times 6 \text{ oranges} = \mathbf{12\ oranges}$
(diagram not required)

18 oranges

$\frac{2}{3}$ were eaten	6 oranges
	6 oranges
$\frac{1}{3}$ were not eaten	6 oranges

10. $\left(\dfrac{3}{9} + \dfrac{3}{9}\right) - \dfrac{4}{9} = \dfrac{6}{9} - \dfrac{4}{9} = \mathbf{\dfrac{2}{9}}$

11. $1 - \dfrac{2}{5} = \dfrac{5}{5} - \dfrac{2}{5} = \mathbf{\dfrac{3}{5}}$

12. $\dfrac{1}{3} \times \dfrac{5}{6} = \mathbf{\dfrac{5}{18}}$

13. $6)\overline{14}$ (with $2\frac{2}{6}$ above) $\quad 2\dfrac{2}{6} = \mathbf{2\dfrac{1}{3}}$
$$\begin{array}{r} \underline{12} \\ 2 \end{array}$$

14. $2\dfrac{4}{5} + 4\dfrac{4}{5} = 6\dfrac{8}{5} = \mathbf{7\dfrac{3}{5}}$

15. $\dfrac{4}{12} + \dfrac{4}{12} = \dfrac{8}{12} = \mathbf{\dfrac{2}{3}}$

16. Multiples of 8: 8, 16, 24, 32, 40, …
Multiples of 6: 6, 12, 18, 24, 30, …
The LCM of 8 and 6 is **24.**

17. Side length = 44 in. ÷ 4 = **11 in.**

18. $\dfrac{4}{5} \times \dfrac{5}{4} = \mathbf{1}$

19. 6, 12, 18, 24, **30, 36, 42**
Each number is 6 more than the preceding number. Note that these numbers are multiples of six.

20. $\angle Z$ (or $\angle XZY$ or $\angle YZX$)

TEST 7

1. **1800**

2. $\mathbf{\dfrac{7}{4}}$

3. Perimeter = 4×12 in. = **48 in.**

4. $8 \times 13 = \mathbf{104\ floor\ tiles}$

5. **5**

6. **0.13**

7. $(6 \times 100) + (9 \times 10)$
$= 600 + 90 = \mathbf{690}$

8. Radius $= \dfrac{\text{diameter}}{2} = \dfrac{26 \text{ in.}}{2} = \mathbf{13\ in.}$

9. $0.53 = \mathbf{\dfrac{53}{100}}$

10. $\dfrac{4}{5} \ <\ \dfrac{5}{4}$
Note that $\frac{4}{5}$ is less than 1 and $\frac{5}{4}$ is greater than 1.

11. $\dfrac{37}{100} = \mathbf{0.37}$

12. $\dfrac{4}{7} + \dfrac{5}{7} = \dfrac{9}{7} = \mathbf{1\dfrac{2}{7}}$

13. $\dfrac{3}{4} \times \dfrac{4}{5} = \dfrac{12}{20} = \mathbf{\dfrac{3}{5}}$

14. $20 \times 22 \div 11 = 440 \div 11 = \mathbf{40}$

15.
$$\begin{array}{r} 14 \\ 4\overline{)56} \\ 4 \\ \hline 16 \\ 16 \\ \hline 0 \end{array} \qquad 3 \times 14 = \mathbf{42}$$

(diagram not required)

$$\begin{array}{c|c} & 56 \\ \hline \left.\begin{array}{c} \\ \dfrac{3}{4} \\ \\ \end{array}\right\{ & \begin{array}{|c|} \hline 14 \\ \hline 14 \\ \hline 14 \\ \hline \end{array} \\ \dfrac{1}{4}\left\{ \vphantom{x}\right. & \begin{array}{|c|}\hline 14 \\ \hline \end{array} \end{array}$$

16. $\dfrac{4}{9} - \dfrac{1}{9} = \dfrac{3}{9} = \mathbf{\dfrac{1}{3}}$

17. $w - 20 = 80$
$$\begin{array}{r} 80 \\ + 20 \\ \hline \mathbf{100} \end{array}$$

18. $4 + N = 5\dfrac{5}{6}$

$N = 5\dfrac{5}{6} - 4 = \mathbf{1\dfrac{5}{6}}$

19. $40\% = \dfrac{40}{100} = \mathbf{\dfrac{2}{5}}$

20.
$$\begin{array}{r} 8 \\ 5\overline{)40} \\ 40 \\ \hline 0 \end{array} \qquad \begin{array}{r} 8 \\ \times 3 \\ \hline \mathbf{24 \text{ answers}} \end{array}$$

TEST 8

1. $\angle BDC$ or $\angle CDB$

2. **3**

3. $\dfrac{12}{20} = \mathbf{\dfrac{3}{5}}$

4. $7106 + 3860 \approx 7000 + 4000 = \mathbf{11{,}000}$

5. $4^2 + \sqrt{25} = 16 + 5 = \mathbf{21}$

6.
$$\begin{array}{r} 80 \\ 86 \\ + 80 \\ \hline 246 \end{array} \qquad \begin{array}{r} 82 \\ 3\overline{)246} \\ 24 \\ \hline 06 \\ 6 \\ \hline 0 \end{array}$$

7. Factors of 20: 1, 2, 4, 5, 10, 20
Factors of 30: 1, 2, 3, 5, 6, 10, 15, 30
The GCF of 20 and 30 is **10**.

8. $\dfrac{80}{100} = \dfrac{8}{10} = \mathbf{\dfrac{4}{5}}$

9. $2\dfrac{1}{4} + 1\dfrac{3}{4} = 3\dfrac{4}{4} = \mathbf{4}$

10. $1 - \dfrac{7}{8} = \dfrac{8}{8} - \dfrac{7}{8} = \mathbf{\dfrac{1}{8}}$

11. $\dfrac{1}{3} \times \dfrac{2}{3} = \mathbf{\dfrac{2}{9}}$

12.
$$\begin{array}{r} 7:6\,5 \\ \cancel{8}:\cancel{0}5 \\ - 5:32 \\ \hline 2:33 \end{array}$$

2 hours 33 minutes

13. Multiples of 4: 4, 8, 12, 16, 20, 24, …
Multiples of 10: 10, 20, 30, 40, 50, 60, …
The LCM of 4 and 10 is **20**.

14. 9 in. $\times$ 9 in. = **81 square-inch tiles**

15. **8**

16. **15.006**

17.
$$\begin{array}{r} 0.6 \\ + 0.8 \\ \hline \mathbf{1.4} \end{array}$$

18.
$$\begin{array}{r} 0.3 \\ 0.48 \\ +\ 6.0 \\ \hline \mathbf{6.78} \end{array}$$

19.
$$\begin{array}{r} 0.18 \\ -\ 0.07 \\ \hline \mathbf{0.11} \end{array}$$

20.
$$\begin{array}{r} 0.17 \\ \times\ 0.5 \\ \hline \mathbf{0.085} \end{array}$$

TEST 9

1. $(8 \times 4) + (5 - 4) = 32 + 1 = \mathbf{33}$

2. $\dfrac{5}{8} \times \dfrac{8}{5} = 1; \ \dfrac{8}{5} = \mathbf{1\dfrac{3}{5}}$

3. Side length $= 48\,\text{cm} \div 4 = \mathbf{12\ cm}$

4. Area $= 20\,\text{in.} \times 15\,\text{in.} = \mathbf{300\ sq.\ in.}$

5. **15.8**

6. $\dfrac{25}{100} = \dfrac{1}{4} \qquad \dfrac{1}{4} \times \dfrac{\$40}{1} = \mathbf{\$10}$

7. Five dozen (5×12) is 60.

$$\begin{array}{r} 12 \\ 5\overline{)60} \qquad 4 \times 12 = \mathbf{48} \\ \underline{5} \\ 10 \\ \underline{10} \\ 0 \end{array}$$

(diagram not required)

```
        60
     ┌ ┌────┐
     │ │ 12 │
  4  ┤ ├────┤
  5  │ │ 12 │
     │ ├────┤
     │ │ 12 │
     └ ├────┤
     ┌ │ 12 │
     │ ├────┤
  1  ┤ │ 12 │
  5  └ └────┘
```

8. D. **90%**

9. **4**

10. $11\% = \dfrac{11}{100} = \mathbf{0.11}$

11. Factors of 15: 1, 3, 5, 15
Factors of 10: 1, 2, 5, 10
The GCF of 15 and 10 is **5**.

12. $5\dfrac{2}{5} + \sqrt{16} = 5\dfrac{2}{5} + 4 = \mathbf{9\dfrac{2}{5}}$

13. $\dfrac{1}{3} \times \dfrac{3}{5} = \dfrac{3}{15} = \dfrac{\mathbf{1}}{\mathbf{5}}$

14.
$$\begin{array}{r} 0.27 \\ 7.3 \\ +\ 9.0 \\ \hline \mathbf{16.57} \end{array}$$

15. $4.32 + n = 8$
$$\begin{array}{r} 8.00 \\ -\ 4.32 \\ \hline \mathbf{3.68} \end{array}$$

16.
$$\begin{array}{r} 0.39 \\ \times\ 0.48 \\ \hline 312 \\ 156 \\ \hline \mathbf{0.1872} \end{array}$$

17. $5\dfrac{6}{7} - 3\dfrac{2}{7} = \mathbf{2\dfrac{4}{7}}$

18. $\dfrac{3}{4} \times \dfrac{6}{6} = \dfrac{\mathbf{18}}{24}$

19.
$$\begin{array}{r} \mathbf{0.225} \\ 4\overline{)0.900} \\ \underline{8} \\ 10 \\ \underline{8} \\ 20 \\ \underline{20} \\ 0 \end{array}$$

20. B. $\overline{SR}$

TEST 10

1. $(0.2 + 0.6) - (0.2 \times 0.6)$
$= 0.8 - 0.12 = \mathbf{0.68}$

2. **2**

3. $8\% = \dfrac{8}{100} = 0.08$

$$\begin{array}{r} \$17 \\ \times\ 0.08 \\ \hline \mathbf{\$1.36} \end{array}$$

4. $4239 + 3763 \approx 4000 + 4000 = \mathbf{8000}$

5.
$$\begin{array}{r} \mathbf{1{,}003} \\ 50\overline{)50{,}150} \\ \underline{50} \\ 0\ 150 \\ \underline{150} \\ 0 \end{array}$$

6.
$$\begin{array}{r} 89 \\ 91 \\ +\ 84 \\ \hline 264 \end{array} \qquad \begin{array}{r} 88 \\ 3\overline{)264} \\ \underline{24} \\ 24 \\ \underline{24} \\ 0 \end{array}$$

7. $\dfrac{17}{35} + \dfrac{4}{35} = \dfrac{21}{35} = \dfrac{\mathbf{3}}{\mathbf{5}}$

8. Three dozen (3×12) is 36.

$\begin{array}{r} 6 \\ 6\overline{)36} \end{array}$ $5 \times 6 = \mathbf{30}$

(diagram not required)

$$\begin{array}{c} 36 \\ \dfrac{5}{6}\left\{\begin{array}{|c|} \hline 6 \\ \hline 6 \\ \hline 6 \\ \hline 6 \\ \hline 6 \\ \hline \end{array}\right. \\ \dfrac{1}{6}\left\{\begin{array}{|c|} \hline 6 \\ \hline \end{array}\right. \end{array}$$

9. $\dfrac{1}{4} = \dfrac{2}{8}$ $\dfrac{2}{8} + \dfrac{1}{8} = \dfrac{\mathbf{3}}{\mathbf{8}}$

10. Multiples of 3: 3, 6, 9, 12, 15, 18, ...
Multiples of 4: 4, 8, 12, 16, 20, ...
Multiples of 6: 6, 12, 18, 24, 30, ...
The LCM of 3, 4, and 6 is **12**.

11. Area $= 80\,\text{cm} \times 35\,\text{cm} = \mathbf{2800\ sq.\ cm}$

12. Circumference $= 3.14 \times 10\,\text{cm} = \mathbf{31.4\ cm}$

13. **23.07**

14. $1 - m = 0.53$

$$\begin{array}{r} 1.00 \\ -\ 0.53 \\ \hline \mathbf{0.47} \end{array}$$

15.
$$\begin{array}{r} 5.0 \\ 4.96 \\ +\ 0.9 \\ \hline \mathbf{10.86} \end{array}$$

16. $0.3 \times 0.4 \times 0.7 = \mathbf{0.084}$

17.
$$\begin{array}{r} \mathbf{0.039} \\ 4\overline{)0.156} \\ \underline{12} \\ 36 \\ \underline{36} \\ 0 \end{array}$$

18.
$$\begin{array}{r} \mathbf{1.8} \\ 0.4\overline{)0.7\,2} \\ \underline{4} \\ 3\,2 \\ \underline{3\,2} \\ 0 \end{array}$$

19.
$$\begin{array}{r} 8\dfrac{1}{5} = 7\dfrac{6}{5} \\ -\ 2\dfrac{4}{5} = 2\dfrac{4}{5} \\ \hline 5\dfrac{\mathbf{2}}{\mathbf{5}} \end{array}$$

20. $13.4 \times 100 = \mathbf{1340}$

TEST 11

1.
$$\begin{array}{r} 3.14 \\ \times\ \ \ 4 \\ \hline \mathbf{12.56\ cm} \end{array}$$

2.
$$\begin{array}{r} 1.3 \\ \times\ 0.028 \\ \hline 104 \\ 26 \\ \hline \mathbf{0.0364} \end{array}$$

3. Perimeter $= 22\,\text{cm} + 33\,\text{cm} +$
$22\,\text{cm} + 33\,\text{cm} = \mathbf{110\ cm}$

4. Area $= 22\,\text{cm} \times 33\,\text{cm} = \mathbf{726\ sq.\ cm}$

5. $\dfrac{\text{length}}{\text{width}} = \dfrac{33}{22} = \dfrac{3}{2}$

6. $70\% = \dfrac{70}{100} = \dfrac{7}{10}$

$\dfrac{7}{10} \times \dfrac{50}{1} = \dfrac{350}{10} = $ **35 questions**

7. $8 - n = 3.2$

$$\begin{array}{r} 8.0 \\ -\ 3.2 \\ \hline \mathbf{4.8} \end{array}$$

8. $0.80 \ \textcircled{=}\ 0.8$

9. **90.02**

10. $7.83\,|56 \longrightarrow $ **7.84**

11.
$$\begin{array}{r} 2\dfrac{1}{4} \\ +\ 3\dfrac{3}{4} \\ \hline 5\dfrac{4}{4} = \mathbf{6} \end{array}$$

12.
$$\begin{array}{r} 7\dfrac{1}{3} = 6\dfrac{4}{3} \\ -\ 3\dfrac{2}{3} = 3\dfrac{2}{3} \\ \hline 3\dfrac{\mathbf{2}}{\mathbf{3}} \end{array}$$

13. $\dfrac{5}{9} \times \dfrac{3}{15} = \dfrac{15}{135} = \dfrac{\mathbf{1}}{\mathbf{9}}$

14. $x = 1\dfrac{1}{5} - \dfrac{3}{5} = \dfrac{6}{5} - \dfrac{3}{5} = \dfrac{\mathbf{3}}{\mathbf{5}}$

15. $2.65 \div 10^2 = 2.65 \div 100 = $ **0.0265**

16.
$$\begin{array}{r} 0.0475 \\ 0.8\overline{)0.0\,3800} \\ \underline{32} \\ 60 \\ \underline{56} \\ 40 \\ \underline{40} \\ 0 \end{array}$$

17. $\sqrt{25} \div 0.5 = 5 \div 0.5 \qquad 0.5\overline{)5.0}^{\,10}$

18.
$$\begin{array}{r} \dfrac{2}{3} = \dfrac{4}{6} \\ +\ \dfrac{1}{6} = \dfrac{1}{6} \\ \hline \dfrac{5}{6} \end{array}$$

19. $\dfrac{\cancel{2}\cdot\cancel{3}\cdot 5\cdot\cancel{7}}{\cancel{2}\cdot 2\cdot\cancel{3}\cdot 3\cdot\cancel{7}} = \dfrac{5}{2\cdot 3} = \dfrac{\mathbf{5}}{\mathbf{6}}$

20. **2.3**

TEST 12

1. Out of six possible outcomes, 4 are greater than 2.

$\dfrac{4}{6} = \dfrac{\mathbf{2}}{\mathbf{3}}$

2. Perimeter $= 6 \times 9$ in. $=$ **54 in.**

3. Area $= 18\text{ cm} \times 3\text{ cm} = $ **54 sq. cm**

4. **3**

5.
$$\begin{array}{r} 6.0 \\ 2.3 \\ +\ 0.54 \\ \hline \mathbf{8.84} \end{array}$$

6.
$$\begin{array}{r} 1.00 \\ -\ 0.48 \\ \hline \mathbf{0.52} \end{array}$$

7.
$$\begin{array}{r} 0.36 \\ \times\ 0.4 \\ \hline \mathbf{0.144} \end{array}$$

8.
$$\begin{array}{r} 3.14 \\ \times\ 40 \\ \hline 125.60 \approx \mathbf{126\ inches} \end{array}$$

9. **B. 6**

10.
$$\begin{array}{r} 28 \\ -\ 12 \\ \hline 16 \end{array} \qquad \dfrac{\text{win}}{\text{loss}} = \dfrac{12}{16} = \dfrac{\mathbf{3}}{\mathbf{4}}$$

11.

$$7 = 6\frac{4}{4}$$

$$-\ 3\frac{1}{4} = 3\frac{1}{4}$$

$$3\frac{3}{4}$$

12. $9\overline{)0.72}$ **0.08**

13. $0.02\overline{)1.40}$ **7 0**

14. $1.48 \times 100 = \mathbf{148}$

15.

$$\frac{2}{3} = \frac{8}{12}$$

$$+\ \frac{3}{4} = \frac{9}{12}$$

$$\frac{17}{12} = \mathbf{1\frac{5}{12}}$$

16.

$$\frac{5}{6} = \frac{10}{12}$$

$$-\ \frac{1}{4} = \frac{3}{12}$$

$$\frac{7}{12}$$

17. $\frac{7}{8} \enclose{circle}{>} \frac{3}{4}$ because $\frac{7}{8} > \frac{6}{8}$

18.

$$4\frac{7}{8} = 4\frac{7}{8}$$

$$+\ 2\frac{1}{4} = 2\frac{2}{8}$$

$$6\frac{9}{8} = \mathbf{7\frac{1}{8}}$$

19.

$$6\frac{1}{3} = 5\frac{4}{3}$$

$$-\ 2\frac{2}{3} = 2\frac{2}{3}$$

$$3\frac{2}{3}$$

20.

$$\begin{array}{r} \$8.59 \\ \times\ \ 0.08 \\ \hline 0.6872 \end{array}$$

The sales tax was **$0.69.**

1. Side length $= 12$ ft $\div\ 4 = 3$ ft
Area $= 3$ ft $\times\ 3$ ft $= \mathbf{9\ sq.\ ft}$

2. $\frac{3}{4} \times \frac{4}{4} = \frac{12}{16}$

3. $\frac{2}{6} = \mathbf{\frac{1}{3}}$

4. $\dfrac{\text{radius}}{\text{diameter}} = \dfrac{19}{38} = \mathbf{\dfrac{1}{2}}$

5. **B.**

6.

$$50 = \mathbf{2 \cdot 5 \cdot 5}$$

$$\begin{array}{r} 1 \\ 5\overline{)5} \\ 5\overline{)25} \\ 2\overline{)50} \end{array}$$

7. $\frac{4}{5} \enclose{circle}{>} \frac{3}{4}$ because $\frac{16}{20} > \frac{15}{20}$

8.

$$\begin{array}{r} \$13.78 \\ \times\ \ \ 0.08 \\ \hline \$1.1024 \end{array} \qquad \begin{array}{r} \$13.78 \\ +\ \ \$1.10 \\ \hline \mathbf{\$14.88} \end{array}$$

9.

$$\begin{array}{r} \$7.00 \\ -\ \$0.07 \\ \hline \mathbf{\$6.93} \end{array}$$

10.

$$\begin{array}{r} 4.8 \\ 14.0 \\ +\ \ 3.76 \\ \hline \mathbf{22.56} \end{array}$$

11.

$$\begin{array}{r} 8.60 \\ -\ 5.36 \\ \hline \mathbf{3.24} \end{array}$$

12.

$$\begin{array}{r} 0.32 \\ \times\ 0.21 \\ \hline 32 \\ 64\ \ \\ \hline \mathbf{0.0672} \end{array}$$

13.

$$8\overline{)0.280}$$ → **0.035**

24
40
40
0

14.

$$0.3\overline{)9.0}$$ → **3 0**

15.

$$5\frac{3}{4} = 5\frac{3}{4}$$
$$-1\frac{1}{2} = 1\frac{2}{4}$$
$$4\frac{1}{4}$$

16.

$$1\frac{3}{4} = 1\frac{6}{8}$$
$$+2\frac{7}{8} = 2\frac{7}{8}$$
$$3\frac{13}{8} = 4\frac{5}{8}$$

17.

$$\frac{1}{2} = \frac{5}{10}$$
$$\frac{3}{5} = \frac{6}{10}$$
$$+\frac{3}{10} = \frac{3}{10}$$
$$\frac{14}{10} = 1\frac{4}{10} = 1\frac{2}{5}$$

18. $m = 4\frac{2}{3} - 1 = 3\frac{2}{3}$

19. 6 faces

20. $3\frac{1}{3} = \frac{10}{3}$ $\quad \frac{10}{3} \cdot \frac{1}{5} = \frac{10}{15} = \frac{2}{3}$

4.

$$\begin{array}{r} 0.12 \\ \times\ 0.34 \\ \hline 48 \\ 36 \\ \hline \mathbf{0.0408} \end{array}$$

5. $\frac{11}{25} \times \frac{4}{4} = \frac{\mathbf{44}}{\mathbf{100}}$

6. $0.34 \div \sqrt{16} = 0.34 \div 4$

$$4\overline{)0.340}$$ → **0.085**

32
20
20
0

7. B.

8. $0.5 + 0.5 \;\textcircled{=}\; 0.5 \div 0.5$
because $1.0 = 1.0$

9.

$$\begin{array}{r} \$22.98 \\ \times\ \ 0.07 \\ \hline \$1.6086 \end{array} \qquad \begin{array}{r} \$22.98 \\ +\ \$1.61 \\ \hline \mathbf{\$24.59} \end{array}$$

10. $5.3 \div 100 = \mathbf{0.053}$

11.

$$0.08\overline{)2.80}$$ → **3 5**

24
40
40
0

12.

$$6\frac{3}{5} = 6\frac{6}{10}$$
$$-4\frac{1}{2} = 4\frac{5}{10}$$
$$2\frac{1}{10}$$

13.

$$3\frac{3}{4} = 3\frac{6}{8}$$
$$1\frac{1}{8} = 1\frac{1}{8}$$
$$+4\frac{1}{2} = 4\frac{4}{8}$$
$$8\frac{11}{8} = 9\frac{3}{8}$$

TEST 14

1. 3

2. Side length $= 52\,\text{m} \div 4 = 13\,\text{m}$
Area $= 13\,\text{m} \times 13\,\text{m} = \mathbf{169\ sq.\ m}$

3. $8 - (0.75 + 2) = 8 - 2.75 = \mathbf{5.25}$

14. $2 \times 2 \times 2 = $ **8 small cubes**

15. Perimeter $= 5 \times 14$ cm $=$ **70 cm**

16. $\dfrac{36}{60} = \dfrac{2 \cdot 2 \cdot 3 \cdot 3}{2 \cdot 2 \cdot 3 \cdot 5} = \dfrac{3}{5}$

17. $2\dfrac{2}{3} \times 1\dfrac{1}{4} = \dfrac{8}{3} \times \dfrac{5}{4} = \dfrac{10}{3} = 3\dfrac{1}{3}$

18. $2\dfrac{1}{2} \div 3 = \dfrac{5}{2} \times \dfrac{1}{3} = \dfrac{5}{6}$

19. A. •———————▶

20. $\dfrac{\text{green}}{\text{total}} = \dfrac{4}{12} = \dfrac{1}{3}$

TEST 15

1. Area $= 4$ cm $\cdot 7$ cm $=$ **28 sq. cm**

2.

$\$5.95$	$\$5.95$	$\$10.00$
$\times\ \ 0.07$	$+\ \$0.42$	$-\ \$6.37$
$\$0.4165$	$\$6.37$	$\mathbf{\$3.63}$

3. $\dfrac{2}{3} \;\boxed{<}\; \dfrac{9}{12}$ because $\dfrac{8}{12} < \dfrac{9}{12}$

4. **Point F**

5. $(2, -3)$

6. **A. hexagon**

7. $\dfrac{56}{72} = \dfrac{2 \cdot 2 \cdot 2 \cdot 7}{2 \cdot 2 \cdot 2 \cdot 3 \cdot 3} = \dfrac{7}{9}$

8. $\dfrac{3}{4}$

$$
\begin{array}{r}
0.75 \\
4\overline{)3.00} \\
2\,8 \\
\hline
20 \\
20 \\
\hline
0
\end{array}
$$

9. $35\% = \dfrac{35}{100} = 0.35 = \dfrac{7}{20}$

10. Multiples of 4: 4, 8, 12, 16, 20, 24, …
Multiples of 12: 12, 24, 36, 48, 60, …
Multiples of 24: 24, 48, 72, 96, 120, …
The LCM of 4, 12, and 24 is **24.**

11. $8 - (4.8 + 0.75) = 8 - 5.55 =$ **2.45**

12.

$$
\begin{array}{r}
3.7 \\
\times\ 0.8 \\
\hline
2.96
\end{array}
$$

13.

$$
\begin{array}{r}
1\,5 \\
0.8\overline{)12.0} \\
8 \\
\hline
4\,0 \\
4\,0 \\
\hline
0
\end{array}
$$

14. $38.5 \div 100 =$ **0.385**

15.

$$
\begin{array}{r}
\dfrac{1}{4} = \dfrac{3}{12} \\[4pt]
\dfrac{2}{3} = \dfrac{8}{12} \\[4pt]
+\ \dfrac{1}{2} = \dfrac{6}{12} \\[4pt]
\hline
\dfrac{17}{12} = 1\dfrac{5}{12}
\end{array}
$$

16.

$$
\begin{array}{r}
5\dfrac{4}{5} = 5\dfrac{12}{15} \\[4pt]
+\ 2\dfrac{1}{3} = 2\dfrac{5}{15} \\[4pt]
\hline
7\dfrac{17}{15} = 8\dfrac{2}{15}
\end{array}
$$

17.

$$
\begin{array}{r}
10\dfrac{3}{8} = 10\dfrac{3}{8} = 9\dfrac{11}{8} \\[4pt]
-\ 2\dfrac{3}{4} = 2\dfrac{6}{8} = 2\dfrac{6}{8} \\[4pt]
\hline
7\dfrac{5}{8}
\end{array}
$$

18. $1\dfrac{1}{3} \times 2\dfrac{1}{4} = \dfrac{4}{3} \times \dfrac{9}{4} = \dfrac{36}{12} = 3$

19. $6 \div 1\dfrac{1}{5} = \dfrac{6}{1} \times \dfrac{5}{6} = \dfrac{30}{6} = 5$

20. $10.8 - n = 4.3$

$$
\begin{array}{r}
10.8 \\
-\ 4.3 \\
\hline
6.5
\end{array}
$$

TEST 16

1. 2 cups

2. $\dfrac{44}{50} = \dfrac{88}{100} =$ **88%**

3. Area $= \dfrac{6 \text{ cm} \cdot 8 \text{ cm}}{2} =$ **24 sq. cm**

4. Perimeter $= 5 + 3 + 5 + 3 =$ **16 units**

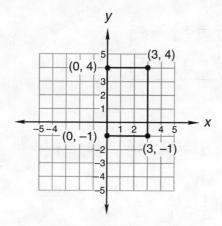

5. Area $= 5 \times 3 =$ **15 sq. units**

6. $\dfrac{3}{5}$ $\bigcirc$ $< \; 0.8$ because $0.6 < 0.8$

7. $\dfrac{2}{3} \times \dfrac{12}{1} = 8$

$\begin{array}{r} 12 \\ -\ 8 \\ \hline \textbf{4 doughnuts} \end{array}$

8. True

9. $40\% = \dfrac{40}{100} = \dfrac{2}{5}$ $\dfrac{2}{5} \times \dfrac{30}{1} =$ **12 boys**

10.
$$\begin{array}{r} 2\,5 \\ 0.4\overline{)10.0} \\ 8 \\ \hline 2\,0 \\ 2\,0 \\ \hline 0 \end{array}$$

11.
$$\begin{array}{r} 8.0 \\ 7.5 \\ +\ 3.21 \\ \hline \textbf{18.71} \end{array}$$

12. $4.4 \div \sqrt{64} = 4.4 \div 8$
$$\begin{array}{r} 0.55 \\ 8\overline{)4.40} \\ 4\,0 \\ \hline 40 \\ 40 \\ \hline 0 \end{array}$$

13. $2.57 \div 10^2 = 2.57 \div 100 =$ **0.0257**

14.
$$\begin{array}{r} \dfrac{5}{6} = \dfrac{5}{6} \\ -\ \dfrac{1}{2} = \dfrac{3}{6} \\ \hline \dfrac{2}{6} = \dfrac{1}{3} \end{array}$$

15.
$$\begin{array}{r} 3\dfrac{3}{5} = 3\dfrac{6}{10} \\ +\ 5\dfrac{7}{10} = 5\dfrac{7}{10} \\ \hline 8\dfrac{13}{10} = 9\dfrac{3}{10} \end{array}$$

16.
$$\begin{array}{r} 6\dfrac{5}{8} = 5\dfrac{13}{8} \\ -\ 3\dfrac{7}{8} = 3\dfrac{7}{8} \\ \hline 2\dfrac{6}{8} = 2\dfrac{3}{4} \end{array}$$

17. $4\dfrac{1}{2} \times 2\dfrac{2}{3} = \dfrac{9}{2} \times \dfrac{8}{3} =$ **12**

18. $\dfrac{9}{20} \times \dfrac{5}{5} = \dfrac{45}{100}$

19. $3 \div 2\dfrac{2}{5} = \dfrac{3}{1} \times \dfrac{5}{12} = \dfrac{15}{12} = 1\dfrac{3}{12} = 1\dfrac{1}{4}$

20. $156.7\,|810 \longrightarrow$ **156.8**

TEST 17

1. $75\% = \dfrac{75}{100} = \dfrac{3}{4} =$ **0.75**

2. $\dfrac{16 \text{ cm} \cdot 6 \text{ cm}}{2} = \dfrac{96 \text{ sq. cm}}{2} =$ **48 sq. cm**

3. $\dfrac{4}{5} = \dfrac{80}{100} =$ **80%**

SOLUTIONS

4. 3.14×200 mm = **628 mm**

5. Area = 7 in. $\times$ 6 in. = **42 sq. in.**

6. Perimeter = 7 in. + 9 in. + 7 in. + 9 in.
= **32 in.**

7. Opposite angles in a parallelogram are equal.
m$\angle B$ = m$\angle D$ = **138°**

8. $\dfrac{\text{cats}}{\text{dogs}} = \dfrac{20}{12} = \dfrac{5}{3}$

9.

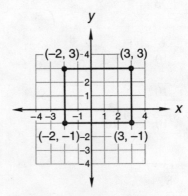

The fourth vertex is at **(3, 3).**

10. $3\overline{)600}$ 2×200 books = **400 books**
(diagram not required)

	600 books
$\frac{2}{3}$ were sold	200 books
	200 books
$\frac{1}{3}$ were not sold	200 books

11. 3 ft = 36 in.   36 in. − 8 in. = **28 in.**

12. $6 \times 3 + 4 \times 5 = 18 + 20 = 38$

13. Volume = (11 in.)(8 in.)(5 in.)
= **440 cu. in.**

14. $\dfrac{6}{n} = \dfrac{10}{15}$
$10n = 6 \cdot 15$
$n = \dfrac{90}{10} = 9$

15. $8 - (0.2 + 0.5) = 8 - 0.7 = 7.3$

16. $4x = 0.48$ $4\overline{)0.48}$ gives **0.12**
$x = \dfrac{0.48}{4}$

17. $6\frac{7}{8} = 6\frac{7}{8}$
$+ 2\frac{1}{4} = 2\frac{2}{8}$
$8\frac{9}{8} = 9\frac{1}{8}$

18. $5\frac{1}{4} = 5\frac{1}{4} = 4\frac{5}{4}$
$- 3\frac{1}{2} = 3\frac{2}{4} = 3\frac{2}{4}$
$1\frac{3}{4}$

19. $3\frac{3}{4} \times 2 = \dfrac{15}{4} \times \dfrac{2}{1} = \dfrac{15}{2} = 7\frac{1}{2}$

20. $4 \div 1\frac{1}{3} = \dfrac{4}{1} \times \dfrac{3}{4} = 3$

TEST 18

1. $\dfrac{1}{4} \times \dfrac{300}{1} = 75$ did not sprout.
$300 - 75 = $ **225 seeds sprouted.**

2. $\dfrac{\text{boys}}{\text{girls}} = \dfrac{4}{3} = \dfrac{b}{15}$
$3b = 60$
$b = \dfrac{60}{3} = $ **20 boys**

3. Area = $\dfrac{12 \text{ in.} \times 9 \text{ in.}}{2} = $ **54 sq. in.**

4. Perimeter = 9 in. + 12 in. + 15 in.
= **36 in.**

5. $7.00 - 4.69 = 2.31$

6. $3.2 \times 0.04 = 0.128$

328   *Saxon Math 7/6—Homeschool*

7. $0.26 \div \sqrt{64} = 0.26 \div 8$

$$\begin{array}{r} 0.0325 \\ 8\overline{)0.2600} \\ 24 \\ \hline 20 \\ 16 \\ \hline 40 \\ 40 \\ \hline 0 \end{array}$$

8. $0.7\overline{)28.0}$ → **4 0**

9. $6^2 = 36$ and $7^2 = 49$
D. 6 and 7

10.
$$\frac{7}{8} = \frac{7}{8}$$
$$+ \frac{1}{2} = \frac{4}{8}$$
$$\frac{11}{8} = 1\frac{3}{8}$$

11.
$$4\frac{1}{4} = 3\frac{5}{4}$$
$$- 1\frac{1}{2} = 1\frac{2}{4}$$
$$2\frac{3}{4}$$

12. $3\frac{3}{5} \times 2\frac{2}{9} = \frac{18}{5} \times \frac{20}{9} = \mathbf{8}$

13. $9 \div 1\frac{1}{2} = \frac{9}{1} \times \frac{2}{3} = \mathbf{6}$

14. $4y = 10$
$y = \frac{10}{4} = 2\frac{2}{4} = \mathbf{2\frac{1}{2}}$

15. Circumference $= 3.14 \times 20\,\text{mm} = \mathbf{62.8\ mm}$

16. Area $= 3.14 \times 10\,\text{mm} \times 10\,\text{mm}$
$= \mathbf{314\ sq.\ mm}$

17. $85\% = \frac{85}{100} = \frac{\mathbf{17}}{\mathbf{20}} = \mathbf{0.85}$

18. Volume $= (10\,\text{cm})(8\,\text{cm})(5\,\text{cm})$
$= \mathbf{400\ cu.\ cm}$

19. **8 vertices**

20. There are three possible answers:
$(-1, -2), (7, -2),$ and $(1, 8).$
One of the possible parallelograms is shown below.

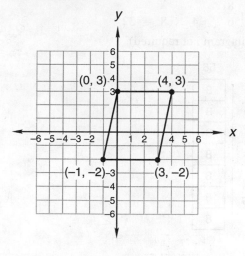

TEST 19

1. $\dfrac{\text{favorable}}{\text{possible}} = \dfrac{3}{6} = \dfrac{\mathbf{1}}{\mathbf{2}}$

2. 80, 80, 85, **85**, 90, 95, 100

3. $\frac{4}{8} = \frac{1}{2} = \frac{50}{100} = \mathbf{50\%}$

4. **6 edges**

5. Area $= \dfrac{12\,\text{cm} \times 5\,\text{cm}}{2} = \dfrac{60\ \text{sq. cm}}{2}$
$= \mathbf{30\ sq.\ cm}$

6. **B. right triangle**

7. $20 - 5 \times 3 + 3^2 = 20 - 15 + 9 = \mathbf{14}$

8. $\left(1\frac{1}{2}\right)^2 = \frac{3}{2} \cdot \frac{3}{2} = \frac{9}{4} = \mathbf{2\frac{1}{4}}$

9. $\frac{5}{9} = \frac{10}{n}$
$5n = 90$
$n = \frac{90}{5} = \mathbf{18}$

10. $6^2 - \sqrt{36} = 36 - 6 = \mathbf{30}$

11. $\dfrac{(18)(16)}{24} = \dfrac{288}{24} = \mathbf{12}$

12. $7\overline{)56}^{\,8}$ $4 \times 8 = \mathbf{32}$

(diagram not required)

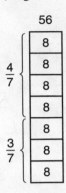

13. $5 - (0.42 + 1.4) = 5 - 1.82 = \mathbf{3.18}$

14.
$$0.5\overline{)6.0}^{\,12}$$
$$\begin{array}{r} 5 \\ \hline 1\,0 \\ 1\,0 \\ \hline 0 \end{array}$$

15.
$$\begin{aligned} 5\dfrac{3}{5} &= 5\dfrac{6}{10} \\ 4 \phantom{\dfrac{3}{5}} &= 4 \\ +\, 2\dfrac{1}{2} &= 2\dfrac{5}{10} \\ \hline 11\dfrac{11}{10} &= \mathbf{12\dfrac{1}{10}} \end{aligned}$$

16.
$$\begin{aligned} 2\dfrac{1}{8} &= 2\dfrac{1}{8} = 1\dfrac{9}{8} \\ -\, 1\dfrac{1}{2} &= 1\dfrac{4}{8} = 1\dfrac{4}{8} \\ \hline & \phantom{1\dfrac{4}{8}} \mathbf{\dfrac{5}{8}} \end{aligned}$$

17. $1\dfrac{3}{5} \div 4 = \dfrac{8}{5} \times \dfrac{1}{4} = \mathbf{\dfrac{2}{5}}$

18. $\dfrac{1}{3} \times 100\% = \dfrac{100\%}{3} = \mathbf{33\dfrac{1}{3}\%}$

19. $3\dfrac{3}{4} = \mathbf{3.75}$

$3.75 + 3.5 = \mathbf{7.25}$

20. Perimeter $= 6 + 5 + 6 + 5 = \mathbf{22\ units}$

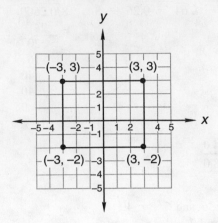

TEST 20

1.
$$\dfrac{4}{5} = \dfrac{16}{Q}$$
$$4Q = 80$$
$$Q = \dfrac{80}{4} = \mathbf{20\ queens}$$

2. $\dfrac{1}{6} \times \dfrac{1}{6} = \mathbf{\dfrac{1}{36}}$

3. (a) $0.8 = \dfrac{8}{10} = \mathbf{\dfrac{4}{5}}$

(b) $0.8 = \dfrac{8}{10} = \dfrac{80}{100} = \mathbf{80\%}$

4. (a) $8\% = \dfrac{8}{100} = \mathbf{\dfrac{2}{25}}$

(b) $8\% = \mathbf{0.08}$

5. $\dfrac{4}{5} \times \dfrac{35}{1} = 28$ kittens finished.

$35 - 28 = \mathbf{7\ kittens\ did\ not\ finish.}$

6. Area $= 10\,\text{mm} \times 15\,\text{mm} = \mathbf{150\ sq.\ mm}$

7. Perimeter $= 12\,\text{mm} + 15\,\text{mm} +$
$ 12\,\text{mm} + 15\,\text{mm} = \mathbf{54\ mm}$

8. $m\angle 5 = m\angle 1 = \mathbf{100°}$

9. $\mathbf{360°}$

10. $\sqrt{1600} = \mathbf{40}$

11. $-2 + -6 = \mathbf{-8}$

12. $-4 - (-7) = -4 + 7 = \textbf{3}$

13. Volume $= 8$ in. $\times 7$ in. $\times 6$ in. $= \textbf{336 cu. in.}$

14. $12 - 6 = \textbf{6}$

15. $4^3 - \sqrt{16} + 3 \times 2 = 64 - 4 + 6 = \textbf{66}$

16. $5\frac{1}{5} = 5.2$ $5.2 - 3.25 = \textbf{1.95}$

17. $\frac{4}{25} = \frac{m}{100}$
$25m = 400$
$m = \frac{400}{25}$
$m = \textbf{16}$

18. $y + 3.4 = 5$
$y = 5 - 3.4 = \textbf{1.6}$

19. $(13°F) - (-5°F) = 13°F + 5°F = \textbf{18°F}$

20. Area $= 3.14 \times 7$ in. $\times 7$ in.
$= \textbf{153.86 sq. in.}$

TEST 21

1.
$$\begin{array}{r} \$3.215 \\ 4)\overline{\$12.860} \quad \textbf{\$3.22} \\ \underline{12} \\ 0\;8 \\ \underline{\;8} \\ 06 \\ \underline{\;4} \\ 20 \\ \underline{20} \\ 0 \end{array}$$

2.

	Ratio	Actual Count
Finished	60%	15
Did not finish	40%	n
Total	100%	s

$\frac{60}{100} = \frac{15}{s}$
$60s = 1500$
$s = \frac{1500}{60} = \textbf{25 runners}$

3. $(2, 4)$

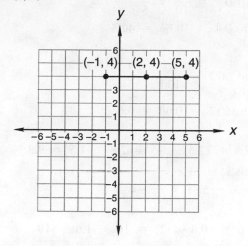

4. $3^2 + \sqrt{36} - 2 \times 10$
$= 9 + 6 - 20 = \textbf{-5}$

5. Volume $= 4\,\text{cm} \times 2\,\text{cm} \times 2\,\text{cm}$
$= 16$ cu. cm
16 sugar cubes would be needed.

6. $8\,\text{ft} - 4\,\text{ft} = 4\,\text{ft}$
$6\,\text{ft} - 4\,\text{ft} = 2\,\text{ft}$
Perimeter $= 8\,\text{ft} + 6\,\text{ft} + 4\,\text{ft} +$
$2\,\text{ft} + 4\,\text{ft} + 4\,\text{ft} = \textbf{28 ft}$

7. Area $= 3.14 \times 20\,\text{ft} \times 20\,\text{ft} = \textbf{1256 ft}$

8. $3\text{ tons} \times \dfrac{2000\text{ pounds}}{1\text{ ton}} = \textbf{6000 pounds}$

9. $\frac{4}{6} = \frac{b}{15}$
$6b = 60$
$b = \frac{60}{6} = \textbf{10}$

10.

	Ratio	Actual Count
Girls	7	g
Boys	8	b
Total	15	45

$\frac{7}{15} = \frac{g}{45}$
$15g = 315$
$g = \frac{315}{15} = \textbf{21 girls}$

11. (a) $1\frac{3}{4} = \frac{7}{4}$ $\quad 4)\overline{7.00}^{\,\textbf{1.75}}$

(b) $1\frac{3}{4} = 1.75 \times 100\% = \textbf{175\%}$

12. (a) $0.4 = \dfrac{4}{10} = \dfrac{2}{5}$

(b) $0.4 \times 100\% = \mathbf{40\%}$

13. (a) $60\% = \dfrac{60}{100} = \dfrac{3}{5}$

(b) $60\% = 0.60 = \mathbf{0.6}$

14. $7.8 + (2.5 - 1) = 7.8 + 1.5 = \mathbf{9.3}$

15. $-3 - 5 + 4 = \mathbf{-4}$

16. $16 \div (0.8 \times 2) = 16 \div 1.6 = \mathbf{10}$

17.
$$7\dfrac{3}{4} = 7\dfrac{9}{12}$$
$$+\ 5\dfrac{2}{3} = 5\dfrac{8}{12}$$
$$\overline{\qquad 12\dfrac{17}{12} = \mathbf{13\dfrac{5}{12}}}$$

18. $3\dfrac{3}{4} \times 1\dfrac{3}{5} = \dfrac{15}{4} \times \dfrac{8}{5} = \mathbf{6}$

19. $5\dfrac{1}{3} \div 2\dfrac{2}{3} = \dfrac{16}{3} \times \dfrac{3}{8} = \mathbf{2}$

20. Area $= \dfrac{6\,\text{ft} \cdot 4\,\text{ft}}{2} = \mathbf{12\ \text{sq. ft}}$

Test 22

1. $180° - (80° + 80°) = \mathbf{20°}$

2.

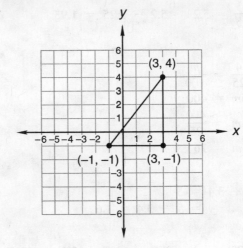

3.
$$5\dfrac{2}{3} = 5\dfrac{4}{6} = 4\dfrac{10}{6}$$
$$-\ 3\dfrac{5}{6} = 3\dfrac{5}{6} = 3\dfrac{5}{6}$$
$$\overline{\qquad\qquad\qquad \mathbf{1\dfrac{5}{6}}}$$

4. $5\dfrac{5}{6} \div 2\dfrac{1}{3} = \dfrac{35}{6} \times \dfrac{3}{7} = \dfrac{5}{2} = \mathbf{2\dfrac{1}{2}}$

5. $\dfrac{1}{8} = 0.125 \qquad 0.125 + 5.25 = \mathbf{5.375}$

6. $\mathbf{-3,\ 0.3,\ \dfrac{3}{4}}$

7.

Area $= \dfrac{4 \times 5}{2} = \mathbf{10\ \text{sq. units}}$

8. $-5 - 3 - 4 = \mathbf{-12}$

9. $\dfrac{5}{16} \cdot \dfrac{4}{15} = \mathbf{\dfrac{1}{12}}$

10. $2^3 + 3^2 - 2 \times 4 = 8 + 9 - 8 = \mathbf{9}$

11. (a) $3\dfrac{1}{2} = \dfrac{7}{2} \qquad 2\overline{)7.0}^{\,3.5}$

(b) $3\dfrac{1}{2} = 3.5 \times 100\% = \mathbf{350\%}$

12. (a) $2.4 = 2\dfrac{4}{10} = \mathbf{2\dfrac{2}{5}}$

(b) $2.4 \times 100\% = \mathbf{240\%}$

13. (a) $70\% = \dfrac{70}{100} = \mathbf{\dfrac{7}{10}}$

(b) $70\% = 0.70 = \mathbf{0.7}$

14. Volume $= 4\,\text{cm} \times 3\,\text{cm} \times 3\,\text{cm}$
$= \mathbf{36\ \text{cu. cm}}$

15. Circumference $= 3.14 \times 28$ mm
$= $ **87.92 mm**

16.
$$\begin{array}{r} 30 \\ -\ 12 \\ \hline 18 \end{array}$$
$\dfrac{\text{girl}}{\text{boy}} \cdot \dfrac{18}{12} = \dfrac{3}{2}$

17. $\dfrac{b}{18} = \dfrac{40}{60}$
$60b = 18 \cdot 40$
$60b = 720$
$b = \dfrac{720}{60} = $ **12**

18. $3n + 3 = 39$
$3n = 36$
$n = \dfrac{36}{3} = $ **12**

19. 6 in. $-$ 1 in. $=$ 5 in.
8 in. $-$ 2 in. $=$ 6 in.
Perimeter $=$ 8 in. $+$ 1 in. $+$ 6 in. $+$
5 in. $+$ 2 in. $+$ 6 in. $=$ **28 in.**

20. Area $=$ (2 in. $\times$ 6 in.) $+$ (6 in. $\times$ 1 in.)
$=$ 12 sq. in. $+$ 6 sq. in. $=$ **18 sq. in.**

TEST 23

1. $\dfrac{98}{4} = 24$ R 2
24, 24, 25, 25

2. $2\dfrac{1}{4}$ billion $= 2.25$ billion $=$ **2,250,000,000**

3. $64 = 2 \cdot 2 \cdot 2 \cdot 2 \cdot 2 \cdot 2 = 2^6$

4. $3 \,\cancel{\text{yd}} \times \dfrac{3 \text{ ft}}{1 \,\cancel{\text{yd}}} = $ **9 ft**

5. $(4 \times 10^3) + (6 \times 10^1)$
$= 4000 + 60 = $ **4060**

6. $\dfrac{3}{6} \cdot \dfrac{2}{5} = \dfrac{1}{5}$

7. Perimeter $= 3 \times 10$ in. $=$ **30 in.**

8. Perimeter $=$ 9 cm $+$ 10 cm $+$ 11 cm
$=$ **30 cm**

9. Diameter $= 2 \times 25$ cm $= 50$ cm
Circumference $= 3.14 \times 50$ cm $=$ **157 cm**

10. 2 ft $+$ 3 ft $=$ 5 ft
7 ft $-$ 3 ft $=$ 4 ft
Perimeter $=$ 7 ft $+$ 3 ft $+$ 4 ft $+$ 2 ft $+$
3 ft $+$ 5 ft $=$ **24 ft**

11. Area $=$ (7 ft $\times$ 3 ft) $+$ (2 ft $\times$ 3 ft)
$=$ 21 sq. ft $+$ 6 sq. ft $=$ **27 sq. ft**

12. (a) $5\overline{)1.0}$ with quotient **0.2**

(b) $\dfrac{1}{5} = 0.2 \times 100\% = $ **20%**

13. (a) $0.05 = \dfrac{5}{100} = \dfrac{1}{20}$

(b) $0.05 \times 100\% = $ **5%**

14. (a) $175\% = \dfrac{175}{100} = \dfrac{7}{4} = $ **$1\dfrac{3}{4}$**

(b) $175\% = $ **1.75**

15.
$$\begin{array}{r} 10.00 \\ -\ 2.65 \\ \hline 7.35 \end{array} \qquad \begin{array}{r} 7.350 \\ +\ 0.385 \\ \hline 7.735 \end{array}$$

16. $2.7 \div (0.6 \times 0.5) = 2.7 \div 0.3 = $ **9**

17. $(-4)(-8) = $ **32**

18. $\dfrac{-24}{+8} = $ **-3**

19. Area $= \dfrac{4 \text{ in.} \times 5 \text{ in.}}{2} = $ **10 sq. in.**

20. $100\% - 80\% = 20\%$
$\dfrac{20}{100} = \dfrac{4}{x}$
$20x = 400$
$x = \dfrac{400}{20} = $ **20 questions**